Ecology of
Arctic Environments

SPECIAL PUBLICATION NUMBER 13 OF THE

BRITISH ECOLOGICAL SOCIETY

EDITED BY

SARAH J. WOODIN

Department of Plant and Soil Science
University of Aberdeen, Aberdeen

AND

MICK MARQUISS

Institute of Terrestrial Ecology
Banchory Research Station
Hill of Brathens, Glassel
Banchory

Blackwell
Science

© 1997 the British Ecological Society
and published for them by
Blackwell Science Ltd
Editorial Offices:
Osney Mead, Oxford OX2 0EL
25 John Street, London WC1N 2BL
23 Ainslie Place, Edinburgh EH3 6AJ
350 Main Street, Malden
  MA 02148-5018, USA
54 University Street, Carlton
  Victoria 3053, Australia

Other Editorial Offices:
Arnette Blackwell SA
  224, Boulevard Saint Germain
  75007 Paris, France

Blackwell Wissenschafts-Verlag GmbH
  Kurfürstendamm 57
  10707 Berlin, Germany

  Zehetnergasse 6
  A-1140 Wien
  Austria

First published 1997

Set by Setrite Typesetters, Hong Kong
Printed and bound in Great Britain
at the University Press, Cambridge

The Blackwell Science Logo is a
trade mark of Blackwell Science Ltd,
registered at the United Kingdom
Trade Marks Registry

DISTRIBUTORS

Marston Book Services Ltd
PO Box 269
Abingdon
Oxon OX14 4YN
(*Orders*: Tel: 01235 465500
         Fax: 01235 465555)

USA
Blackwell Science, Inc.
Commerce Place
350 Main Street
Malden, MA 02148-5018
(*Orders*: Tel: 800 759 6102
         617 388 8250
         Fax: 617 388 8255)

Canada
Copp Clark Professional
200 Adelaide Street West, 3rd Floor
Toronto, Ontario M5H 1W7
(*Orders:* Tel: 416 597-1616
         800 815-9417
         Fax: 416 597-1617)

Australia
Blackwell Science Pty Ltd
54 University Street
Carlton, Victoria 3053
(*Orders*: Tel: 3 9347 0300
         Fax: 3 9347 5001)

A catalogue record for this title
is available from the British Library

ISBN 0-632-04218-4

Library of Congress
Cataloging-in-publication Data

Ecology of Arctic environments / edited by Sarah J.
  Woodin, Mick Marquiss.
         p.   cm.—(Special publication number 13
  of the British Ecological Society)
    Includes bibliographical references and index.
    ISBN 0-632-04218-4
    1. Ecology—Arctic regions—Congresses.
    2. Nature—Effect of human beings on—Arctic
  regions—Congresses.
    I. Woodin, Sarah J. (Sarah Jane), 1961–.
    II. Marquiss, Mick.
    III. Series: Special publication ... of the British
  Ecological Society; no. 13.
  QH84. 1. E26 1997
  574. 5'2621–DC20   96-43822
                                                 CIP

Contents

Preface

The ecosystems of the Arctic are inherently fragile, and thus susceptible to destabilization. Once thought of as a pristine environment, it is now all too apparent that the Arctic is a sink for pollutants transported northwards over long distances in the atmosphere and oceans. Furthermore, predictions in recent years have indicated that the Arctic will be subject to major climatic change as a result of global warming. It is this in particular which has prompted a great increase in interest in the Arctic. Not only is it quite possible that the first ecological changes attributable to global warming will be detected in this region, but the presence of a significant proportion of the entire world's soil carbon in arctic soils means that any destabilization of arctic ecosystems could have feedback effects on global climate systems. Many ecologists are currently seeking to further our understanding of how arctic ecosystems function, and to predict and detect anthropogenic changes which may occur within them.

Research in the Arctic by British ecologists was stimulated by a Natural Environment Research Council (NERC) funded research programme on arctic terrestrial ecology based at Ny-Ålesund, Svalbard, and similar national and international programmes have been established in other countries. A recent increase in availability of information on, and access to, arctic regions of the former USSR has provided a particular spur to international collaboration. This high level of interest and research activity in the Arctic, and the completion of the first phase of the NERC research programme in 1994, prompted the organization of a British Ecological Society Special Symposium on the Ecology of Arctic Environments. This was held at the University of Aberdeen on 27–29 March 1995, as part of the University's Quincentenary year celebrations.

This volume contains 12 of the 20 papers which were presented at the symposium. The first is a background chapter providing an introduction to the soils and periglacial processes and features of the Arctic: an essential starting place for ecologists. Current knowledge about the composition and role in ecosystem function of microbial communities and lower plant communities in the Arctic is reviewed in the next two chapters. Scaling up, this is followed by a discussion of how entire arctic plant communities influence ecosystem, regional and global processes, particularly in the context of climatic change. Relationships between individual arctic species and their environment are considered in Chapters 5–7, which also have climatic variation as a common theme. Chapter 5 suggests that the character traits which enable arctic plant species to survive in a very fragile and variable environment preadapt them to the consequences of climate change. Chapter 6

v

describes the environments, physiological adaptations and life-cycle biology of arctic terrestrial arthropods, and reports population responses to simulated climatic change. Moving up the trophic levels, Chapter 7 summarizes data collected over nearly 30 years showing associations between the reproductive variability of polar bears, ringed seal populations and environmental fluctuations, including climate. Completing the theme of community and species ecology is an elegant chapter which draws together terrestrial and aquatic ecosystems, and all trophic levels, in a consideration of trophic cascades. This culminates with the application of the trophic cascade concept to the ecology of lesser snow geese feeding on arctic coastal marshes.

The volume ends with four chapters concerning man's impacts on the arctic environment. The sources, pathways and effects of radionuclides, toxic metals and persistent organochlorines in the Arctic are reviewed. The effects of acid deposition on arctic vegetation are discussed, concentrating particularly on nitrogen deposition which is potentially an increasing problem. Early results of experiments on the effects of UV-B radiation on subarctic vegetation are reported, and the final chapter describes the effects of increasing atmospheric CO_2 concentration and changes in climate on the net CO_2 flux of arctic ecosystems.

We are grateful to all the authors who have made this volume possible, and to all the referees who contributed to the quality of the finished product by their comments. Special thanks are due to Dr Ian Stirling who, in addition to his research paper, presented an inspiring University public lecture on the natural history of polar bears. The symposium would not have happened without much organization by Dr Mark Young, and Dr Clare Woolgrove and Christine Payne contributed enormously to its smooth running and cheerful atmosphere. We also thank 'The Gathering' and 'Spitting Guinness' for excellent music, the delegates for heroic attempts at ceilidh dancing, and Professor and Mrs Chapin for joining in on the fiddle and tin whistle!

S. J. Woodin

1. Arctic soils and permafrost

EWART A. FITZPATRICK

Department of Plant and Soil Science, University of Aberdeen, Meston Building, Meston Walk, Aberdeen AB24 2UE, UK

INTRODUCTION

One of the most striking features of the arctic environment is the occurrence of a unique range of soils and surface features. Probably the most important feature in terms of soil development and geomorphology of the Arctic is the presence of permafrost or permanently frozen subsoil, which in places like Svalbard extends to depths of hundreds of metres. This continuously frozen barrier frequently leads to impeded drainage, reducing conditions, salinity and efflorescences of salts on soil surfaces. This chapter provides an introduction to the soils and permafrost of the Arctic. It describes the processes which lead to formation of the unique surface and subsurface features generally known as *periglacial features*. This is followed by a brief description of the major soils found in the region, and their relationships to the periglacial features and vegetation.

The Arctic is generally considered to be the region north of the 14°C July isotherm and is known as a periglacial environment. The term periglacial was introduced by Lozinski (1912). It has no specific definition, although it has come to mean those Arctic and Antarctic areas that have an *active layer* at the surface and subsurface *permafrost*. However, some maritime areas that are generally regarded as a part of the Arctic do not have permafrost. The active layer freezes every winter and thaws every summer, while the permafrost remains permanently frozen. The term periglacial also includes those high mountain areas that have a very cold climate, even though permafrost may be absent.

Precipitation in the Arctic varies from 50 to 1200 mm year^{-1} with mean annual temperatures of < 0°C, and minimum winter temperatures < −40°C. Areas with a maritime climate have higher rainfall which is evenly distributed throughout the year, while the cold continental climates have low rainfall with most of it occurring during the short cool summers (Tedrow 1977). The classical works about arctic geomorphology include those of Geikie (1894), Högbom (1914), Taber (1929, 1930, 1943), Leffingwell (1919) and Lachenbruch (1962). General accounts of the periglacial environment have been given by Embleton and King (1975), French (1996) and Washburn (1973).

In the Arctic, Tedrow (1977) recognizes three main climatic subzones: polar desert, subpolar desert and the tundra, each of which has a number of important distinguishing features.

1   The polar desert subzone is characterized by barren rocky surfaces, intensely cold, strong winds and low rainfall. Algae and lichens are the principal components of the plant life. The soils are unleached and carbonates are often present. Salt efflorescences and desert varnish also occur. The main soils are Leptosols, Regosols and Solonchaks.

2   The subpolar desert is dominated by mosses, lichens, sedges and grasses, with occasional shrubs, birch and dwarf willows. The main soils are Leptosols, Cambisols, Gleysols, some Histosols and rarely Solonchaks.

3   The tundra subzone is characterized by mosses, sedges, grasses, shrubs, birch, black spruce, white spruce, pine, larch and willows. Salt accumulations occur in the very dry areas. The dominant soils are Cambisols, Gleysols and Histosols, which occur mainly in oceanic areas and depressions. Podzols and Luvisols are rare, shallow, and show clear evidence of winter freezing. Phaeozems and Andosols are very rare. Throughout these subzones there are Fluvisols associated with the rivers.

There is some variability in the climate of these subzones, particularly in the amount of precipitation which increases towards the oceanic areas.

PERIGLACIAL PROCESSES IN THE ARCTIC

The change in state of water to ice upon freezing, and the accompanying change in density with its 10% increase in volume, are very important processes. With a further reduction in temperature, ice behaves like a normal solid and decreases in specific volume but it has a very high coefficient of linear expansion (51×10^{-6} K^{-1}), thus with a 30°C fall in temperature, cracks at least 1.5 mm wide will form in the soils. These unique contraction–expansion properties of water result in many of the processes and features developed in the Arctic.

Physical processes in arctic soils

Freeze–thaw processes

The results of these processes depend upon both the amount of water in the soil system, and the rate of decrease in temperature. When there is an adequate supply of water, and the temperature of the soil falls rapidly overnight to less than 0°C, the surface freezes, forming a massive structure in which there is an intimate mixture of ice and soil. When cooling is slow, ice crystals with a characteristic elongate form, or piprake, more than 5 cm long, grow out from the surface carrying a small capping of soil or gravel. Alternatively, they may grow beneath stones heaving them to the surface, where a continuous layer of stones and boulders may form as in the polar desert subzone.

When freezing takes place over a number of days beneath the surface or under controlled laboratory conditions a variety of patterns develop, as determined by the speed of freezing, particle size distribution and water content. In very sandy material the ice fills the large pores giving a massive structure. In loams and silts a distinctive lenticular structure develops, with lenses of soil surrounded by continuous more or less horizontally bifurcating sheets of pure ice. In silty clays, ice segregates mainly vertically to give a prismatic structure. In clays, ice segregates in both vertical and horizontal layers to isolate subcuboidal blocks of frozen soil (Kokkonnen 1926; Taber 1929). In some situations distinctive subspherical aggregates are formed (Morozova 1965; Van Vliet-Lanoë, Coutard & Pissart 1984).

The thickness of the ice segregations is a function of the speed of freezing as well as the water content. When the water content is low there is no ice segregation. When freezing is very slow and progressive over months or years and there is a plentiful supply of water, thick sheets of ice form, since water is steadily drawn up to the freezing front. The thickness of the ice sheets and soil lenses increases gradually with depth; they may be 1–2 mm thick at the surface and increase to over 1 cm at about 50 cm deep. However, sheets > 10 m thick occur in parts of Siberia and northern Canada. This segregation and growth of ice causes the soil to lift and is known as frost heaving. The amount of heaving varies from site to site with about 25 mm in mud circles to 26 cm in frost boils (Czeppe 1960; Everett 1965; Washburn 1967; Benedict 1970). This is one of the most important processes in the Arctic and in many areas immediately to the south. It is the mechanism that is responsible for the break-up of many roads and is therefore of immense economic importance.

The water in soils contains dissolved air, and usually some extra CO_2 derived from respiring roots and decomposing organic matter. These gases are expelled from solution when the water freezes, and form gas bubbles or vesicles in both the ice and the soil. When the soil thaws the vesicular structure in the soil is maintained, while some of the air bubbles in the ice may coalesce to form larger vesicles in the recently melted, and almost fluid, soil. Laminar pores are also created by the melting of the ice sheets.

Ice segregates as the result of the formation of ice crystals that grow and elongate about their c-axes in a manner similar to quartz or calcite. The elongation is in the direction of heat loss which is usually vertically. Thus, in thin sections of permafrost cut normal to the surface it is possible to see vertically elongated ice crystals (E.A. FitzPatrick unpublished data). Ice segregations also cause a reorganization of the soil by withdrawing water and compacting the lenticular or subcuboidal soil units. This may cause some of the fine sand and silt grains to become orientated parallel to the surfaces of the soil aggregates. These units are so well developed and compressed that they maintain their form when the soil thaws. Lenticular and subcuboidal peds released by melting permafrost at the end of the last glaciation are still present in the soils of a great many areas (FitzPatrick 1956, 1993). The amount of segregation depends also on the texture of the material with little taking place in sands, some in clays and considerable segregation in silts. The materials that develop large segregations and hence considerable heaving of the surface are said to be frost susceptible (Penner 1974).

In polar areas freezing of the surface soil during the winter and thawing during the summer creates an annual cycle which, coupled with ice segregations, cause profound disturbance of the soil surface and the formation of a number of characteristic features. Since stones have a lower specific heat than the surrounding soil, they cool down and heat up more quickly, thus attaining 0°C before the surrounding soil and therefore form loci for the formation of ice. This ice can become quite thick around the stone and displaces the surrounding soil. Repeated freezing and thawing, resulting in the formation and disappearance of ice, can cause the reorientation of the stones which in flat situations develop a vertical orientation of their long axes and simultaneously are gradually forced to the surface. Thus, typical permafrost will be composed of bifurcating sheets of ice surrounding compact lenses of soil, both of which will have vesicles of gases. If stones are present they will have a complete sheath of ice which will also have gas vesicles.

Frost shattering (frost wedging)

The occurrence in polar and alpine regions of extensive surfaces strewn with angular rock fragments is generally interpreted as being due to freeze–thaw processes. As water freezes, pressures of over $500\,kg\,cm^{-2}$ are created and it is assumed that these pressures exist in the pores of the rocks causing them to split apart, leaving a number of loose fragments as it thaws. Therefore the amount of fragmentation will increase as the number of freeze–thaw cycles increases. This explanation, however plausible, does not stand up to careful analysis since many areas in the middle latitudes that have more frost–thaw cycles than polar areas do not have the same extensive occurrence of angular rock debris. Everett (1961) suggested that the active agent is the steady growth of ice crystals in the fine pores of the rocks during prolonged periods of low temperatures. This creates high pressures which severely weaken the rocks causing them to split apart. Probably the main mechanism for shattering is the very wide temperature fluctuations that take place, from about +50°C at the surfaces of rocks on a hot summer day to −50°C during the winter. This 100°C variation causes great thermoelastic stress and probably leads to the development of a contraction crack system in the rocks that can be exploited by the formation of ice crystals (McGreevy 1981; Hallet 1983; Hallet, Walder & Stubbs 1991). Rock fragments in soils and other unconsolidated materials can also be frost shattered. Further, it has been shown that over a long period rocks can be reduced by freeze–thaw processes to accumulations of silt- and clay-size particles (Lautridou 1982; Lautridou & Ozouf 1982; Fukuda 1983).

Solifluction

This is the process whereby material is moved down the slope, and is now used as an umbrella term to include gelifluction and soil creep. Gelifluction is the rapid movement of material downslope, and creep is the slower and more insidious process in the active layer (Baulig 1956; Harris 1981). However, neither of these is used for the flow of permafrost down the slope, although this could be included in creep since this is a very slow process. Thus, there is gelifluction, active-layer creep and permafrost creep as described below. In addition there is movement of material by surface wash, mud-flows and land slides.

On slopes, repeated freezing and thawing cause mass movement of material downslope forming stratified slope deposits. The exact nature of the mechanism involved is gradually being understood through the use of laboratory experiments (Harris, Gallop & Coutard 1993; Harris, Davies & Coutard 1995, 1996). Different soils behave differently, and mass movement rates are strongly related to the content of silt. Coarse materials tend to move by creep due to the formation of needle ice, whereas with fine material gelifluction tends to take place due to the melting of the segregated ice. Creep takes place at an annual rate of 4–10 mm at dry sites, and 20–30 mm at wet sites, but there is considerable annual variation. Gelifluction takes place at an annual rate of 10–40 mm at dry sites and 30–70 mm at wet sites. Gelifluction can be accelerated in the presence of permafrost which is impermeable and causes water to accumulate within the surface horizons. Jahn (1960) has shown that the rate of solifluction in Spitsbergen is about 1–3 cm year^{-1} on slopes of 3–4°, and 5–12 cm year^{-1} on slopes of 5–7°. In all the measurements that have been made on solifluction it has been shown that the rate of movement is greatest at the surface and decreases with depth attaining zero at about 1 m. When the movement of material downslope is gradual, the stones tend to become orientated with their long axes parallel to the slope and normal to the contour. In addition, the surfaces of such areas are characterized by a number of lobes and terraces produced by differential movement.

The characteristic arcs at depths of 1–3 m in the permafrost on slopes in Arctic areas are attributed to movement resulting from freezing and thawing, and therefore regarded as normal solifluction. It is more likely that these arcs are due to plastic deformation and slow flow of the permafrost downslope. Since the depth of summer thawing in the Arctic seldom exceeds 1 m and the arcs occur below this depth, it is difficult to explain their presence as due to freezing and thawing. However, the temperature in the upper 1 m of the permafrost in the summer is only a few degrees below freezing, therefore it could flow by plastic deformation during this period of the year. The warmer and less viscous upper part would flow more rapidly than the cooler and deeper more viscous part, thus accounting for the arcs. In addition, these arcs are destroyed when they are incorporated in the active layer with its annual freeze–thaw cycle. Therefore, on slopes there is an upper part of the soil that has an annual freeze–thaw cycle and can move slowly or rapidly and disruptively, and a lower frozen part that moves very slowly by plastic deformation to form the arcs (FitzPatrick 1971, 1983; Thompson & Sayles 1972).

Wind action

Average wind speeds of about 18 km h^{-1} with maxima up to 72 km h^{-1} occur in the Arctic (Filstrup 1953). These winds deflate the surfaces of alluvium and other deposits, leading to the formation of sand dunes and loess that accumulate to over 30 m, as in parts of central Alaska. The particles carried by the very strong winds cause abrasion of surfaces with the formation of wind-faceted stones or ventifacts, which are very common on strand flats and raised beaches.

Chemical processes in arctic soils

The main chemical processes that take place in soils generally, also occur in the Arctic and include the following: solution, hydration, hydrolysis, oxidation, reduction, carbonate accumulation, salinization, salt wedging, cavernous weathering, case hardening, desert varnish formation and biological weathering.

Solution, hydration and hydrolysis

Solution is the process that dissolves the soluble minerals in soils. It affects particularly salts, carbonates and primary minerals such as apatite, but these are usually present in very small amounts. Thus, this process is generally of minor importance except where the rocks contain much carbonate or soluble salts. It should be pointed out that quartz weathers by solution, but it is very sparingly soluble; hence, it accumulates in soils while the other minerals steadily decline in volume. However, there is usually enough silica in solution in the Arctic to support populations of diatoms in many wet situations.

Hydration is the process whereby minerals take up water from their surroundings. This applies particularly to primary minerals, such as biotite, and to the secondary oxides and oxyhydroxides of iron and aluminium. This is also of minor importance in the Arctic.

Hydrolysis (proton donation) is the most important process affecting the primary minerals in soils generally. During this process, dissociated hydrogen ions in the soil solution replace basic cations such as calcium, sodium and potassium on the surface of a mineral. These basic ions may form an integral part of the mineral or they may form part of the exchange complex with clay minerals. The lower the pH of the soil solution, the more rapid the decomposition of the minerals. The pH of arctic soils varies from about 5 to 7 and in a few situations it may be as low as 3.0. It is determined in part by the CO_2 concentration derived mainly from the atmosphere, which may be the only source of CO_2 in polar desert areas where there is little or no vegetation and very low microbial activity. However, it is supplemented considerably from decomposing organic matter and respiring roots where vegetation is present. Since some arctic soils contain large amounts of organic matter it would be expected that the rate of hydrolysis would be high, but this is not the case because of the low temperatures which reduce both the rate of organic matter decomposition and the reaction rate between the hydrogen ions and the minerals.

Other elements, such as iron and aluminium, are also released by the hydrolysis of the primary minerals and may stay in solution, or may be complexed with organic compounds or precipitated as hydrated oxides on surfaces of rock fragments or soil aggregates. These three processes are active throughout the Arctic, but at a much reduced rate as compared with areas immediately to the south. This is due to the lower temperatures and to the small amounts of water in some areas, particularly in the polar deserts subzone. However, there are sufficient dissolved acids and water movement in many places in the subpolar desert and tundra subzones to cause most of the soils to be acid and for a range of ions to be translocated to waterways as shown in Figs 1.1 and 1.2 (Brown *et al.* 1962). Figure 1.1 shows that the concentrations of the main cations in the waterways in northern Alaska are low and seldom exceed $5 \, mg \, l^{-1}$. The streams at Barrow have high concentrations of calcium and magnesium and, as expected, those influenced by calcareous materials have the highest, up to $30 \, mg \, l^{-1}$. Figure 1.2 shows that the concentrations of the main microelements are also generally low, except for that of iron, which

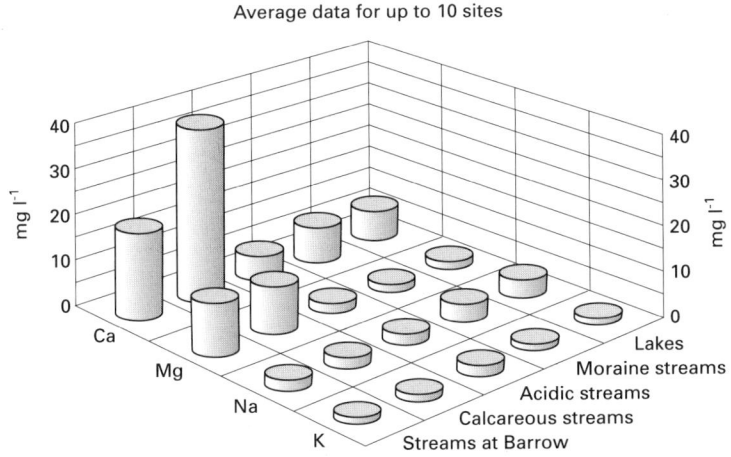

FIG. 1.1. The major cations occurring in the streams and lakes of northern Alaska.

Average data for up to 10 sites

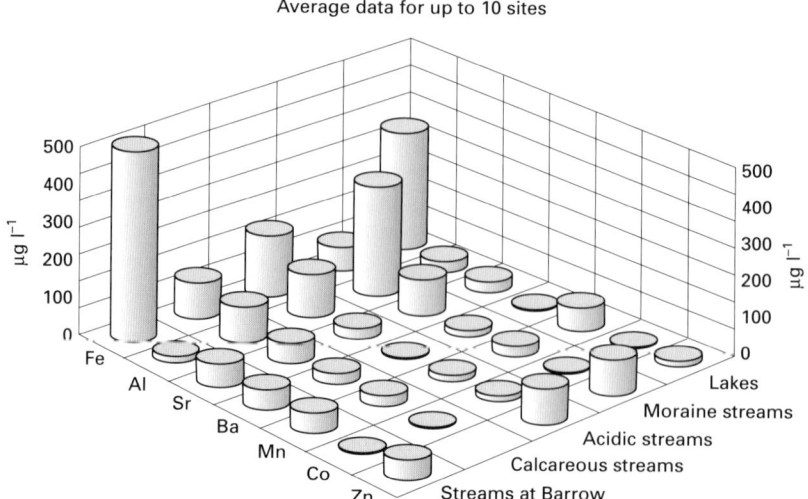

FIG. 1.2. The microelements occurring in the streams and lakes of northern Alaska.

is high in the streams at Barrow, the acidic streams and in the lakes. The aluminium concentration is highest in the streams from the moraines, probably because of the high surface area of this relatively fresh glacial material containing many feldspars. These low figures for most elements may represent the low degree of weathering that is taking place. On the other hand, they could indicate that there is little tendency for ions to accumulate in the environment in that part of Alaska. Ugolini, Stoner and Marrett (1987) identify a number of factors which severely restrict the rates of chemical weathering in the Arctic.

Oxidation and reduction

Iron is the main mineral that undergoes oxidation and reduction in soils. In aerobic soils, such as Cambisols in upland and moderate to strongly sloping situations, the small amount of iron that is released by hydrolysis is usually oxidized to give yellow and brown oxides and oxyhydroxides in the upper and middle parts of the soil. It should be mentioned that most of the primary minerals containing iron are greenish in colour, indicating that the iron is in the reduced, Fe^{2+} form. When it is released by hydrolysis in an aerobic environment, it is almost immediately oxidized to give yellow and brown oxides, which give the coloration to many of the freely draining soils, including the Cambisols. In the tundra subzone, iron exists mainly in a reduced form in the soils on flat and gently sloping situations that are wet or saturated with water for most of the summer. The middle and lower horizons in these soils are grey and blue, and may have yellow and brown mottles. In these situations most of the iron released by hydrolysis stays in the grey or blue Fe^{2+} state because of the restricted entry of oxygen into pores filled with water. The yellow mottling is produced in the short periods of aeration and oxidation of the iron to lepidocrocite during the latter part of the summer.

Carbonate deposition

Carbonate accumulation is very common in soils of the polar desert subzone and extends into the subpolar desert subzone. It may form a continuous horizon or pendents under stones and boulders. In many of the Cambisols of the drier areas, there is some calcium carbonate deposition within the soil, particularly where the permafrost is deep and there is very low precipitation (Courty *et al.* 1994).

Salinization

In dry areas, evaporation during the later part of the summer often leads to the formation of efflorescences of salts on soil surfaces. There may be a gradual growth and accumulation on the surfaces or there may be periodic drying out of surface water that has accumulated in shallow depressions or around tussocks. Where this process has been continuous for a long period there may be over 10 cm of accumulated salts, with thenardite (Na_2SO_4) being the principal component. The salts originate from a variety of sources: they may accumulate from weathering, be present in geological strata or be derived from sea spray.

Salt wedging takes place when salts crystallize in the pores and joints of rocks creating high pressures similar to those caused by freezing, and results in the physical break down of the rocks. It is of greatest importance in the polar subzone with sulphates producing more disintegration than chlorides or carbonates. The outer surface of many rocks is *case hardened*, caused by the deposition of material by evaporation of solutions containing minerals. Many rocks, particularly in the polar subzone, develop a cavernous appearance as a result of *salt fretting*. The effect is sometimes enhanced when the original rock has been case hardened. The loosened material may accumulate around the base of the rocks or it may be removed by wind.

Desert varnish formation

Desert varnish is a very dark, most often black, coating on the surfaces of rocks and rock fragments. It seems to be composed mainly of varying amounts of manganese, iron and aluminium and forms by the migration of material from inside the rock fragment to its surface. Desert varnish formation takes place only on very old land surfaces and thus seems to require a considerable period of time to develop.

Further, it only forms in the very driest sites, and thus is present mainly in the polar desert subzone where the coated stones and boulders may form a continuous layer or hamada at the surface.

Biological processes

The main biological processes associated with soil formation are the accumulation of litter, nitrogen cycling and biochemical weathering, all of which are associated with plants. The plant communities in the Arctic vary from a sparse covering of lichens on rock surfaces in the polar desert subzone to well-developed forest communities of spruce, pine and larch in the southern part of the tundra subzone. Thus, the contribution of litter is equally variable but it seldom exceeds $2\,t\,ha^{-1}\,year^{-1}$ in the forested areas. This is relatively small as compared with a tropical rainforest which contributes about $25\,t\,ha^{-1}\,year^{-1}$. On the other hand, the soils of the tundra subzone may contain several centimetres of free organic matter at the surface while a tropical soil may only contain $<5\%$ organic matter mixed with the mineral soil. The difference between these sites is due to differences in the rate of decomposition which is much more rapid in the tropics. The type of organic matter is also important. In the cold areas of the world the organic material is usually acid to very acid which helps mineral weathering, but the low temperatures tend to retard weathering and soil formation generally. Thus, in the Arctic there is a relatively small production of organic matter but a relatively large amount in soils. Figure 1.3 shows the amount of carbon in a transect from the polar desert subzone to the southern boreal forest (Giblin *et al.* 1991). It shows an almost steady increase in the amount of biomass with a slight peak in the tussock areas. However, the dead organic matter shows two distinct peaks, one in the area of tussock vegetation in the tundra subzone and a second much higher peak in the northern boreal forest. The high amounts of dead organic matter are related both to the rate of addition as well as the slow rate of decomposition. As the rate of addition increases so does the amount of dead organic matter, indicating that decomposition is less than addition. This seems related to the increase in wetness and decrease in decomposition rate from the polar desert to the wet tussock vegetation. The rapid fall in dead organic matter in the tall shrub vegetation is due to drier conditions which reduce plant growth but increase the rate of decomposition. There is then a steady increase in the addition of organic matter, but a marked jump in the amount of dead organic matter in the northern boreal forest again associated with wetter soils. Thereafter, the amount of dead organic matter decreases as the rate of decomposition increases commensurate with the increase in temperature. This transect clearly demonstrates the varied relationships between organic-matter accumulation, rate of decomposition, temperature, soil water and plant growth in the Arctic.

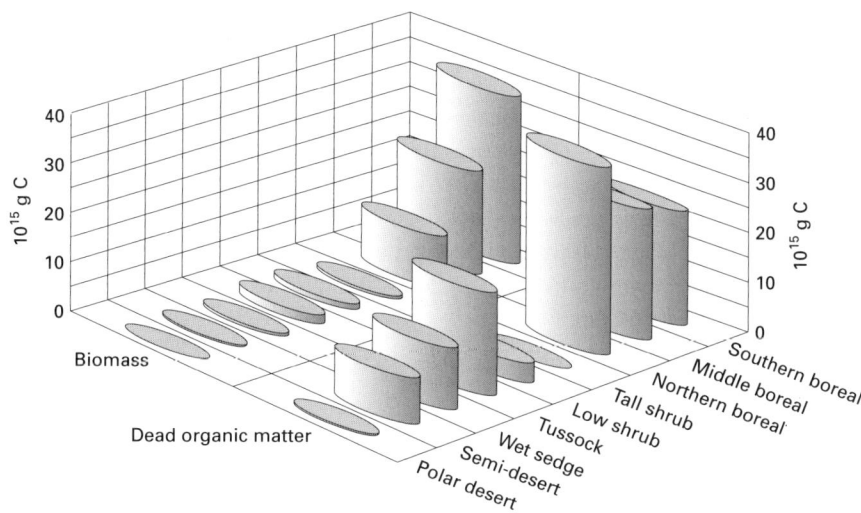

FIG. 1.3. A transect showing the changes in biomass and dead organic matter in soils from the polar desert to the boreal forest.

The Arctic is probably the most important part of the Earth's surface for direct biological weathering which is brought about mainly by lichens and mosses. Lichens are common through the Arctic and take on a dominant role in the polar desert subzone where in some places they are the only form of vegetation and thus the only contributors to the soil organic matter. They are the first colonizers and produce the so-called lichen acids which initiate weathering of the rocks (Parfenova & Yarilova 1965). This is followed by moss colonization which can be more efficient in weathering since mosses trap much larger amounts of water. When the litter from plants decomposes, it produces organic acids which help in the weathering process. Although this is not strictly biological weathering it does make a considerable contribution. These decomposition products include simple acids such as oxalic acid, and also the all-important fulvic acid (Schnitzer & Vendette 1975).

Profile formation

The processes responsible for the differentiation of horizons include all of those mentioned above as well as the distinctive and almost unique processes of translocation. Nearly all of the various types of materials that occur in soil can or have been translocated by a variety of processes. Translocation by water and frost churning are the two main processes responsible for translocation in arctic soils. Elements that are brought into solution by hydrolysis and solution can easily be translocated by water, especially when there is a large amount entering the soil. Thus, in the high rainfall areas of the Arctic much of the material in solution is washed completely out of the soil. Whereas in areas outside the Arctic percolation is usually vertically downwards, in the Arctic the permafrost acts as a very efficient barrier causing lateral percolation over its surface. This can lead to flushing in the middle and lower parts of the landscape and in some situations the dissolved salts may form an efflorescence where the water comes to the surface and evaporates. As stated above some soluble materials such as calcium bicarbonate may be transformed to calcium carbonate and deposited to form pendents beneath stones or to cement the whole horizon.

Iron released by hydrolysis in the upper parts of the soil may follow one of at least three different pathways: (i) it may stay in solution and be lost from the system; (ii) it may be oxidized *in situ* to form brown horizons that include the middle cambic horizon in Cambisols and the upper horizon beneath the hamada in the Regosols of the polar desert; and (iii) it may be chelated or complexed with organic compounds and travel a short distance down the soil profile and then deposited as in the case of Podzols. There are many theories advanced for this deposition of iron. Recently, Lundström (1994) demonstrated that the deposition is probably due to bacterial activity.

Lysimeter studies by Ugolini *et al.* (1987) have demonstrated that there is sufficiently active leaching in the tundra subzone to produce Podzols. Leaching is also active in the subpolar desert zone but not as active in the tundra subzone; nevertheless, some soils have chemical criteria that qualify them to be Podzols.

The incorporation of organic matter into the soil is normal in areas with vigorous faunal activity, but with the absence of soil fauna in much of the Arctic other methods of organic matter incorporation operate. One method is for the soluble decomposition products produced by fungi and bacteria in the surface organic matter to be translocated in solution by percolating water. The second method is by frost churning which incorporates organic matter that ranges from small particles of partially humified material to large lumps of the litter from the surface, down to over 1 m as in the case of some Cryic Gleysols (see Fig. 1.14 below).

A very distinctive feature of arctic soils is the translocation of silt. In most soils with free percolation there are silt cappings on the upper surfaces of soil aggregates and stones as first described by FitzPatrick (1956, 1971) and shown in Fig. 1.4. Subsequently, this feature has been described for virtually all locations in the Arctic. When ice melts in the active layer during summer it releases a large volume of water that can travel very quickly through the relatively large spaces left as it melts. Any silt dislodged during freezing and thawing of the active layer will be transported in the percolating water which may also erode some of the surfaces of the aggregates, thereby collecting more silt. This silt is deposited in some of the spaces as they are formed and also in the vesicles produced by freezing. It is clear from thin sections

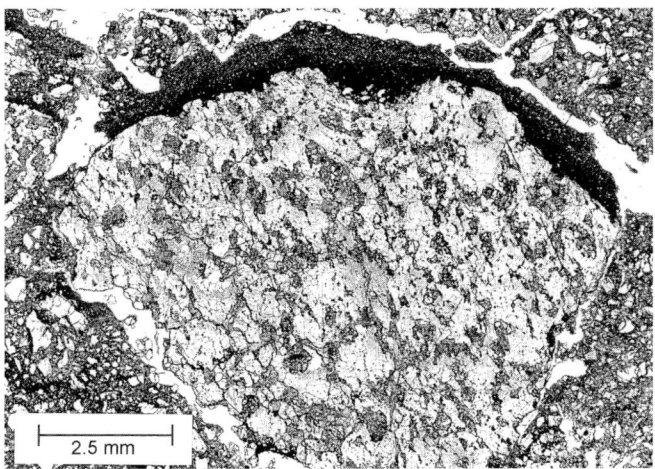

FIG. 1.4. A silt capping on the upper surface of a stone; note the absence of any coarse grains of sand indicating that the silt accumulated in a pre-existing space created by ice.

that the silt cappings on the surfaces of stones have formed by the filling of a pre-existing space and not the filling of pores in the matrix because of the absence of coarse material. Likewise, the silt cappings on the surfaces of soil aggregates show minimal impregnation of their surfaces.

Clay translocation also takes place in arctic soils, but not on the same scale as in some other environments (McKeague, MacDougall & Miles 1973; Sokolov *et al.* 1976). This process forms clay coatings in the deep subsoil that is little influenced by deep freezing. Their general absence near the surface suggests that if they do form they are destroyed by the following winter freezing.

PERIGLACIAL FEATURES

The arctic landscape is characterized by a number of unique surface features that have been produced by freeze–thaw processes. They vary widely in their morphology and other characteristics and have been difficult to classify adequately. The system produced by Washburn (1956) has been the most popular. He divides surface features into two groups: sorted and non-sorted. The sorted features are those that have very distinct patterns formed by stones and boulders. The non-sorted features are no less distinct and may also have patterns, but are not formed by the arrangement of stones and boulders. However, this classification scheme is not all-embracing and does not include important features such as talus cones and tors. Although its use is somewhat restricted, it will be used as the basis for the description of arctic periglacial features as given in Fig. 1.5.

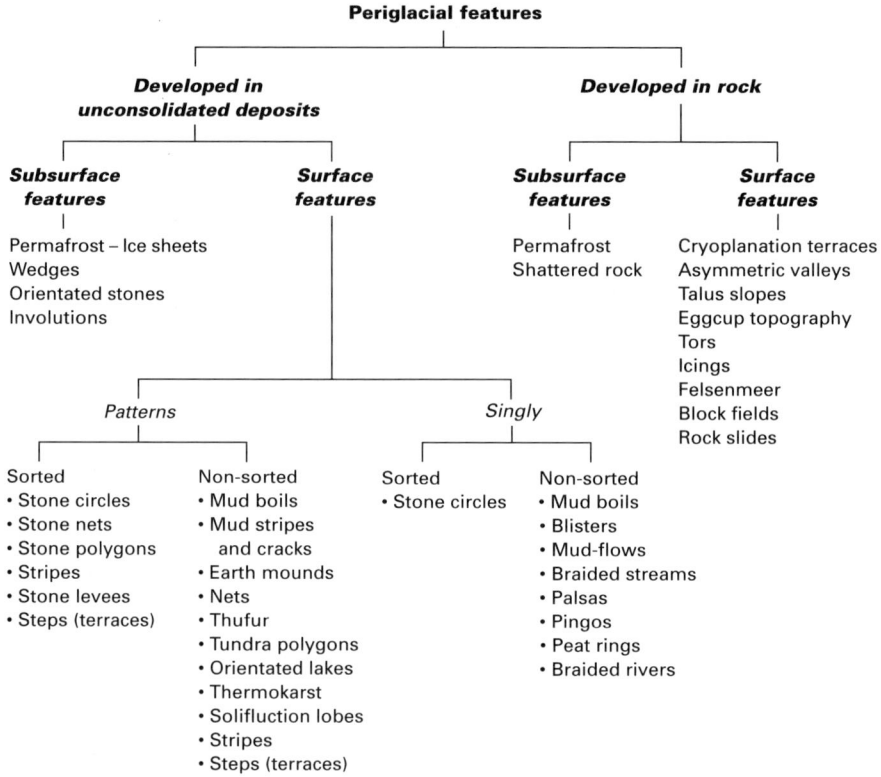

FIG. 1.5. The range of periglacial features occurring in the Arctic.

Features developed in unconsolidated deposits

Subsurface features

Permafrost. This is the permanently frozen subsoil (Müller 1947) and is mainly of two types: wet and dry. Dry permafrost contains insufficient water to form a frozen mass whereas wet permafrost has a variety of types of segregated ice as described above. Permafrost develops at temperatures just below 0°C and therefore can form or is forming at present in parts of the Arctic. The thickest wet permafrost of 1500 m is in Siberia at Schalagonzy (Baranov & Kudfryavtsev 1963) while depths of 600 m occur in Alaska. In many situations there is an unfrozen layer – talik – sandwiched between frozen layers.

In the lower part of the Arctic in both North America and Siberia where the mean annual temperature is higher, permafrost is present as a relic feature because of the insulating effect of forests that considerably reduces the amount of heat entering the soil. If the forest is removed rapidly, the permafrost melts leading to the formation of thermokarst as described below and may also lead to severe soil erosion. Thus, very strict control methods have to be introduced before the forest can be removed, either for timber or for cultivation as in some of the more southerly areas in Siberia.

Wedges. Ice wedges are the most conspicuous type of wedge but there are also sand wedges and soil wedges. *Ice wedges* are wedge-shaped masses of ice that extend vertically downwards from the top of the permafrost as shown in Fig. 1.6, and develop where the mean annual temperature is $< -6°C$. During the polar winter when the temperature falls to -40 to $-60°C$ and the active layer cools to -15 to $-20°C$, the whole soil contracts sufficiently to cause the ground, including the permafrost, to crack into hexagonal or rectangular patterns forming tundra polygons, about 10–30 m in diameter as shown in Fig. 1.7 (Mackay 1993a,b, 1995). Hoar frost and spring melt water accumulate in these cracks and freeze, so that during the following summer when the ground expands a thin vertical sheet of ice is trapped

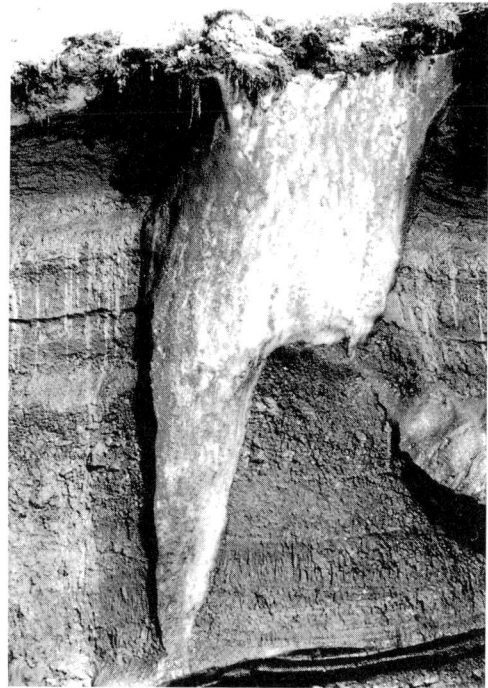

FIG. 1.6. An ice wedge 4 m deep from Garry Island in the Mackenzie Delta, northern Canada.

FIG. 1.7. Landscape at Point Barrow with well-developed tundra polygons about 30 m in diameter. The area in the distance was recently occupied by an orientated lake that has emptied into the sea and now has the beginnings of ice-wedge formation.

within the permafrost (Lauriol, Ducheshe & Clark 1995). The part of this thin sheet of ice in the active layer melts during the summer allowing the material to reunite. This process is repeated many times, possibly not annually, but there may be a biennial or triennial cycle, ultimately forming wedges which are usually 1–1.5 m wide at the top and about 3–4 m deep. In parts of Siberia they can be up to 10 m wide and 50 m deep (Figs 1.6 & 1.7) (Lachenbruch 1962). It is estimated that the rate of growth is about 0.5–1 mm year^{-1} and the tendency is for the largest wedges to occur in the tundra subzone and for the size to decrease in the polar desert areas (Black 1952). Thus, the soil profile at wedge sites is one of mineral soil overlying ice.

Ice wedges are ubiquitous on flat or very gently sloping sites throughout the Arctic, being largest in the old landscapes such as northern Alaska and Siberia that were not glaciated during the last glaciation and thus have been continuously cold for a very long time. In some cases organic matter such as leaves and lemming droppings fall into the cracks and become trapped and can be used as a means of dating. In most situations ice wedges form after the deposition of some type of material, be it loess, alluvium or a glacial deposit. Such wedges are known as epigenetic wedges. In some cases the wedges develop as the sediment is being deposited. This leads to the formation of tiers of wedges – such wedges are syngenetic. Some of the old permafrost in Alaska, northern Canada and Siberia that is no longer forming may contain ice wedges and sheets which are contorted as a result of subsidence or plastic flow.

In the very dry environments, particularly in Antarctica, the cracks become filled with wind-blown material to form *sand wedges* (Péwé 1959; Ugolini & Walters 1973). In other situations, the cracks become filled with soil to form *soil wedges* but little is known about their formation.

Orientated stones. The repeated freezing and thawing that causes stones to move steadily up and be heaved to the surface also causes them to become vertically orientated on flat sites. On slopes, as stated above, the progressive movement due to creep causes the stones to become orientated normal to the contour and parallel to the surface. When movement is fast they are also orientated in this manner but upon meeting an obstruction such as a large boulder they are turned at right angles.

Involutions. Involutions are vertical distortions of materials as a result of differential growth of ice in the active layer. They usually develop in stratified materials with different frost susceptibilities and occur in most alluvial deposits. They are best developed where some of the strata have a high content of silt with its high frost susceptibility. Different types of deformation develop depending upon whether the most frost susceptible material is at the top or bottom of the sequence of strata. It is suggested that this process is ultimately responsible for the formation of stone circles and mud boils. If, for example, silt overlies stones, the stones are heaved to the surface and a stone circle is formed but if stones overlie silt a mud boil is formed with surrounding stones (Van Vliet-Lanoë 1985). However, French (1986) has warned that distortions in sediments produced by slumping are virtually indistinguishable from involutions.

Surface features

Patterns – sorted. Stone circles are composed of a raised, circular border of large stones surrounding an area of finer material (Fig. 1.8). They occur singly or in groups and vary in size from about 1 to 3 m with the permafrost being slightly deeper at the centre of the circle. In many cases some of the stones have been converted into ventifacts by strong winds. *Stone polygons* are similar to stone circles but have a distinctly polygonal outline while *stone nets* form a transitional stage between stone circles and stone polygons. On slopes of about 25° there are longitudinal *stripes* of alternating bare soil and vegetation or bare soil and stones. As the slope angle decreases the stripes become lobes or small steps (terraces). On moderate slopes between the bottoms of mountains and the main rivers there are small deeply incised valleys continuous with gullies in the mountains. These valleys have *stone levees* caused by the rapid flow in the spring of the heavily laden melt waters from the mountain gullies.

FIG. 1.8. Large stone circles 2–3 m wide on Baffin Island. (Photograph by F.M. Synge.)

Patterns – non-sorted. Mud boils are bare areas of soil surrounded by vegetation (Fig. 1.9). They occur on flat to very gently sloping situations and are of two different types. One type has a slightly convex surface, is about 1 m in diameter and surrounded by mosses, sedges and grasses that occupy a slight depression between the boils. The second type also has a flat to slightly convex surface but is surrounded by a narrow ridge of vegetation which may include *Salix polaris*. A section through these features shows that the permafrost is closer to the surface beneath the vegetation which acts as a very efficient insulator. These features can occur over large areas and form the spotty tundra of Siberia. If mud boils have a surround of stones as suggested above then they are classified as sorted boils. *Mud stripes* are elongate mud boils that form on low-angle slopes, often they have associated linear cracks between the stripes.

Earth mounds are raised areas about 1–2 m in diameter and up to 1 m high. They have an ice core during the early part of the summer and are thought to have formed by the growth of the ice. They seem to be the mineral counterpart of palsas that form in peat.

Thufurs are hummocks of soil and vegetation with a height and diameter that vary from 10 cm to over 1 m (Fig. 1.10). Thufurs occur in the tundra subzone and also in areas without permafrost such as in Iceland.

In many of the flat areas of the Arctic, such as the northern part of Alaska and the Ob river area of Siberia, there are a large number of shallow *orientated lakes*. They are generally orientated in a north–south direction and may cover more than 30% of the land surface. The distinctive feature of these lakes is that they are migratory and steadily move across the land surface, eroding along one side and

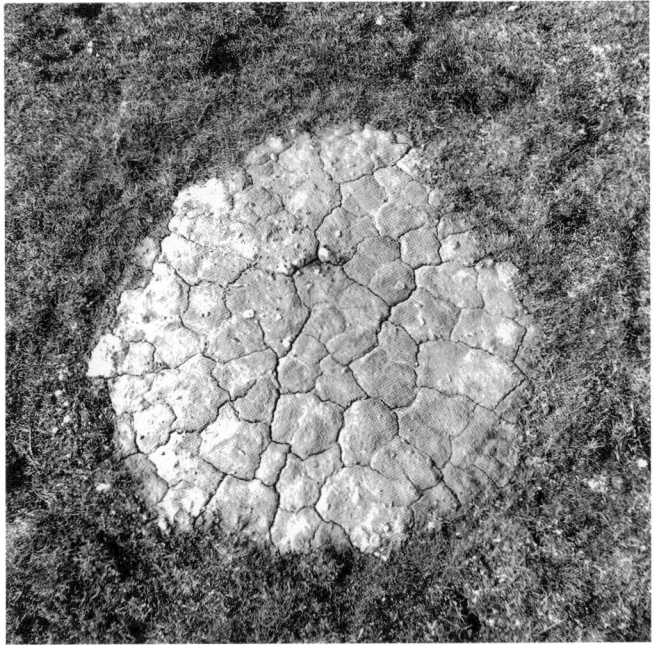

FIG. 1.9. A mud boil 120cm wide in northern Alaska with desiccation cracks.

FIG. 1.10. Thufur in Iceland, they vary in size from 20 to 40cm in length.

filling up along the opposite side. In many cases they migrate to the sea where they empty, exposing a ground surface where new features can form, particularly ice wedges (see Figs 1.6 & 1.7). These are possibly a special form of *thermokarst* which is hummocky terrain formed by the uneven melting of the permafrost. It can form naturally by a change in an environmental factor such as an increase in temperature, or the death of the vegetation following a fire or flooding. Often in the forested tundra, thermokarst is induced by humans due to deforestation. In central Yakutia in Siberia over 40% of the land surface is affected (Kachurin 1962; Solov'ev 1969). A distinctive type of thermokarst is the alas which is a large, deep depression. Often alases become filled with water to form lakes, which in turn become filled with sediment in which a new generation of periglacial features will develop including ice wedges.

On gentle slopes, the movement of material tends to be relatively slow and it is here that *steps* (terraces) form. They may have a stony riser (sorted steps) but they may have a riser of vegetation (non-sorted steps) (see Fig. 1.13 below).

On gentle to moderate slopes *solifluction lobes* run parallel to the contour and vary from a few metres to over 100 m in length. They are very conspicuous in the tundra subzone and form by the slow movement of material down the slope. The movement is so slow that there is usually a soil profile – Cryic Gleysol. A section normal to the contour through these lobes shows that they have moved down over the soil surface and that a buried profile can be traced beneath them up the slope for many metres. Where there are forests the trees are tilted but they attempt to continue to grow vertically. Generally, solifluction lobes develop outside the forested areas.

Soil blisters form during the autumn when the soil freezes gradually down from the surface, and upwards from the permafrost, causing a layer of unfrozen material to be trapped between the two layers. As the temperature continues to fall, the trapped unfrozen material expands causing a blister which may rupture forming a geyser, and an outflow of material that quickly freezes on the side of the blister. These features have been reported from the forested tundra of Siberia where they cause the trees to be tilted to form what has been called 'drunken forest' (Nikiforoff 1928).

Mud flows are small areas of very rapid movement of material down the slope. During the spring the active layer on moderate to steep slopes may become saturated with water causing a sudden movement of material down the slope. These features are very common in hilly areas without vegetation.

Palsas are irregularly shaped peat hummocks containing ice cores which grow annually thus causing a progressive increase in height that ranges from 1 to 10 m. Their diameter may extend laterally for up to 1 km, forming a plateau dominated by black spruce with a ground vegetation of *Sphagnum*. These 'forest plateaux' are higher than the surrounding treeless areas, which are colonized by sedges which make up the main body of the peat itself. These plateaux are unstable with the edges tending to slip while the removal of the trees can cause the plateaux to collapse into hummocky thermokarst due to the melting of the permafrost (Åkerman 1987).

FIG. 1.11. A pingo from Spitsbergen showing the characteristic conical form and rupture of the sides due to the updoming as the ice grew. It is about 20 m high.

A *pingo* is a small circular hill with an ice core. A large number are about 10–20 m high, but they can vary from <5 m to over 60 m high and up to 300 m in diameter (Fig. 1.11). Pingos form by slow growth of the ice core which is fed from a source of water. It is believed that there are two types of pingos – open system and closed system. Open-system pingos occur as isolated features and are common in the tundra subzone where the permafrost is thin or discontinuous. Under such conditions, water moves upwards through openings in the permafrost and freezes, causing an updoming of the surface.

Closed-system pingos occur almost exclusively in areas of continuous permafrost, and in contrast to open-system pingos they often occur in clusters. They form in alluvium or old lake beds which were previously unfrozen but freeze upon drainage, then large ice segregations form and the surface is forced upwards. One such conspicuous cluster is on the island of Tuktoyaktuk in northern Canada (Mackay 1979). There is still some uncertainty about the rate of growth of pingos but evidence from northern Canada suggests that initially they grow at a rate of 1.5 m year^{-1} and then at slower rates continuing to grow for up to 1000 years (Mackay 1979). The upward growth of pingos causes cracks to form in the upper part while at the top a small crater develops and may become filled with water during the summer.

A distinctive feature of the rivers and streams in the Arctic is their multiple channels that tend to occupy the whole of the valley floor (Fig. 1.12). These are known as *braided rivers*. Their courses are very shallow and intertwining, often changing direction on a daily basis depending upon the amount of water entering the system from melting snow and ice. This results in a large area of exposed river bed from which material can be removed by wind to form loess and sand dunes. There are other stream forms in the Arctic such as *beaded streams*.

FIG. 1.12. The braided Tanana river near Fairbanks, Alaska. Note the ephemeral nature of the islands that show destruction of the forest as the river periodically changes it course. Such large exposed areas are excellent sources for wind-blown material, hence the thick accumulations of loess in the area.

Features developed in rock

Surface features

Cryoplanation terraces (altiplanation terraces) are large terraces up to tens of metres wide and hundreds of metres long. They have slopes of 1–12° and risers of 5–7 m high. Their surfaces are covered with a thin layer of frost-shattered rock which is moving down the slope and there may be frost circles and polygons. They seem to form by freeze–thaw processes acting on the rock and by snowmelt water removing any fine material (Demek 1969; Czudek 1995).

Many of the valleys in the Arctic have steeper north- than south-facing slopes. These are known as *asymmetric valleys*, and it is suggested that they form by more vigorous weathering and erosion on the warmer south-facing slopes. This type of landscape is by no means universal in the Arctic thus there is still some controversy as to whether it is truly an arctic phenomenon.

Throughout the mountainous regions of the Arctic, a dominant process is the accumulation of rock debris at the base of slopes to form *talus slopes*. It is the result of the combination of very active frost shattering and steep slopes. In some areas, such as Spitsbergen, the very fissile rocks with horizontally bedded strata give the best examples in the form of talus cones (Fig. 1.13). *Eggcup topography* is characterized by large, deeply incised, v-shaped gullies cut in rock from which material has been removed by freeze–thaw processes to form talus cones below

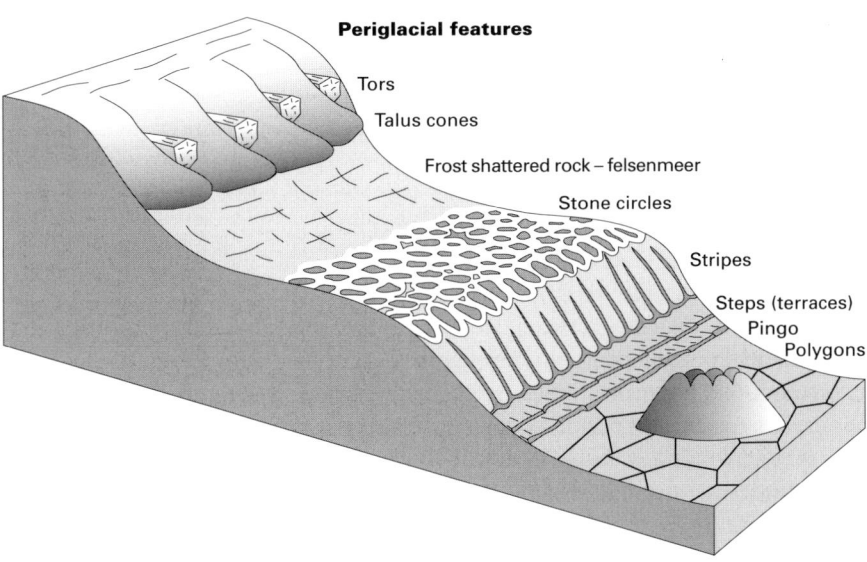

FIG. 1.13. Spatial relationships between some of the periglacial features and slope.

(Fig. 1.13). Such features are very widespread in the Arctic with Templefjellet in Spitsbergen being the most impressive.

Tors (arctic-type) are projections of rock on the tops or sides of hills and mountains, formed by the differential physical weathering and movement of material downslope (Fig. 1.13). They range in size from 2 m to over 25 m and are very common and conspicuous in some landscapes such as Spitsbergen.

Icings (frosting) are accumulations of ice as result of seepage of water on to a surface. The water has many origins including normal springs, thermal springs and seepage from pingos and blisters.

Felsenmeer are very large, flat areas with a shallow depth of angular frost-shattered material (Fig. 1.13). They are largely devoid of vegetation and may have a little patterned ground such as stone circles, and where there is a slope, stripes of coarse and fine material may form. There are also large areas of stony material caused by wind deflation of sediments, particularly alluvium.

Block fields are concentrations of large boulders at the surface, which usually occupy relatively small areas and form *in situ* or have travelled only a short distance. On many steep slopes beneath rock outcrops, large boulders that break off fall on to the surface and slide downslope forming *rock slides*. In so doing they plough through the active layer creating grooves and accumulations of soil against their downslope faces.

Subsurface features

Permafrost and shattered rocks are discussed in an earlier section.

There is a fairly good general relationship between the various features and slope as shown in Fig. 1.13. On the steep upper slopes there are tors and talus cones giving a type of eggcup topography. This changes into a less steeply sloping area with frost-shattered material and there may be many rock slides in such areas. Next comes a gentle slope covered with stone circles that form when the material is very stony. Where the slope increases again, the stone circles become lobes and then stripes of soil and vegetation occur. These change into steps on approaching the lowest part of the landscape where there is vegetation, primarily because of the accumulation of water over the permafrost which is quite close to the surface. On the very lowest part of the landscape there are pingos and tundra polygons with ice wedges. There may also be mud boils. This sequence shows that there is a general relationship between angle of slope, degree of stoniness, depth to permafrost, water conditions and vegetation. However, this is not the only possibility; for example, stone circles may occur in the lowest part of the landscape on stony materials, while mud boils may be present at quite high elevations where there is wet, loamy, soil material.

SOILS

Soils with permafrost occupy about one-fifth of the Earth's land surface and probably deserve more attention than they have received. A brief description of the most important types is given according to the *World Reference Base for Soil Resources* (WRB) (Spaaragen 1994), but adequate provision has not been made in this system for all of the soils, and there are inconsistencies. The approximate equivalent terms in the unpublished supplement to *Soil Taxonomy, Agricultural Handbook No. 436* (USDA 1975) are given in Table 1.1, while good accounts of the soils of cold areas are given by Fedoroff (1966), Tedrow (1977) and Rieger (1983). According to the WRB, most of the soils of the Arctic are Cryosols within which there are two main divisions – histic and thizotropic. The histic soils have histic horizons while the thizotropic soils show thizotropic properties.

A classification of the soils of the Arctic must take into consideration the many periglacial features which create situations in which horizons are not continuous for long distances as found in soils of many other areas. For example, an area with stone circles does not show one but at least two different soils: the soils beneath the circles and the soils within the circles. Similar situations arise with many other features such as mud boils, stripes and steps and, on a larger scale, with tundra polygons. The soil that forms the continuous polygonal pattern has an ice wedge as the underlying material while the soils away from the wedges have

TABLE 1.1. Names for soils of the Arctic according to the *World Reference Base for Soil Resources* (Spaaragen 1994) and their equivalents in the unpublished supplement to *Soil Taxonomy, Agricultural Handbook No. 436* (USDA 1975).

World Reference Base for Soil Resources	Soil Taxonomy
Andosols	Gelisols
	Andic Umbrihaplels
Cambisols	Gelisols
Gelic Cambisols	Typic Orthohaplels
Cryosols	Gelisols
Histic Cryosols	Spagnic Fibristels
	Typic Fibristels
Thixotropic Cryosols	Glacic Haploturbels
	Typic Haploturbels
Fluvisols	Gelisols
Dystric (Gelic) Fluvisols	Glacic Psammihaplels
Gleysols	Gelisols
Cryic Gleysols	Typic Aquaturbels
Histosols	Gelisols
Gelic Histosols	Spagnic Fibristels
	Typic Fibristels
Leptosols	Gelisols
Cryic Leptosols	Lithic Anhyturbels
	Lithic Haploturbels
Luvisols	Gelisols
Albic (Gelic) Luvisols	Typic Argihaplels
Phaeozems	Gelisols
Haplic (Gelic) Phaeozems	Typic Molliturbels
Podzols	Spodosols
Gelic Podzols	Pergelic Cryorthods
Regosols	Gelisols
Gelic Regosols	Calcic Anhyturbels
	Lithic Anhyturbels
	Typic Anhyturbels
Solonchaks	Gelisols
Haplic Solonchaks	Salic Anhyhaplels
Umbrisols	Gelisols
Gelic Umbrisol	Typic Umbriturbels
	Typic Umbrihaplels

permafrost that is composed largely of mineral material (see Fig. 1.14). So far no attempt has been made to produce a comprehensive system for these very complex situations.

The parent materials of arctic soils are very variable including different types of rock, glacial drift, alluvium, marine deposits, volcanic tephra and especially loesses of varying thickness. However, the concept of parent material as used for, say, a Podzol cannot be applied to the soils of the Arctic, since the lowest part of the soil is permafrost which for many workers is another horizon. Nearly all soils in the Arctic are shallow and seldom exceed 75 cm in thickness above the permafrost.

In this review an attempt is made to relate the plant communities and periglacial features to the various soils. However, the precise plant community will to a large extent be determined by the nutrient status (Haag 1974; Chapin & Shaver 1985). In particular the Andosols have special nutrient characteristics since they have a high capacity to fix phosphorus due to their high content of allophane.

Andosols

These soils are developed exclusively on volcanic tephra. They have a loose litter resting on a dark-coloured upper horizon, that grades into a fluffy brown middle horizon, and then into the underlying permafrost at about 75 cm. They are of minor significance in the Arctic and only occur in the upper part of the Kamchatka peninsula where they are common, and in Iceland just south of the Arctic Circle (Arnalds, Hallmark & Wilding 1995). However, many soils of the Arctic have varying amounts of volcanic material particularly in eastern Siberia and western Alaska.

Cambisols

Cambisols show a small to moderate degree of development. They are mainly absent from the higher parts of the Arctic but become progressively more common towards the south and occupy about 30% of the arctic landscape. *Gelic Cambisols* occur beneath grass–shrub, forest and tundra vegetation and are the most widespread type of Cambisol. They may have a very thin litter resting on a thin (5–10 cm) dark brown to grey, upper, ochric horizon containing fragments of undecomposed organic matter due to frost churning. This changes to a weakly weathered brown or yellowish-brown cambic horizon which may change sharply into permafrost at about 50 cm. Most of these soils have a lenticular structure in the horizons above the permafrost. This is formed by ice segregation during the winter and is largely maintained after thawing in the summer. The upper surfaces of the lenses have a thin coating of silt that has been translocated during thawing in the early part of the summer. These soils are generally acid with pH values of about 4.5 at the surface increasing to 6.5 in the lower horizon, but when these soils are developed on low ridges and are deep, there may be an accumulation of calcium carbonate (Courty *et al.* 1994). Although widespread, they are restricted to areas of relatively low rainfall and the drier, slightly elevated sites. In the southern part of the tundra subzone thin sections of the upper horizons of these soils show faecal material of enchytraeid worms and larvae, reflecting the longer and warmer summers. Here the upper horizon is a mixture of lenticular peds and spheroidal faecal material. Some Cambisols also have a single or multiple thin horizons at about 15–20 cm from the surface with well-formed clay coatings. If stones are present they will be vertically orientated on flat sites. The periglacial features associated with these soils include stone circles, stone polygons and stone stripes.

Cambisols are common in the tundra subzone and there are many intergrades to Fluvisols, Luvisols, Podzols and Regosols. When Cambisols are cultivated, vigilance must be exercised to prevent the permafrost from melting rapidly and creating thermokarst.

Cryosols

There are two units of Cryosols – Histic Cryosols and Thixotropic Cryosols. Since it is virtually impossible to separate Histic Cryosols from Gelic Histosols they are all included under the former heading.

Histic Cryosols (Gelic Histosols) are very common on the flat or gently sloping areas particularly in the tundra subzone, where they occupy about 5% of the landscape. Their presence is due largely to the permafrost which inhibits drainage, leading to a vegetation dominated by *Carex* spp., *Eriophorum* spp., and mosses, particularly *Sphagnum* spp. This, coupled with the low temperatures, means that organic matter from the plants is not rapidly decomposed and can easily accumulate to form Histosols. Where the accumulation is thick (> 1 m) the permafrost table is present within the Histosol itself. Thin sections show that the plant material is little decomposed due to the wet and cold conditions and it may be involuted due to the differential growth of ice. Many deposits have varying amounts of diatoms. Tundra polygons and palsas are usually associated with Histosols and sometimes thermokarst is present.

Thixotropic Cryosols are soils of the earth hummocks and mud boils. They may have a thin litter, followed by a brown, strongly frost churned horizon which changes state markedly during the year. The change is from hard and massive when frozen to almost fluid during the thaw, and then to hard when the soils dry during the summer. These changes are said to be thixotropic. Although these soils are wet and even saturated with water during the spring they do not show any mottling or reduction.

Fluvisols

Dystric (Gelic) Fluvisols are the soils of the alluvial areas. In the Arctic, the valleys tend to be very wide because they have braided streams that have aggraded over a long period of time and thus there is a relative abundance of alluvium with these soils occupying about 2% of the arctic landscape (see Fig. 1.12). In the early stages of development the profile consists of stratified material of varying texture with permafrost in the lower part. With time, some of the sites become vegetated and then there is a thin organic–mineral mixture at the surface overlying the unaltered clearly stratified alluvium. Often the vegetation is temporary and may be covered by a further deposition of alluvium or destroyed by erosion. If the vegetation continues to develop, the soil will evolve in one of a variety of directions depending upon the nature of the environmental conditions. Usually they develop into Cryic Gleysols because of their low topographic position. The soil surfaces will develop a number of patterns including stone circles and tundra polygons with their associated ice wedges.

Gleysols

Cryic Gleysols are very common soils in the Arctic particularly in the subpolar subzone due mainly to the presence of the impermeable permafrost. They form on flat, gently sloping sites and depressions under mosses, sedges, willows, black spruce and lodgepole pine, and occupy about 20% of the arctic landscape. They are common within tundra polygons and in areas of mud boils. At the surface

FIG. 1.14. Cryic Gleysol (Typic Aquaturbels) with involuted organic matter in the lower part of the active layer and top of the permafrost.

there is a thin (2–10cm) accumulation of organic matter resting on a grey or greyish-brown, weak to moderately mottled horizon (15–20cm) that grades down into a strongly mottled or blue-grey completely reduced horizon (15–20cm). Below is the blue-grey permafrost at 30–40cm or there may be an ice sheet or an ice wedge. There are some variations in the morphology of these soils. The very wet ones have very strong mottling up to the surface. A very common type has discrete, involuted areas of organic matter both above and within the upper part of the permafrost. This suggests frost churning at some time in the past since the permafrost table is now high enough to incorporate organic matter in the permafrost (Fig. 1.14). Thin sections show that the mineral horizons are fairly massive with frequent to abundant undecomposed plant fragments. These soils grade into Histosols (peat) as the middle horizon becomes wetter and the organic matter horizon at the top increases in thickness. They are acid, with pH values of about 4.5 at the surface increasing to about 6.5 and even > 7 at depth. Like most soils in the Arctic they show a low rate of organic matter decomposition with C/N ratio values of about 15.

Histosols

See *Histic Cryosols,* page 27.

Leptosols

Leptosols are soils dominated by angular frost-shattered rock fragments. In some cases it is possible to recognize fragments that can be fitted together as seen in Fig. 1.15. The whole soil is very open and porous allowing free drainage of the spring melt water, hence the absence of surface vegetation. These soils vary from being quite shallow (*Lithic Leptosols*) to over 1 m thick (*Cryic Leptosols*) and are underlain by permafrost that is often developed in the rock. There may be stone circles or polygons on flat sites and stripes on sloping sites. These soils occupy about 10% of the Arctic, and are dominant in the polar deserts where lichens are the main vegetation. They also occur in the subpolar desert subzone on moderate to steep slopes. Surface stones on flat or gently sloping sites may have desert varnish indicating a relatively long period of stability.

Luvisols

Albic (Gelic) Luvisols occur beneath forest vegetation in the southern continental part of the tundra subzone and occupy about 3–4% of the arctic landscape. They have litter at the surface resting on a thin (2–5cm) organic–mineral mixture with lenticular or granular structure, followed by a distinctive leached horizon (2–5cm) also with lenticular or granular structure. There is then a brown argic

FIG. 1.15. Cryic Leptosol (Lithic Haploturbel) 110 cm deep formed by continual freezing and thawing. Note that some of the fragments can be fitted together showing that they were once parts of larger boulders.

horizon (20–30 cm) with a small increase in the clay content with lenticular, prismatic or subcuboidal structure, beneath which is the permafrost.

It would appear that many of these soils do not have clay coatings in the argic horizon which is recognized mainly on the basis of the clay increase. There may be clay translocation but the deep winter freezing seems to prevent coatings from developing although there is not much frost churning. Some of these soils may have mottling above the argic horizon. In some situations solodic soils develop where there is high exchangeable sodium. Luvisols intergrade to Cambisols, Podzols and Regosols.

Phaeozems

Haplic (Gelic) Phaeozems have a restricted distribution and are developed under grassland in the southern part of the tundra subzone and cover < 1% of the arctic landscape. They have a deep (18–25 cm) mollic horizon (dark-coloured mixture of organic and mineral material) at the surface changing to slightly altered material and then into permafrost which is usually < 1 m from the surface. These soils need fairly stable conditions to develop; where there has been a change to cooler conditions some show frost churning with tonguing of the mollic horizon down to the permafrost.

Podzols

Gelic Podzols occupy about 5% of the arctic landscape and develop under both coniferous forest and heath vegetation. They have an accumulation of organic matter (5–10 cm) at the surface followed by a bleached horizon (2–5 cm) that changes to a spodic horizon of variable thickness with an accumulation of humus, iron and aluminium. At depths of over 1 m there is permafrost. Typical Podzols of this type occur mainly in the tundra subzone on coarse material or in shallow depressions (Ugolini 1966). In the subpolar desert subzone there are areas of shallow Podzols with permafrost within 1 m of the surface. The solum in these soils seldom exceeds 15 cm in depth but they do have all of the characteristic horizons of Podzols. Like many Cambisols they also have a well-developed lenticular structure with silt coatings. Thus the horizons in these shallow soils are good examples of compound horizons, namely, those that have processes of leaching and translocation during the summer period and the structural features of winter freezing. In some cases these shallow Podzols have been churned by frost and the horizons have been disrupted. This indicates the presence of a previous warmer period in which Podzols formed and now a return to colder conditions and frost churning. Podzols intergrade to Cambisols, Gleysols, Luvisols and Regosols.

Regosols

Gelic Regosols are weakly developed soils on loess, medium-textured materials and sand dunes. They occupy about 15% of the arctic landscape and are the dominant soils of the polar desert subzone where less than 50% of the surface is covered by vegetation. They have a weakly developed brown upper horizon (5–10 cm) with a small amount of organic matter that seems to have been derived from algae and lichens. This changes into unaltered material with permafrost occurring at about 50–100 cm. They form also where low rainfall and strong winds inhibit the development of vegetation. In such situations deflation results in the formation of stone pavements or hamada over a yellowish-brown horizon with permafrost beneath. The stone pavement may have desert varnish and carbonate pendents beneath stones. Regosols intergrade to Cambisols, Gleysols, Luvisols and Podzols.

Solonchaks

Haplic Solonchaks occur through the Arctic in many different situations and occupy about 1% of the arctic landscape. On dry knolls, water may evaporate leaving an efflorescence of salts; this can be enhanced when the rocks contain salts as in the Mackenzie area of Canada and in Spitsbergen. In the dry Yakutia region of eastern Siberia there is widespread accumulation of salts in the soils as well as in alases. In some situations solodic soils (high in exchangeable sodium) develop. Some of the hummocky (thufur) soils of the subpolar desert subzone develop efflorescences during the end of the summer period as the soil dries out.

Umbrisols

Gelic Umbrisols occur beneath grass–shrub forest and tundra vegetation. They have a thin litter and a thick (10–20 cm), dark umbric horizon that changes into a brown cambic horizon and then into permafrost. They are differentiated from the Greyzems by being much more acid. They require a long period to develop and occur only where frost churning is at a minimum; but there are places where they have been disrupted by frost action subsequent to their formation, indicating a fluctuation in the climate or a removal of the vegetation. The periglacial features and distribution are similar to the Gelic Cambisols.

Other soils occurring in the Arctic include Anthrosols, Arenosols, Glossisols, Planosols and Stagnosols but these are of relatively low frequency in the lower part of the Arctic and are continuous with similar soils without permafrost to the south.

The general relationships between some of these soils, slope, climate and parent material are shown in Figs 1.16 and 1.17. Figure 1.16 shows a topographic relationship between Leptosols, Cambisols, Gleysols and Histosols in an undulating landscape. On the crest, rock comes to the surface and there are shallow Leptosols, most often with stone circles. The middle and lower parts of the slope are covered with a soliflucted loessic material. Below the crest Cambisols have formed under the freely draining and aerobic conditions. They have a dark-coloured ochric horizon at the surface overlying a brown cambic horizon followed by the relatively unaltered material with permafrost. These Cambisols grade into Gleysols that have a very dark, upper, organic rich, umbric horizon. This rests on a strongly mottled horizon with lumps of organic matter that have been incorporated by frost churning. At the bottom of the slope there is peat with ice lenses and an ice wedge.

Figure 1.17 shows a latitudinal relationship between the soils of dry sites and climate. There are Regosols in the polar desert subzone grading to Cambisols of the subpolar desert subzone and then to Podzols in the tundra subzone. The Regosols are very shallow soils with rock containing permafrost. At the surface there is a

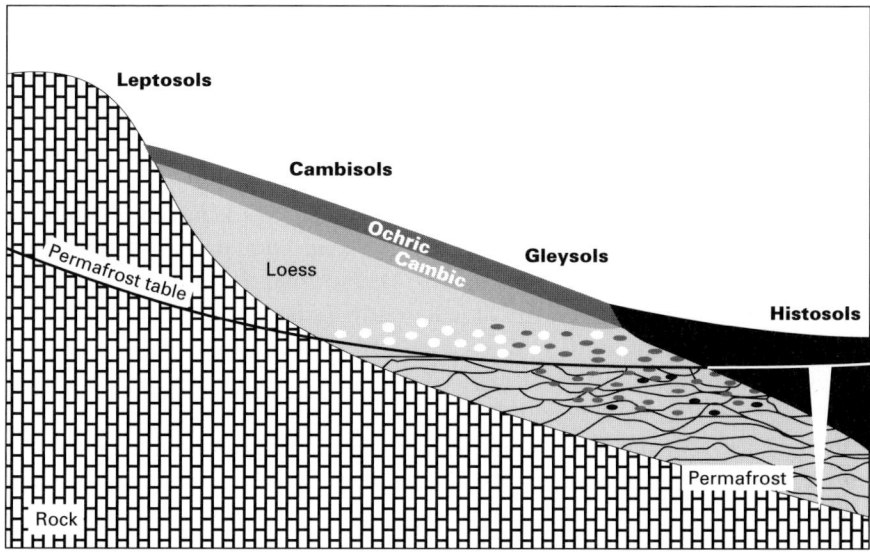

FIG. 1.16. A spatial relationship between Leptosols, Cambisols, Gleysols and Histosols in the tundra subzone.

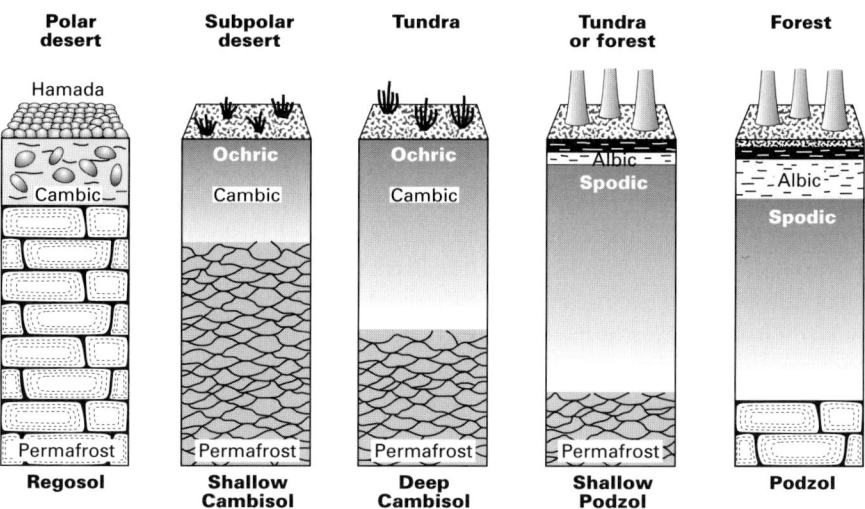

FIG. 1.17. An idealized soil sequence from the polar desert subzone to Podzols south of the tundra subzone.

stone pavement or hamada with desert varnish. In many situations the stones will
have pendents of calcium carbonate. The Cambisols are shallow in the subpolar
desert subzone but increase in thickness with deeper permafrost towards the south.
In the tundra subzone, where there is an increase in mean annual temperature as
well as precipitation, Podzols form and get progressively better developed in the
more southerly parts of the tundra subzone.

One of the unique features of arctic soils is their rapid change from one to
another due to small topographic changes that have been caused by periglacial
processes. Smith (1956) has described a 10-m transect through a number of mud
boils in Spitsbergen. The permafrost is deeper beneath the bare muddy area but
there is little difference in the details of the soil profile which is a Gelic Gleysol
throughout. Douglas and Tedrow (1960) have described a transect from northern
Alaska to show the distribution of the involuted organic matter in a Gelic Gleysol
(see Fig. 1.14). Bunting and Fedoroff (1974) have described a complex transect
through the raised margins of an ice wedge in the polar desert subzone of northern
Canada. The highest part near to the wedge has an efflorescence of gypsum on the
surface overlying an organic–mineral mixture. This is followed by a brown horizon
and then a grey, frost churned horizon containing lumps of peat resting on buried
peat with permafrost. This profile grades downslope through a distance of about
2 m into a Humic Gleysol. Even more complex transects are described for Vaygach
Island off the coast of northern Russia by Ignachenko (1967).

CONCLUSIONS

It should be clear from the above that the Arctic is a unique and complex area
with a temperature range of > 100°C between summer and winter, and a range of
precipitation of < 50 to > 1200 mm year^{-1} between the polar desert subzone and
some oceanic sites. These ranges in climatic conditions have led to the development
of the unique range of periglacial features. In addition, it is a very fragile environment
since it is virtually impossible in the short term to re-establish vegetation on the
tundra if it is destroyed by track vehicles or pollution by oil spillage. The widespread
formation of thermokarst following deforestation in the tundra subzone is clear
evidence for its fragility.

It should be evident that the most important processes for geomorphology
and soil formation in the Arctic are freezing and thawing. When accompanied by
varying amounts of water they lead to the wide range of periglacial features that
are encountered. The very low temperatures are important since they are especially
necessary for the formation of frost shattering, ice wedges and desert varnish which
is also quite common in Antarctica where temperatures are likewise very low.

Mass movement on slopes is not unique to the Arctic but when accompanied
by freeze–thaw processes the speed is much greater. This process dominates all of

the slopes, thus the whole landscape is constantly on the move – all of the material on the slopes is steadily moving into the valleys giving the Arctic a special status in this respect.

Probably permafrost is the unique horizon in the soils of the Arctic. It acts as an impermeable barrier to the vertical movement of water causing a high degree of wetness in the soils so as to produce the largest area of Gleysols in the world. The restriction of water movement also leads to other features such as stone circles and mud polygons, both with their laterally discontinuous horizons. Apart from Vertisols in the tropics no other area of soils has such a variety of irregularities and discontinuities.

The historical approach to the study of periglacial features has largely been through observations in the field, but a few workers such as Kokkonen (1927), Taber (1929, 1930) and Corte (1962, 1966) attempted some experimental work. Serious monitoring of the rate of formation and change in periglacial features was conducted by Washburn (1967) in Greenland. The tempo of research in these areas has increased significantly in recent years: for example, monitoring the changes in palsa peatland in northern Québec by Laberge and Payette (1996). Also elaborate laboratory experiments have been designed to investigate the formation of involutions (Pissart 1982) and to measure the rate of solifluction (Harris *et al.* 1993, 1995, 1996). A recent innovation has been the introduction of isotope studies, as in the investigation of the rate of carbonate accumulation in Spitsbergen (Courty *et al.* 1994) and the accumulation of ice in ice wedges (Lauriol *et al.* 1995). These new developments mean that progress is being made with regard to determining the mechanism and rate of formation of many of the features.

There is a great need to study the Arctic for its own sake as well as to obtain an insight into the conditions that prevailed many hundreds of kilometres to the south during the maximum of the glaciations. There is scarcely a soil in the periglacial zone of the last glaciation that does not show clear evidence of periglacial processes. Former ice wedges, pingos, solifluction deposits and permafrost are everywhere to be seen; for example in north-east Scotland the majority of soils are developed in solifluction deposits that were associated with permafrost. A similar situation occurs in a broad swathe from northern France to eastern Siberia and northern China, and across central North America. Also within these areas, the presence of silt cappings on the surfaces of stones and boulders in the subsoil is unequivocal evidence of relic permafrost (FitzPatrick 1956, 1971; Van Vleit-Lanoë 1985; Paton 1992).

A very significant contribution has been made by Haesaerts and Van Vleit-Lanoë (1981); they state that features can be used as climatic indicators which, coupled with sedimentological and pedological events, give climatic curves that are almost as accurate as those given by pollen analysis. There are not many soil situations that are as dynamic as these. The nearest are Vertisol landscapes but they do not seem to be as complicated.

REFERENCES

Åkerman, J. (1987). Periglacial forms of Svalbard: a review. *Periglacial Processes and Landforms in Britain and Ireland* (Ed. by J. Bordman), pp. 9–25. Cambridge University Press, Cambridge.

Arnalds, O., Hallmark, C.T. & Wilding, L.P. (1995). Andisols from four different regions of Iceland. *Soil Science Society of America Journal*, **59**, 161–169.

Baranov, I.J. & Kudfryavtsev, V.A. (1963). Permafrost in Eurasia. *Proceedings of the Permafrost International Conference Indiana (Lafayette, 1963)*. National Academy of Sciences, National Research Council Publication, **1287**, 98–102.

Baulig, H. (1956). Pénéplaines and pédiplanes. *Societe Belge d'Etude Geographiques Bulletin*, **25**, 25–58.

Benedict, J.B. (1970). Downslope soil movement in a Colorado alpine region: rates, processes and climatic significance. *Arctic and Alpine Research*, **2**, 165–226.

Black, R.F. (1952). Growth of ice-wedge polygons in permafrost near Barrow, Alaska. *Geological Society of America Bulletin*, **63**, 1235–1236 (abstract).

Brown, J., Grant, C.L., Ugolini, F.C. & Tedrow, J.C.F. (1962). Mineral composition of some drainage waters from arctic Alaska. *Journal of Geophysical Research*, **67**, 2447–2453.

Bunting, B.T. & Fedorof, N. (1974). Micromorphological aspects of soil development in the Canadian High Arctic. *Soil Microscopy* (Ed. by G.K. Rutherford), pp. 350–365. The Limestone Press, Kingston, Ontario, Canada.

Chapin, F.S. III & Shaver, G.R. (1985). Individualistic growth responses of tundra plant species to environmental manipulation in the field. *Ecology*, **66**, 564–576.

Corte, A. (1962). Vertical migration of particles in front of a moving freezing plane. *Journal of Geophysical Research*, **67**, 1085–1090.

Corte, A. (1966). Particle sorting by repeated freezing and thawing. *Biuletyn Peryglacjalny*, **15**, 175–240.

Courty, M.-A., Marlin, C., Dever, L., Tremblay, P. & Vachier, P. (1994). The properties, genesis and environmental significance of calcite pendents from the High Arctic (Spitsbergen). *Geoderma*, **61**, 71–102.

Czeppe, Z. (1960). Thermic differentiation of the active layer and its influence upon frost heave in periglacial regions (Spitsbergen). *Academie Polonaise Sciences Bulletin Series Science. Géographie*, **8**, 149–152.

Czudek, T. (1995). Cryoplanation terraces – a brief review of some remarks. *Geografiska Annaler Series A – Physical Geography*, **77A**, 95–105.

Demek, J. (1969). Cryogene processes and the development of cryoplanation terraces. *Biuletyn Peryglacjalny*, **18**, 115–125.

Douglas, L.A. & Tedrow, J.C.F. (1960). Tundra soils of Arctic Alaska. *Transactions of the 7th International Congress of Soil Science*, **iv**, 291–304.

Embleton, C. & King, C.A.M. (1975). *Periglacial Geomorphology*. Edward Arnold, London.

Everett, D.H. (1961). The thermodynamics of frost damage to porous solids. *Transactions of the Faraday Society*, **57**, 1541–1551.

Everett, K.R. (1965). Slope movement and related phenomena. *Environment of the Cape Thompson Region, Alaska* (Ed. by N.J. Wilimovsky), pp. 175–220. US Atomic Energy Commission, Division of Technical Information.

Fedoroff, N. (1966). Les sols du Spitsberg occidental. Spitsberg, 1964. *Centre National de la Recherches Scientifique Recherches Cooperation Programme 42 (Lyon)*, pp. 111–228.

Filstrup, B. (1953). Wind erosion in the Arctic. *Geografisk Tidsskrift*, **52**, 51–65.

FitzPatrick, E.A. (1956). An indurated soil horizon formed by permafrost. *Journal of Soil Science*, **7**, 248–254.

FitzPatrick, E.A. (1971). *Pedology: A Systematic Approach to Soil Science*. Oliver & Boyd, Edinburgh.

FitzPatrick, E.A. (1983). *Soils, their Formation, Classification and Distribution*, Longman. London.

FitzPatrick, E.A. (1993). *Soil Microscopy and Micromorphology*, John Wiley, Chichester.

French, H.M. (1986). Periglacial involutions and mass displacement structures, Banks Island, Canada. *Geografiska Annaler*, Series A, **68A**, 167–174.

French, H.M. (1996). *The Periglacial Environment*, 2nd edn. Longman, London.

Fukuda, M. (1983). The pore water pressure profile in porous rocks during freezing. *Proceedings of the 4th International Conference on Permafrost*, pp. 322–327. National Academy Press, Washington, DC.

Geikie, J. (1894). *The Great Ice Age and its Relation to the Antiquity of Man*. Edward Stanford, London.

Giblin, A.E., Nadelhoffer, K.J., Shaver, G.R., Laundre, J.A. & MacKerrow, A.J. (1991). Biochemical diversity along a riverside toposequence in Arctic Alaska. *Ecological Monographs*, **61**, 415–435.

Haag, R.W. (1974). Nutrient limitations to plant productivity in two tundra communities. *Journal of Botany*, **52**, 103–116.

Haesaerts, P. & Van Vliet-Lanoë, B. (1981). Phénomènes periglaciares et sols fossiles observés à Maisières-Canal, à Harmignies et à Rocourt. *Biuletyn Peryglacjalny*, **28**, 291–325.

Hallet, B. (1983). The breakdown of rocks due to freezing: a theoretical model. *Proceedings of the 4th International Conference on Permafrost*, pp. 433–438. National Academy Press, Washington, DC.

Hallet, B., Walder, J.S. & Stubbs, C.W. (1991). Weathering by segregation ice growth in microcracks at sustained subzero temperatures: verification from experimental study using acoustic emissions. *Permafrost and Periglacial Processes*, **2**, 283–300.

Harris, C. (1981). *Periglacial Mass-wasting: A Review of Research*. British Geomorphological Research Group, Research Monograph Series 4, 204 pp. Geo Abstracts, Norwich.

Harris, C., Davies, M.C.R. & Coutard, J. (1995). Laboratory simulation of periglacial solifluction – significance of porewater pressures, moisture contents and undrained shear strengths during soil thawing. *Permafrost and Periglacial Processes*, **6**, 293–311.

Harris, C., Davies, M.C.R. & Coutard, J.P. (1996). An experimental-design for laboratory simulation of periglacial solifluction processes. *Earth Surface Processes and Landforms*, **21**, 67–75.

Harris, C., Gallop, M. & Coutard, J.P. (1993). Physical modelling of gelifluction and frost creep – some results of a large-scale laboratory experiment. *Earth Surface Processes and Landforms*, **18**, 383–398.

Högbom, B. (1914). Über die geologische bedeutung des frostes. *Bulletin of the Geological Institute, University of Uppsala*, **12**, 257–389.

Ignachenko, V.I. (1967). Soil complexes of Vaygach Island. *Soviet Soil Science*, 1216–1229.

Jahn, A. (1960). Some remarks on evolution of slopes on Spitsbergen. *Zeitschrift fur Geomorphologie*, Supplement 1, 49–58.

Kachurin, S.P. (1962). Thermokarst within the territory of the U.S.S.R. *Biuletyn Peryglacjalny*, **11**, 49–55.

Kokkonen, P. (1926). Boebachtungen über die struktur des bodenfrostes. *Acta Forestalia Fennica*, **30**, 1–55.

Laberge, M.J. & Payettes, S. (1996). Long term monitoring of permafrost change in palsa peat lands in Northern Quebec, Canada, 1983–1993. *Alpine and Arctic Research*, **28**, 169.

Lachenbruch, A.H. (1962). *Mechanism of Thermal Contraction Cracks and Ice-wedge-polygons in Permafrost*. Geological Society of America Special Paper No. 70, 69 pp.

Lauriol, B., Ducheshe, C. & Clark, I.D. (1995). Systematic water infiltration in ice wedges – results of an O-18 and deuterium study. *Permafrost and Periglacial Processes*, **6**, 47–55.

Lautridou, J.P. (1982). La fraction fine des débris de gélifraction expérimentale. *Biuletyn Peryglacjalny*, **29**, 78–85.

Lautridou, J.P. & Ozouf, J.C. (1982). Experimental frost shattering: 25 years of research at the Centre de Géomorphologie du CNRS. *Progress in Physical Geography*, **6**, 215–232.

Leffingwell, E. de K. (1919). The Canning River region, northern Alaska. *United States Geological Survey Professional Papers*, **109**, 1–251.

Lozinszi, W. (1912). Die periglazaile fazies der mechanischen verwitterung. *Transactions of the Eleventh International Geological Congress*, **2**, 1039–1053.

Lundström, U.S. (1994). Significance of organic acids for weathering and the podzolization process. *Environment International*, **20**, 21–30.

McGreevy, J.P. (1981). Some perspectives on frost shattering. *Progress in Physical Geography*, **5**, 56–75.

Mackay, J. (1979). Pingos of Tuktoyaktuk peninsula area, Northwest Territories. *Géographie Physique et Quaternaire*, **33**, 3–61.

Mackay, J.R. (1993a). The sound and speed of ice-wedge cracking, Arctic Canada. *Canadian Journal of Earth Sciences*, **30**, 509–518.

Mackay, J.R. (1993b). Air temperature, snow cover, creep of frozen ground, and time of ice wedge cracking, western Arctic coast. *Canadian Journal of Earth Sciences*, **30**, 1720–1729.

Mackay, J.R. (1995). Ice wedges on hillspoles and landform elevation in the late Quaternary, western Arctic Coast, Canada, *Canadian Journal of Earth Sciences*, **32**, 1093–1105.

McKeague, J.A., MacDougall, J.I., & Miles, N.M. (1973). Micromorphological, physical, chemical and mineralogical properties of a catena of soils from Prince Edward Island in relation to their classification and genesis. *Canadian Journal of Soil Science*, **53**, 281–295.

Morozova, T.D. (1965). Micromorphological characteristics of pale yellow permafrost soils in Central Yakoutia in relation to cryogenesis. *Soviet Soil Science*, 1333–1342.

Müller, S.W. (1947). *Permafrost or Perennially Frozen Ground and Related Engineering Problems*. Edwards Brothers, Ann Arbor, Mich.

Nikiforoff, C.C. (1928). The perpetually frozen subsoil of Siberia. *Soil Science*, **26**, 61–81.

Parfenova, Y.I. & Yarilova, E.A. (1965). *Mineralogical Investigations in Soil Science*. Israel Program for Scientific Translation, Jerusalem.

Paton, R.W. (1992). Fragipan formation in argillic brown earths (Fragiudalfs) of the Milfield Plain, north-east England. I. Evidence for a periglacial stage of development. *Journal of Soil Science*, **43**, 621–644.

Penner, E. (1974). The mechanism of frost heaving in soils. *Highway Research Board Bulletin*, **225**, 1–22.

Péwé, J.L. (1959). Sand-wedge polygons (teselations) in the McMurdo Sound Region, Antarctica – a progress report. *American Journal of Science*, **257**, 545–552.

Pissart, A. (1982). Déformations de cylindres de limon entourés de graviers sous l'action d' alternaces gel-dégel. Experiences sur l'origine des cryoturbations. *Biuletyn Peryglacjalny*, **26**, 275–285.

Rieger, S. (1983). *The Genesis and Classification of Cold Soils*. Academic Press, London.

Schnitzer, M. & Vendette, E. (1975). Chemistry of humic substances extracted from an Arctic soil. *Canadian Journal of Soil Science*, **55**, 93–103.

Smith, J. (1956). Some moving soils in Spitsbergen. *Journal of Soil Science*, **7**, 10–21.

Sokolov, I.A. & Belousova, N.I. (1964). Organic matter in Kamchatka soils and certain problems of illuvial-humic soil formation. *Soviet Soil Science*, 1026–1035.

Sokolov, I.A., Naumov, Y.M., Gradusov, B.P., Tursina, T.V. & Tsyurupa, V.V. (1976). Ultra continental taiga soil formation on calcareous loams in Central Yakutia. *Soviet Soil Science*, 144–160.

Solov'ev, P.A. (1969). The Lena–Amga watershed; thermokarst phenomena and forms associated with cryogenous frost heaving. *Guide to a Trip Round Central Yakutia: Paleography and Periglacial Phenomena* (Ed. by E.M. Katasonov & P.A. Solov'ev), pp. 29–36. Soviet Academy of Science, Yakutsk. (In English)

Spaaragen, O.C. (Ed.) (1994). *World Reference Base for Soil Resources*. International Soil Reference and Information Centre, Wageningen.

Taber, S. (1929). Frost heaving. *Journal of Geology*, **37**, 428–461.

Taber, S. (1930). The mechanics of frost heaving. *Journal of Geology*, **38**, 303–317.

Taber, S. (1943). Perennially frozen ground in Alaska: its origin and history. *Geological Survey of America Bulletin*, **54**, 1433–1548.

Tedrow, J.C.F. (1968). Pedogenic gradients of the polar regions. *Journal of Soil Science*, **19**, 197–204.

Tedrow, J.C.F. (1977). *Soils of the Polar Landscapes*. Rutgers University Press, New Brunswick, NJ.

Thompson, E.C. & Sayles, F.H. (1972). *In situ* creep analysis of room in frozen soil. *ASCE Journal of Soil Mechanics and Foundation Division*, **98**, 899–916.

Ugolini, F.C. (1966). Soils of the Mesters Vig district, Northeast Greenland: I. The Arctic Brown and related soils. *Meddelelser om Grønland*, **165**, 196 pp.

Ugolini, F.C., Stoner, M.G. & Marrett, D.J. (1987). Arctic pedogenesis: 1. Evidence for contemporary podzolization. *Soil Science*, **144**, 90–100.

Ugolini, F.C. & Walters, J. (1973). Pedological survey of the Noatak River Basin. *The Environment of the Noatak River Basin, Alaska* (Ed. by S.B. Young), pp. 86–156 and pp. 526–580. Contribution from the Centre for Northern Studies No. 1.

USDA (1975). *Soil Taxonomy, Agricultural Handbook No. 436*. US Government Printing Office, Washington, DC.

Van Vliet-Lanoë, B. (1985). Frost effects in soils. *Soils and Quaternary Landscape Evolution* (Ed. by J. Boadman), pp. 117–158. Wiley, Chichester.

Van Vliet-Lanoë, B., Coutard, J.P. & Pissart, A. (1984). Structures caused by repeated freezing and thawing in various loamy sediments. A comparison of active fossil and experimental data. *Earth Surface Processes and Landforms*, **9**, 535–565.

Washburn, A.L. (1956). Classification of patterned ground and review of suggested origins. *Geological Society of America Bulletin*, **67**, 823–866.

Washburn, A.L. (1967). Instrumental observations of mass-wasting in Mesters Vig district, Northeast Greenland. *Meddelelser om Grønland*, **166**, 318 pp.

Washburn, A.L. (1973). *Periglacial Processes and Environments*. Edward Arnold, London.

2. Microbial ecology, decomposition and nutrient cycling

CLARE H. ROBINSON* AND PHILLIP A. WOOKEY†

*Sheffield Centre for Arctic Ecology, Department of Animal and Plant Sciences,
The University, 26 Taptonville Road, Sheffield S10 5BR, UK and
†Department of Biological Sciences, The University of Exeter,
Exeter EX4 4PS, UK

INTRODUCTION

Environmental and biological characteristics of arctic soils are not clearly defined; gradients exist from dry to water-logged soils, from pH values of 3.5 to 8.0 in surface soils, from mineral to peat soils, from 100% vegetation cover to less than 5%, with meadow, tussock, dwarf-shrub, sparse woodland and taiga vegetation (after Heal 1981). However, pedogenesis in arctic environments has usually occurred 'under some of the worst climatic conditions in the world' (Cragg 1981; Table 2.1), and soils often are permanently frozen with only a shallow (30–75 cm) thaw layer (Bliss 1981). These widely different tundra soils together contain approximately 11% of the world's soil carbon pool (Melillo *et al.* 1990), and hold 95% of the organically bound nutrients in the tundra ecosystem (Jonasson 1983). Owing to slow decomposition rates in cold, wet or dry soil environments, plant growth is often limited by nutrient availability (e.g. Haag 1974; Chapin & Shaver 1985; Wookey *et al.* 1993, 1994, 1995; Parsons *et al.* 1995). Since saprotrophic fungi are primary decomposers of organic matter, and mycorrhizae of polar dwarf shrubs facilitate plant nutrient uptake, fungal activity is a major influence on the nutrition of (arctic) plants through uptake, immobilization (S. Jonasson *et al.* unpublished data) and mineralization of nutrients (Frankland, Dighton & Boddy 1990). In wet tundra, bacteria may play a more important role in the availability of plant nutrients, especially as the *Carex* and *Eriophorum* spp., which are often predominant in such areas, are non-mycorrhizal (Tester, Smith & Smith 1987). In all ecosystems, some processes of considerable ecological significance (e.g. nitrification, denitrification, methanogenesis and methanotrophy) are performed predominantly by soil bacteria.

Within the past three decades, studies of the role of micro-organisms in decomposition and nutrient cycling processes have taken two almost completely divergent

*Present address: Division of Life Sciences, King's College, University of London, Campden Hill Road, Kensington, London W8 7AH, UK.
†Present address: Department of Geography, Royal Holloway College, University of London, Egham, Surrey TW20 0EX, UK.

TABLE 2.1. Soil characteristics at selected Arctic sites.

Location	Habitat	Lat./Long.	Elevation (m)	Parent material	Mean ann. temp. (°C)	Air temp. >0°C	Soil temp. at −5 cm >0°C	Mean ann. soil temp. (°C)	Mean ann. pptn (mm)	Depth organic matter (cm)	pH	Moisture (% dry wt)
Ny-Ålesund, Svalbard	Polar semi-desert	78°56′N 11°50′E	22	Carboniferous limestone	−6.0				371	0–50	6.9–8.0	47.0 ± 12.3
Barrow, Alaska	Wet meadow	71°18′N 156°40′W	4	Marine sediment	−12.5	276	330	−5	124	0.5	5.3	
Abisko, Sweden	Dwarf-shrub heath	68°21′N 18°49′E	400		−0.8	1267		1.9	304	0–40	4.0–4.5	153.8 ± 29.8
Fairbanks, Alaska	Spruce forest	64°45′N 148°15′W	167–470	Loess	−3.3		654		300	>90		351 ± 177
Moor House, England	*Calluna* bog	55°8′N 2°45′W	550–575		5.4	1097	1963	5.3	1981	0–37+	4.9	Est. 400/100 max./min.

Mean annual air temperature and precipitation values for Ny-Ålesund and Abisko were provided by the Norwegian and Swedish Meteorological Institutes, respectively. The soil characteristics of these two sites are taken from Robinson *et al.* (1995). Accumulated air temperature and mean annual soil temperature at the Abisko site were kindly provided by Mr M.E. Tjus of the Abisko Naturvetenskapliga Station. Other data are abstracted from Van Cleve and Alexander (1981).

paths. According to Swift (1985), microbial ecologists (particularly mycologists) have been concerned in the past to demonstrate the diversity of decomposer species and their patterns of distribution and abundance (community structure). An alternative approach has been orientated toward the description of broader soil processes (equivalent to microbial community function) and has paid little attention to roles of individual species or to the variation between them. This approach has met with a degree of success in increasing understanding of the patterns of organic decay and of mineral nutrient dynamics in soil. In the latter approach the decomposer organisms are pictured as 'driving variables' (Swift, Heal & Anderson 1979), and at most are divided into a small number of trophic groups. This chapter is therefore divided into two parts: (i) microbial ecology, mainly relating to microbial community structure; and (ii) decomposition and nutrient cycling (the broad function of the microbial community) in arctic environments, reflecting the nature of the results collected to date. Finally, we discuss how microbial community structure and function could be linked in the context of future research in arctic ecosystems.

MICROBIAL ECOLOGY

A great proportion of the work carried out on the microbial ecology of arctic environments was performed during the 'Tundra Biome' study of the International Biological Programme (IBP; see Holding *et al.* 1974; Bliss, Heal & Moore 1981). Little work has been carried out on microbial ecology in the Arctic since this programme, perhaps due to the highly selective methods of detecting fungi and bacteria in soil and litter and the difficulty in drawing general conclusions concerning the ecological roles of these organisms. However, recent international interest in environmental change in arctic environments (Chapin *et al.* 1992), together with biodiversity (Chapin & Körner, 1995) plus the development of new techniques, for example, characterization of the phospholipid fatty acids in the microbial community (Frostegård, Bååth & Tunlid 1993) and extraction and analysis of microbial DNA (e.g. Gill *et al.* 1994), should lead to a rekindled interest in arctic microbial ecology.

Microbial community structure

Fungal species spectra and biomass in soil and plant litter

The fungal flora of polar regions is reported to consist of cold-tolerant isolates of cosmopolitan species which have been selected or become adapted to low temperatures (1–5°C; Latter & Heal 1977; Widden & Parkinson 1978; Kerry 1990; Finotti *et al.* 1993). From the limited data available it cannot be stated that certain genera or species of fungi are restricted to tundra areas (Holding 1981). Compared with the extensive literature from non-tundra sites, the genera *Chrysosporium* and *Scytalidium* seemed to occur more frequently on plant litter collected from tundra, whilst *Aureobasidium, Trichoderma* and *Mortierella* were less common on this substratum (Holding 1981). *Penicillia*, particularly *P. decumbens* and *P. frequentans*, together with pink yeasts and *Phoma exigua*, were isolated frequently from litter of five plant species from Stordalen (Dowding & Widden 1974) in Swedish Lapland. Also at Abisko, *Penicillium brevicompactum, Mucor hiemalis* and *Truncatella truncata* were frequently isolated (isolation frequency > 5%) from soil-inoculated *Vaccinium* litter collected from a subarctic dwarf-shrub heath which was incubated at 15°C (Gehrke *et al.* 1995).

In soils, rather than litter, of about 100 genera of fungi isolated in 33 tundra areas from seven regions of North America, Antarctica and the British Isles, only the species *Chrysosporium* and sterile mycelia were observed in all seven regions, and the genera *Cladosporium, Mortierella* and *Penicillium* in six regions (Dowding & Widden 1974). Sterile mycelia were also predominant in fungi isolated from a range of soils in northern Russia (Syzova & Panikov 1995). The above groups are common in temperate soils, but in such soils sterile mycelia tend to make up a smaller proportion of the mycoflora. Conversely, *Trichoderma* spp. are common in temperate situations, but are found infrequently in most tundra areas (Dowding & Widden 1974). The frequency of *Chrysosporium pannorum* was greater at lower temperatures on a mixture of coniferous and deciduous leaf litter inoculated with a soil suspension and incubated over a 45-month period at 10, 4 and 1°C. Ivarson (1973) therefore believed cold temperature to have a selective effect on this species. In a different area of research, the study of spoilage micro-organisms on chilled and frozen food, Gunderson (cited in Farrell & Rose 1967) found that *Aureobasidium pullulans* was the most frequently isolated species, and isolates from 21 genera could grow at 0–5°C. Dowding and Widden (1974) used multiple regression analysis of soil fungal counts on up to 11 soil and climatic variables, but no clear results were obtained, although temperature, moisture and pH appeared to be consistently important, while soil N and frost-free days were not.

At Point Barrow, Alaska, basidiomycete ectomycorrhizal associations with *Dryas, Salix, Vaccinium* and *Cassiope* were commonly formed by species of *Hebeloma, Cortinarius, Laccaria, Russula* and *Lactarius* (Miller & Laursen 1974). In situations where few or no ectomycorrhizal associations occurred, species of *Coprinus, Galerina* and *Hypholoma* were the common saprotrophic basidiomycetes (Miller & Laursen 1974). Water moulds of the genera *Achlya, Saprolegnia* and *Pythium* occurred in troughs in sites with polygonal ridges (Flanagan & Scarborough 1974).

Total fungal biomass is a measure of absolute fungal presence, and has been widely used in soil perturbation studies (e.g. Bardgett 1991) and site comparisons (review by Kjøller & Struwe 1982). The length of fungal mycelium in 21 tundra sites was presented by Dowding and Widden (1974), and ranged from a mean of 4 mg^{-1} in a new moraine at Signy Island, Antarctica, to 7000 mg^{-1} in a birch wood at Hardangervidda, Norway. In temperate deciduous forest, estimates of

1900–4432 mg^{-1} were given for the litter layer (Kjøller & Struwe 1982). Fungal hyphal length cannot only differ extremely between sites, but values from individual samples can be over a 100-fold different within the same soil horizon at the same site (see Kjøller & Struwe 1982). Widden and Parkinson (1979) found that fungal biomass in raised beach soils at Truelove Lowland, Devon Island, Canada, was positively correlated with presence of *Dryas integrifolia*, and at 10 cm horizontal distance from the cushions there was a noticeable drop in fungal biomass. This could reflect the ectomycorrhizal nature of this plant species and/or a concentration of saprotrophs in the organic matter developed under the vegetation, but it is clear that any detailed studies of the microbiology of polar desert soils, or soils under tundra cushion plant communities, must consider the patchiness of the higher plant community (Widden & Parkinson 1979). Total mycelial length has been correlated with 11 soil and climatic parameters, and the 'best' regression included pH, minimum soil moisture and maximum temperature (i.e. fungal hyphal length increased with increasing soil moisture and temperature, and decreasing soil pH), but though significant ($P = 0.05$), it only accounted for 46% of the variance (Dowding & Widden 1974). Lengths of fungal mycelia were observed at seven summer sampling times in five plots at Point Barrow (Miller & Laursen 1974) and the greatest lengths around 20 June and 11 August appeared to be related to high soil moisture.

To facilitate comparisons between different groups of micro-organisms, it is advantageous to convert hyphal lengths to biomass. This is carried out by converting the hyphal volume to dry weight per gram dry weight of soil or per square metre; the problems associated with this conversion have been discussed by Frankland *et al.* (1990). Values for tundra range from 0.1 g m^{-2} (polar semi-desert, 0–3 cm depth; Robinson *et al.* 1996) to 22.2 g m^{-2} (a hummock at Stordalen mire, 0–5 cm depth; Hayes 1973, cited in Kjøller & Struwe 1982) and 416 g m^{-2} (ombrotrophic bog, Tver region of northern Russia; Syzova & Panikov 1995).

In summary, the fungal species composition in arctic environments is very similar to those in temperate latitudes, although the isolates are often adapted to low temperatures. Fungal biomass may be lower than in temperate ecosystems in extreme soils, for example polar semi-desert lithosols, and is markedly seasonal. Thus, fungal community structure on a broad scale has been characterized to some degree in a range of arctic soils, given the selective nature of isolation, the problems associated with measurement of fungal biomass (Frankland *et al.* 1990; Stahl, Parkin & Eash 1995) and the very heterogeneous nature of tundra soil environments. However, saprotrophic fungal community structure has been elucidated to a lesser degree than in temperate ecosystems. To give three examples, no detailed information has been gained on the interactions within and between species *sensu* Rayner and Webber (1984), the mechanisms of resource capture, or patterns of resource occupation at the fine scale. The species lists and biomass values available are almost exclusively products of the research of the Tundra Biome group of the IBP, and as such, are from the period 1964 until 1974 (Bliss *et al.* 1981).

Fungal function is related to fungal isolate

Fungi were isolated from *Eriophorum* and *Carex* spp. from Point Barrow and other Alaskan sites and used in experiments to assess their physiological characteristics (Flanagan & Scarborough 1974) – a fascinating attempt to relate fungal community structure to function at a microscale. Strains of *Cladosporium herbarum*, which appeared to be unable to survive the winter on vegetation, were capable of decomposing starch, gallic and humic acids (the latter two were considered indicative of lignin decomposition by these authors) at both 2–5°C and at approximately 25°C. By contrast, a wide range (80–90%) of fungi from five sources decomposed starch and pectin at 25°C, but about 50% of these organisms, especially the sterile mycelia, lost this activity at 2–5°C. The data of Flanagan and Scarborough (1974) suggest that sterile mycelia are particularly important in the Alaskan soils, but their potential physiological ability varies considerably. Sterile mycelia B2 and B18 could decompose a wide range of substrates, whilst strain S11 was able to decompose starch at 25°C only (Table 2.2). Compared with fungal strains from Alaska, Flanagan and Scarborough (1974) observed that very few strains of fungi representative of Devon Island soils decomposed cellulose, gallic or humic acids, but pectinolytic and amylolytic strains were common. It is worthwhile remembering, however, that the ability to decompose these substrates *in vitro* is not necessarily related to the abilities of the organism to decompose plant tissues (Swift 1976).

The psychrophilic nature of the cellulose decomposition process was also investigated by Flanagan and Scarborough (1974). Under laboratory conditions, decomposition commenced at −7°C and 20% of the strains tested showed an optimum decomposition rate at approximately 6°C. Pectin decomposition was shown to be initiated at about 1°C and reached an optimum rate at 18°C or higher. The psychrophilic properties of cellulase activity correspond with the description of psychrophilism proposed by Deverall (1968), who suggested that a suspected psychrophile should be shown to have a growth optimum near 10°C or below. It is clearly important for psychrophilic micro-organisms not to experience temperatures above 10°C during sampling, isolation and storage if their activity is to be maintained. Although experimental data were not presented, Flanagan and Scarborough (1974) stated that the optimum pH for cellulose decomposition by 60% of the Alaskan strains was 4.5–5.0, while pH 6.0 was the optimum for pectin and starch decomposition. These authors considered that, in general, fungi from the Alaskan sites increased their rate of litter decomposition up to moisture/substrate levels of 300–400% (dry wt). A few sterile mycelia and certain aquatic phycomycetes grew at moisture levels in excess of 2000% (dry wt). A complication to understanding

TABLE 2.2. Percentage occurrence of fungi on vegetation and in soil at Point Barrow in relation to substrate decomposition at ~25°C and 2–5°C.*

Fungus	*Eriophorum* spp.			*Carex* spp.			Substrate decomposition				
	Green leaves	1 year after death	Soil†	Green leaves	1 year after death	Soil†	Pectin	Starch	Gallic acid	Humic acid	Cellulose
Sterile mycelia											
B2	2		7	1	15	5	R	R	+	+	+
B18			38	5	9	1	+	+	+	+	–
B300			1			21	+	R	+	–	–
B302		3	11				R	R	+	–	–
B359	11	1	1				–	R	–	–	–
B221			2	23	3	9	–	+	–	–	–
S11	6			2	51	1	–	R	W	W	–
Cladosporium herbarium	53	78		57	7	2	+	R	W	W	–
Mucor microsporus	2	5	17	2	1		+	+	–	–	–
Phialophora hoffmannii			1			15	R	+	–	+	+
Cylindrocarpon magnusianum						3	R	+	+	–	+
Chrysosporium prunosum	3						+	+	–	–	+

* Abstracted from Flanagan and Scarborough (1974).

† Soil sampled beneath plant species.

W, very weak positive reaction at ~25°C; +, positive at both ~25°C and 2–5°C; R, positive only at ~25°C; –, negative at both temperatures.

the effect of soil moisture status on fungi is that hyphae of the same mycelium may be subject simultaneously to many different environmental regimes, and therefore unfavourable moisture conditions may to some extent be offset by translocation from elsewhere (Boddy 1986).

Thus, the substrates utilized are also used by temperate fungal isolates (see Domsch, Gams & Anderson 1993). Swift and Heal (1984) state 'the minimum requirement for decomposition of a complex resource is that the requisite set of enzymes is brought into contact with their substrates'. The data in Table 2.2 and above on substrate utilization, and therefore enzymatic capacity, show that although some isolates are functionally equivalent (i.e. several different isolates possess the same enzymes), several are more specialized. Therefore, in the soils studied at Devon Island, where there were found to be few cellulolytic and ligninolytic strains, 'empty niches' could occur where stages in decomposition are precluded by the lack of appropriate isolates. It is also apparent that this spectrum in fungal enzymatic activity is overlain by physical controls. For example, not all saprotrophic fungal isolates from arctic soils are strictly psychrophilic, and it is obvious that the combination of temperature and moisture is important to their optimum activities. The major point to emphasize is that under the same set of conditions, resources occupied by different decomposer fungi may decompose at different rates and degrees due to differences in their enzymatic capabilities. Fungal community structure is indeed related to fungal community function at the microscale.

Soil bacteria

Taxonomically, bacterial populations in tundra areas show few differences from populations found in other regions and no types unique to tundra regions have been recognized (Holding 1981). Information on the anaerobic bacterial population of tundra soils is lacking. Bacterial colony counts for the 0–5 cm layer of 38 tundra areas have been related to selected characteristics of soil, vegetation and climate. Of eight characteristics examined, pH (less than or greater than 4.75) appeared to be the most consistently linked site variable and available calcium (less than or greater than 4000 ppm) appeared to have a true nutrient effect not linked to pH or to the plant species being decomposed. The counts did not seem to be related to air temperature, site moisture or geographical location. The effect of available potassium (less than or greater than 400 ppm) or P (> or < 50 ppm) or N (> or < 1%) was difficult to determine. When a principal component and cluster analysis of soil and climatic data (French 1974) was used with maximum and minimum counts per gram oven-dried soil, Holding et al. (1974) observed three major site groupings from a visual examination of the clustering. There was a tendency for the counts to decrease from the 'warm bogs' group (e.g. British Isles) to the

'Fennoscandian' group (e.g. Finland, Norway, Sweden) and then to the 'polar' group (e.g. northern North America, north Russia and Antarctica).

Bacterial biomass estimates are strongly influenced by the unreliability of both plate and direct microscopic counts, and the difficulty in determining whether cells are alive, dead or in a non-culturable but viable state (Colwell *et al.* 1985). From direct counts, bacterial biomass ranged from $2.26\,\mu g\,g^{-1}$ oven-dried soil at Stordalen to 8200 in a horizon at Moor House. Total bacteria biomass estimated by plate counts was $0.16\,g\,m^{-2}$ for two Devon Island sites (Bliss *et al.* 1973, cited in Holding 1981). Bacterial biomass increases with decreasing latitude (Swift *et al.* 1979).

The synthesis studies of Dunican and Rosswall (1974) and Rosswall and Clarholm (1974) emphasized the apparent rarity of bacterial properties restricted only to arctic environments. Although the four main processes involved in the cycling of N compounds are of considerable interest, very few population or process data are available except for N fixation (see below). Most probable number dilution counting methods have failed to show the presence of nitrifying bacteria at three tundra peat sites (Glenamoy, Stordalen and Moor House), although denitrifying bacteria occurred in tundra areas examined in the USA, Sweden and Norway (Holding 1981).

To conclude, the bacterial communities of arctic environments appear to be similar in composition to those at temperate latitudes, but again there are methodological problems in their detection, identification and enumeration. There is little information on community function, and in extreme tundra soils bacterial counts are lower than in temperate ecosystems. In the Ross Desert, Antarctica, diverse combinations of cyanobacteria, green algae, filamentous fungi, yeasts and lichens, together with associated heterotrophic bacteria occur in cryptoendolithic communities (see review by Wynn-Williams (1990)). The microbes are orientated in layers inside rock, parallel to the surface, defined by physical and physiological constraints. No such communities have been discovered in arctic environments yet.

DECOMPOSITION AND NUTRIENT CYCLING, OR THE BROAD FUNCTION OF THE MICROBIAL COMMUNITY

Since the microbial species found in arctic environments are also found in less extreme situations, it seems likely that the main difference between tundra and other ecosystems is the rate of decomposition and nutrient cycling processes rather than the types of processes. The information from process studies is more directly comparable between sites and therefore easier to interpret than that concerning microbial community structure, and laboratory measurements are also more easily related to field conditions.

Decomposition

Decomposition is defined here as the sum of physical, chemical and biological processes occurring in dead organic matter, which results in the reduction of particle size and the transformation of organic compounds into inorganic ones (after Heal *et al.* 1981). Although decomposition of organic matter in arctic ecosystems is generally much slower than at temperate latitudes, it is regulated by the same suite of factors, which are the physicochemical environment, resource quality and the activity of decomposer organisms (Swift *et al.* 1979). The mass loss of a resource provides an integrated measurement of decomposition through the processes of microbial and faunal respiration, leaching and loss of particulate organic matter.

Soil temperature has been shown to be the most important control on litter decomposition in boreal forest ecosystems (Van Cleve & Yarie 1986; Sparrow, Sparrow & Cochran 1992), and increases in soil temperature usually result in increased litter mass loss in such ecosystems (Berg, Ekbohm & McClaugherty 1984; Van Cleve & Yarie 1986; Dyer, Meentemeyer & Berg 1990). Studies by Moore (1984) on the decomposition of *Picea mariana* needles in a subarctic forest in eastern Canada illustrate the effects of surface conditions on heat balance. Decomposition rate constants (k) were 0.20 in litter, 0.17 beneath the lichen mat and 0.24 on the floor of a site which had been subjected to fire; that is, decomposition was slowest under the lichen mat where the annual heat sum was lowest.

The effect of moisture regime on decomposition is demonstrated in the work of Robinson *et al.* (1995), where a 58% increase in summer precipitation was simulated by water additions at both a polar semi-desert ecosystem in Svalbard and a dwarf-shrub heath in Abisko. Although the watering treatment had no effect on mass loss of *Salix polaris* litter at the Svalbard site after 13.5 months, at Abisko the mass loss from *Vaccinium uliginosum* litter was significantly greater ($P < 0.05$) under the watering treatment. The increased litter decomposition in the watered compared to the unwatered treatment at Abisko may have been due to enhanced leaching of soluble materials, or to increased microbial activity. However, there was no significant effect of the watering treatment on bulk soil moisture at either site.

Temperature and moisture act together to determine decomposition rates (Heal & French 1974; Bunnell *et al.* 1977; Meentemeyer 1978). Using a site classification based on these two factors, Heal *et al.* (1981) found that, given approximately optimal conditions of other factors, typical first year litter mass loss rates would be of the order 100% for warm oceanic, 50% for warm continental, 30% for cold oceanic and 20% for cold continental tundra environments.

In arctic environments, topography also exerts an important influence on decomposition rate through interactions with soil temperature and moisture directly, and indirectly through increased or decreased litter inputs due to differences in primary production. For example, within the same climatic zone on Devon Island, a sedge meadow and a beach ridge plateau show wide variation in the accumulation of organic matter (51 and 9 kg m^{-2} respectively; Heal *et al.* 1981). In the latter case, significant quantities of dead organic matter occurred only in the immediate vicinity of scattered vascular plants.

Resource quality of litter material is an expression of the degree to which its chemical constituents and physical structure affect the nutritional requirements of saprotrophs. Deciduous shrub and graminoid leaf litters from arctic environments decompose faster than evergreen leaf litters, which in turn decompose faster than mosses, lichens and woody stems (Heal & French 1974). Heal, Latter and Howson (1978) following the decomposition of 14 litter types in a peat bog found that roots of *Eriophorum vaginatum* (0.5% N, 4.5% soluble carbohydrate, 34% lignin and 18% tannin) had the slowest decomposition rate and leaves of *Rubus chamaemorus* (1.31% N, 8.2% soluble carbohydrate, 27.0% tannin and 6% lignin) the fastest decomposition rates. The mass loss (y) could be explained (96% of the variance) by the equation:

$$y = 25.3 \, (P + Ca) - 0.53 \, (\text{lignin} \times \text{tannin}) \qquad r^2 = 0.977, \, P < 0.001$$

Data from a number of tundra sites showed that differences in mass loss between litters are inversely related to initial lignin concentration (Van Cleve 1974), and various other chemical components have been shown to act as controls on decomposition rates, for example for grass litter and crop residues, N or C/N ratio (Koenig & Cochran 1994) or lignin/N ratio (Aber & Melillo 1980; Melillo, Aber & Muratore 1982) and the lignocellulose index (lignin/(cellulose + lignin); Melillo *et al.* 1989, cited in Nadelhoffer *et al.* 1992) for low-quality tree litters.

The microbial decomposers and their psychrophilic nature in relation to the activity of decomposer organisms have been considered above. Soil invertebrates can, however, regulate mineral C (Anderson *et al.* 1985) or nutrient fluxes (Robinson *et al.* 1992) through the way their feeding and burrowing activities affect microbial populations and physicochemical conditions in temperate ecosystems. Tundra invertebrates are, in general, feeders on microbial tissue, and Nematoda, Enchytraeidae, Acari, Collembola and Diptera are the dominant groups (Ryan 1981). Tundra faunas generally lack large, litter-dwelling invertebrates which actually feed on litter, a role played by isopods, millipedes and earthworms in temperate forest ecosystems. The influence of tundra invertebrates on decomposition rates has usually been assumed to be minor, although O'Lear and Seastedt (1994) found decomposition of fruiting bodies of *Agaricus bisporus* was greater in mesic sites in alpine tundra where soil invertebrates were not excluded by naphthalene applications.

An indication of the rate of organic matter decomposition can be obtained by measuring mass loss of a resource, or by 'soil' respiration. Only soil respiration is associated with the catabolism of organic compounds and is often used as a sensitive, short-term measurement of the influence of particular factors on the rate of decomposition and in detecting changes in decomposer activity over time. No method of measuring soil respiration currently distinguishes between C derived from the breakdown of detritus and that produced by root respiration, although various studies indicate that between one-third and two-thirds of the total respiration originates from living roots (in Anderson 1991). In chernozems, abiotic sources of CO_2 from carbonates may be as much as 30–60% of the flux from the whole profile (Zlotin 1975, cited in Coleman & Sasson 1980), and therefore abiotic release of CO_2 from polar desert lithosols developed over limestone could be a potentially important component of the C balance of these arctic ecosystems.

Thus the controls on decomposition measured by mass loss outlined above also operate on soil respiration. The temperature response of microbial respiration is often expressed by the Q_{10} coefficient, which is the factorial increase in response to a 10°C rise in temperature. The Q_{10} is often assumed to be *c.* 2.0 for biological systems, but Q_{10} values tend to be higher under cold regimes and lower under hot regimes; so, whereas de Boois (1974) recorded a Q_{10} = 2 response over 5–15°C for microbial respiration in temperate woodland leaf litter, the temperature response for litter from the Arctic was 3.7 between 0 and 10°C (Bunnell *et al.* 1977) and 4 in tundra soils averaged over 10°C increments from −10 to 25°C (Flanagan & Bunnell 1976). These authors concluded that, in this way, the psychrophiles and cold-tolerant mesophiles in taiga and tundra were able to maintain comparable decomposition rates to temperate species operating at higher temperatures over a longer season. By contrast, there were no differences in temperature response of microbial respiration over a −15 to 25°C range in glucose-amended and non-amended soils from the Alaskan tundra and a temperate woodland in Virginia (Linkens, Melillo & Sinasbaugh 1984). They showed that rates of endocellulase activity may become more temperature limited below 10°C. Since both experimental approaches involve treatments in which available C was not limiting to microbial activity, the question of physiological adaptations of the microbial community to low temperatures is unresolved (Anderson 1991).

Soil moisture is also an important factor controlling soil respiration. Experiments by Billings *et al.* (1982) and Funk *et al.* (1994) showed that lowering the water table to 5 or 20 cm below the surface of tundra or taiga peats dramatically increased CO_2 efflux from these vegetated cores by increasing the aerobic depth of the soil. An excellent demonstration of the relationship between soil respiration, air temperature and soil moisture has been provided by Flanagan and Veum (1974) for a dry meadow tundra site at Hardangervidda. At a polar semi-desert site in Svalbard, we have measured the amount of CO_2 evolved over 24 h from areas of mainly unvegetated ground, with little organic matter, using 10-ml gas samples collected from static chambers (10 cm diameter, pushed 5 cm into the 'soil' with 10 cm standing above the soil, mean head-space volume 744 cm^3) and injected into a portable infrared gas analyser (EGM-1, PP Systems, Hitchin, Herts, UK). One chamber was inserted on 12 July 1994 in each of four out of six treatment replicates. Sampling was performed on 15 July, 21 July and 21 August 1994. The mean of three replicate syringes from each chamber minus the mean ($n = 15$) ambient CO_2 concentration was used to calculate μmol CO_2 per chamber at standard temperature and pressure (s.t.p.). On all sampling occasions, there was no effect on the amount of CO_2 evolved of an increase in mean 'summer' soil temperature of 0.8°C, produced using polythene tents, but soil respiration was significantly increased by an interaction between the application of fertilizer (5, 5 and 6.3 g m^{-2} year^{-2}, N, P and K) and a 50% increase in summer precipitation (Fig. 2.1; Robinson unpublished data; see Wookey *et al.* 1993 for a description of the experimental site and treatments). This suggests that nutrient availability plus low soil moisture contents together are limiting to decomposition on the unvegetated ground (which comprises approximately 70% of the area) at this polar semi-desert site, although our results do not conflict with those above that show large temperature effects on soil respiration, since the temperature increase produced here was relatively small.

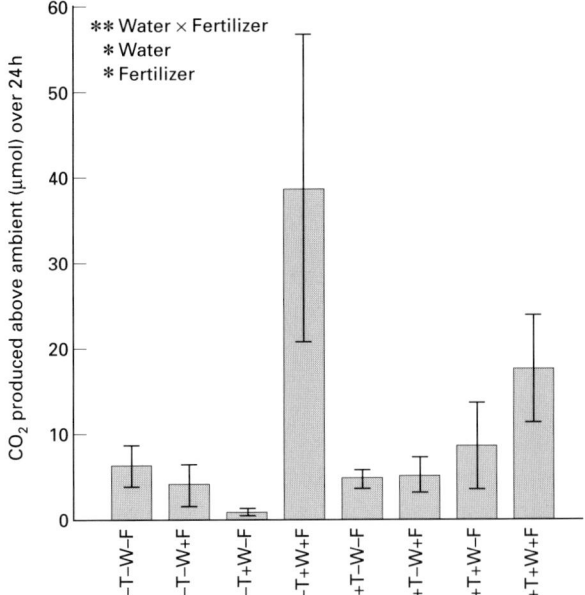

FIG. 2.1. Mean ($\pm$ SE) CO_2 produced above ambient over a 24-h period from chambers inserted into unvegetated ground at a high arctic polar semi-desert, Ny-Ålesund, $n = 4$ (for main effects statistics, $n = 16$).*, **, $P < 0.05$, 0.01, treated vs. untreated, General linear model (GLM). T, tented treatments; W, watered treatments (i.e. a simulation of 58% extra summer rainfall); F, fertilized treatments (5, 5 and 6.3 g m^{-2} N, P and K respectively).

Methane production is significant in some waterlogged organic soils (Gorham 1991). Whalen and Reeburg (1989) suggest that global methane releases from tundra and taiga are in the order of 38 and 15 Tg annually, respectively, which together are 10% of the global atmospheric input. Methane fluxes show high diel, seasonal, intra- and intersite variability (Gorham 1991). Of the IBP sites, methane production was maximal in waterlogged minerotrophic depressions on Stordalen mire, where it accounted for 45% of the total C output, while at Moor House about 12% of the C loss was methane in wet *Sphagnum* areas but only 1% in the drier hummocks. At Point Barrow, artificially heated meadow soils evolved 46–65% of their gaseous C as methane (Heal *et al.* 1981). However, if the soil is aerobic at the surface, methane (CH_4) is absorbed and oxidized by methanotrophic bacteria (Whalen & Reeburg 1990; Torn & Chapin 1993).

Currently, long-term C storage in undrained and unmined boreal and subarctic peatlands is between 76 and 96 Tg C year^{-1} (Gorham 1991), but it is clear that the processes of CO_2 and CH_4 release from soils in arctic environments under the strong climatic controls discussed could be affected extremely by climate change scenarios (Oechel *et al.* 1993). Under these predicted changes for increased C accumulation to take place in ecosystems, more C must be stored in plants and/or soils, although this would require additional nutrients that are limiting in arctic soils (chiefly N and P) to be made available.

To conclude, the decomposition processes of catabolism, leaching and comminution which occur in tundra ecosystems are the same as at more southerly latitudes, but they are usually slower, governed by a suite of often autocorrelated factors that include low soil temperature, exceptionally high or low soil moisture contents, organic matter of low resource quality, strong seasonality and microclimates unfavourable to decomposition generated by changes in topography. The role of soil invertebrates in decomposition in arctic environments has been largely neglected to date.

Nitrogen cycling

The nature of environmental constraints on primary productivity in tundra has been a research focus subsequent to IBP; N limitation to plant growth is common in both high arctic (Henry, Freedman & Svoboda 1986) and subarctic ecosystems (Shaver & Chapin 1980). Excellent reviews of N cycling in tundra and boreal ecosystems have been provided by Rosswall and Granhall (1980), Van Cleve and Alexander (1981) and Nadelhoffer *et al.* (1992).

Nitrogen fixation

Alexander (1974) drew particular attention to the contribution of biological N_2

fixation to the total N input of tundra ecosystems. In general, fixation is responsible for about one-half of the external N input, with precipitation, rather than dry deposition, accounting for the remainder (Chapin & Bledsoe 1992). The range of fixation levels at the different sites at Point Barrow ($7-380\,mg\,N\,m^{-2}\,year^{-1}$), Kevo (136–349) and Stordalen (180–11 500) are clearly important when compared with the quantities added in rainfall of 23, 75 and 41 $mg\,N\,m^{-2}\,year^{-1}$ (Alexander 1974). Cyanobacteria, whether as components of lichens, free-living or intracellular symbionts or epiphytes on mosses, are the most important agents of N_2 fixation (Holding 1981; Chapin & Bledsoe 1992). Chapin, Bliss and Bledsoe (1991) found the rate of cyanobacterial N_2 fixation at Truelove Lowland generally increased during early season, then decreased through mid to late season, was most highly correlated with soil moisture, and that the temperature optimum for N_2 fixation was 20°C. Controls on N_2 fixation are discussed further in a review by Chapin and Bledsoe (1992). Free-living bacteria, in general, do not seem to be important in both lowland (Granhall & Lid-Torsvik 1975) and high arctic tundra (Henry & Svoboda 1986), although Nosko, Bliss and Cook (1994) have found free-living N_2-fixing bacterial genera late in the growing season in association with graminoids from the Canadian High Arctic. Low levels of both aerobic and anaerobic fixation by free-living bacteria have been reported by Granhall and Lid-Torsvik (1975) in Fennoscandian sites, ranging from $1.4\,mg\,N\,m^{-2}$ at the birch wood of Kevo, to $179\,mg\,N\,m^{-2}$ in the drier areas of Stordalen.

Symbiotic bacteria in association with legumes or other higher plants are not usually significant N_2 fixers in arctic environments, although they may be important locally (Chapin & Bledsoe 1992). *Astragalus alpinus* at a dry meadow site in Hardangervidda was estimated by Granhall and Lid-Torsvik (1975) to fix N_2 at rates comparable to free-living bacteria. *Dryas drummondii* (actinorrhizal) was able to fix N_2 in mid- to late-seral stages at the forelands of the Athabasca Glacier, Canada (Kohls *et al.* 1994), and *D. octopetala* is nodulated at some sites, but not at others (Chapin & Bledsoe 1992).

Soil nitrogen availability

Net mineralization represents the balance between the key processes of gross mineralization and immobilization in decomposer biomass (6–7% of total soil N was immobilized in microbial biomass in a subarctic podsol; S. Jonasson *et al.* unpublished data) or through adsorption of ammonium (NH_4^+) on to soil (Nömmik 1981). In most arctic environments, the predominant form of mineral N is NH_4^+, but in high arctic soils of neutral/basic pH, N may be present mainly as nitrate (see below). Simple organic forms of N, such as amino acids, have been shown to be important in tundra N cycles (Kielland & Chapin 1992; Chapin, Moilanen & Kielland 1993).

Nitrogen mineralization is strongly regulated in the Arctic by low soil temperatures (Marion & Black 1987), high or low soil moisture contents and the recalcitrant nature of organic matter (Van Cleve & Yarie 1986). In their review, Van Cleve and Alexander (1981) stated that an indication of the importance of below-ground processes as a 'bottleneck' for N release in arctic ecosystems is that total N mineralization is only 0.4–1.3% of total soil N. The analogous range in temperate ecosystems is 3–7% (Van Cleve & Alexander 1981). Soil N capital values are 1722, 909 and 342 g m^{-2} for a hummocky sedge meadow at Devon Island (75°33′N), a wet meadow at Barrow (71°18′N) and Stordalen mire (68°22′N) respectively, compared to 360 g m^{-2} for temperate deciduous forest at Hubbard Brook (43°56′N). Nitrogen turnover times are therefore increased in high-latitude ecosystems (9200 years) compared with up to 2000 years in ecosystems at temperate latitudes (Van Cleve & Alexander 1981). Although net N mineralization is lower in arctic soils (0.1–0.6gm^{-2} year^{-1}) than soils of temperate regions (6–7gm^{-2} year^{-1}; Van Cleve & Alexander 1981), N release varies widely amongst the different types of arctic ecosystems (Nadelhoffer et al. 1992). Hart and Gunther (1989) reported large differences in annual N mineralization measured in dry and moist tundra ecosystems at a high elevation site in southern Alaska. In the foothills of Alaska's North Slope, five-fold differences were found in N mineralization among dry, moist and wet ecosystems using in situ soil incubations. These differences were mainly due to the quality of the soil organic matter and microclimate (Giblin et al. 1991; Nadelhoffer et al. 1991). Similarly, Robinson et al. (1995) found that the net N mineralization rate differed markedly between a high arctic polar semi-desert ecosystem (0.1 g m^{-2} year^{-1}) developed over a calcareous lithosol and a subarctic dwarf-shrub heath soil (0.4 g m^{-2} year^{-1}) with podsolic characteristics.

Seasonal variations in N availability are also apparent. Extractable NH$_4^+$-N in organic soils was greater just after the soil thaw and before freezing of the soil in autumn rather than in mid-season (Gersper et al. 1980). Despite a flush of N on soil thaw, N mineralization in arctic soils during the winter may be a more important component of annual N mineralization rates than recognized hitherto, since Clein and Schimel (1995) have shown that N was mineralized in tundra and taiga soils kept in the laboratory at −2 and −5°C. Nitrification occurs in some arctic ecosystems, especially with a high soil pH (Fig. 2.2b; Nadelhoffer et al. 1991), although NH$_4^+$ is the dominant form of inorganic N in cold, wet tundra soils (Fig. 2.3a; Giblin et al. 1991; Jonasson et al. 1993). Even so, Giblin et al. (1991) measured nitrification in moist tussock tundra ecosystems on Alaska's North Slope, but found no evidence of this process in nearby wet sedge and dry heath ecosystems. Nitrate accounted for about half the annual net N mineralized, but in contrast to ammonification, nitrification occurred mostly in mid-summer. This may be because nitrification is more sensitive to increases in temperature than ammonification (Paul & Clark 1989).

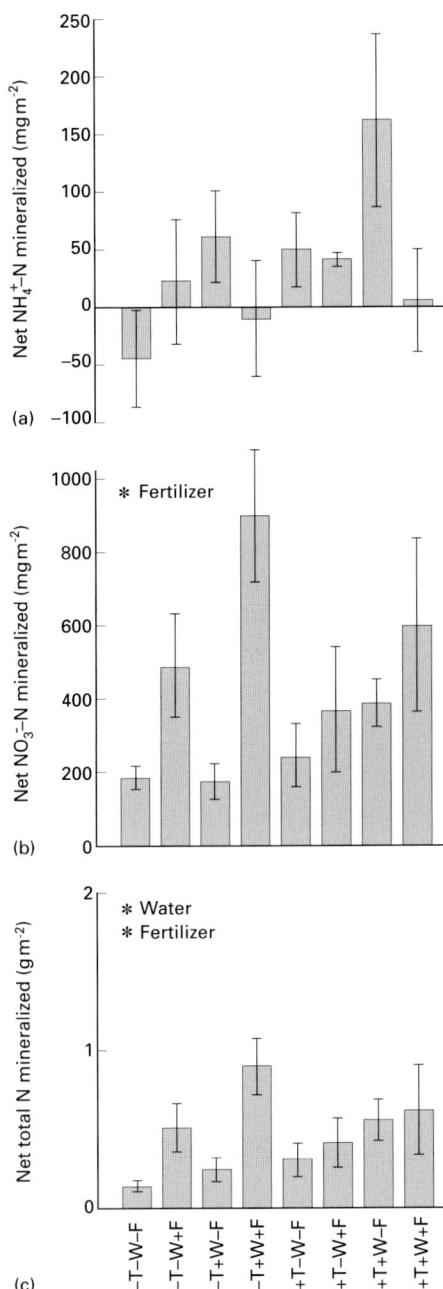

Fɪɢ. 2.2. Mean (± sᴇ) net seasonal (a) NH_4^+–N; (b) NO_3^-–N; and (c) total inorganic N mineralized at a high arctic polar semi-desert, Ny-Ålesund, $n = 6$ (for main effects statistics, $n = 24$). *, $P < 0.05$, treated vs. untreated, GLM. Key to treatments as in Fig. 2.1. Note the predominance of nitrate in this high pH (6.9–8.0) soil. (After Robinson *et al.* 1995.)

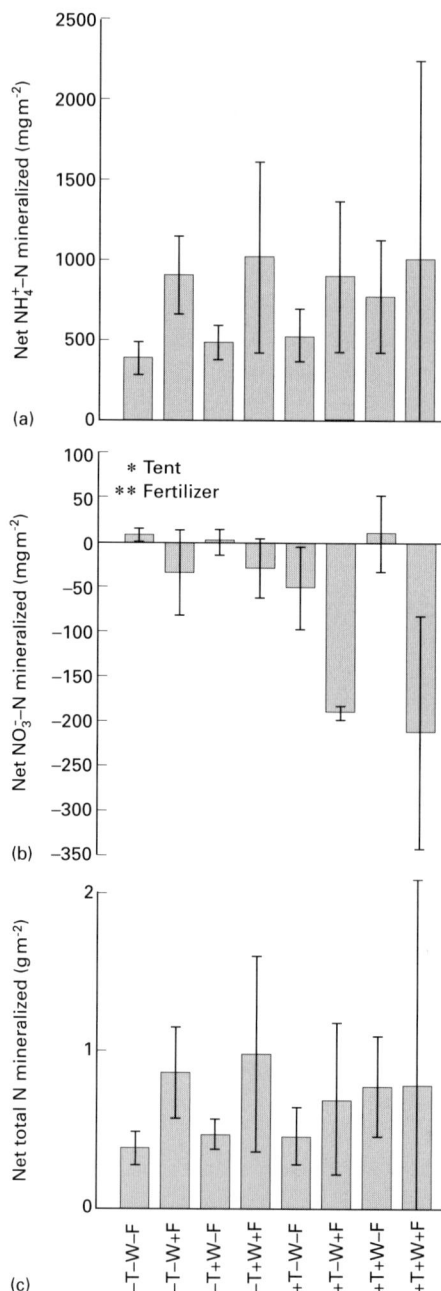

Fig. 2.3. Mean (± SE) net seasonal (a) NH_4^+–N; (b) NO_3^-–N; and (c) total inorganic N mineralized at a subarctic dwarf-shrub heath at Abisko, $n = 4$ (for main effects statistics, $n = 16$), *, **, $P < 0.05$, 0.01, treated vs. untreated, GLM. Key to treatments as in Fig. 2.1. Note the predominance of ammonium in this soil of pH 4.0–4.5. (After Robinson et al. 1995.)

Increases in soil temperature (Van Cleve *et al.* 1983, + 9°C; Nadelhoffer *et al.* 1991, 3–15°C; J.M. Melillo *et al.* unpublished data, +5°C) have been shown to stimulate N availability and mineralization strongly. Nadelhoffer *et al.* (1991) have shown, however, that there was no difference in the amount of N mineralized between 3 and 9°C in moist tundra or wet sedge tundra soils, but mineralization was increased by up to 10 times in the range 9–15°C. Analogous results were evident when no greater net N mineralization was found at a high arctic site in Spitsbergen (mean soil temperatures at 5 cm depth were increased by 0.7°C to 6.8°C (Fig. 2.2c; Robinson *et al.* 1995). By contrast, Jonasson *et al.* (1993) found in a fellfield and treeline ecosystem at Abisko, when mean soil temperatures were increased to give soil temperatures of *c.* 10°C at 3–5 cm depth,there was little effect on N mineralization. Although bulk soil moisture content was not significantly increased, greater net seasonal total N mineralization was found in treatments where a 58% increase in summer precipitation was made at the polar semi-desert study site by Robinson *et al.* (1995; Fig. 2.2c).

The role of recalcitrant organic matter in controlling rates of N mineralization in six tundra soils was deemed to be more important than differences in soil temperature in the field (Nadelhoffer *et al.* 1991). In taiga ecosystems in Alaska studied by Van Cleve and Yarie (1986), the chemistry of soil organic matter was important, but secondary to soil temperature as a control on N cycling. The interaction of temperature with moisture on rates of N mineralization in arctic soils has been identified as a research priority (Nadelhoffer *et al.* 1991).

In conclusion, the net N mineralization rate is low in arctic soils compared with that in temperate ecosystems, but different types of tundra soil exhibit a range of rates. Nitrogen mineralization is seasonal, although little is known about rates during winter, and nitrate and organic N can be as important in plant nutrition as NH_4^+. Below a threshold of approximately 10°C, net N mineralization responds relatively weakly to increases in soil temperature. The chemistry of soil organic matter and soil moisture content are also likely controls on N availability, as they are on C mineralization, but their effects on N mineralization are currently poorly understood.

Denitrification and ammonia volatilization

Gaseous loss of N by biological denitrification has been estimated for a wet meadow at Barrow to reach $3.4 \, mg \, m^{-2} \, year^{-1}$ (Gersper *et al.* 1980), which is 3.4% of the annual N input (Van Cleve & Alexander 1981). Lack of denitrification on drier sites may reflect the absence of anaerobic conditions. At Stordalen denitrification did not occur, probably because nitrate did not occur in any appreciable quantities (Rosswall & Granhall 1980). Loss of N by NH_3 volatilization (an abiotic process) has not been measured at arctic sites, but may be of some importance on drier sites where soil pH rises to the alkaline range (Van Cleve & Alexander 1981).

Soil phosphorus availability

Phosphorus is generally tightly conserved in arctic ecosystems, even when it is not the most limiting element (Nadelhoffer *et al.* 1991), and soluble and exchangeable phosphate pools are typically very small and turn over rapidly (Nadelhoffer *et al.* 1992). For example, in wet tundra soils in Barrow, Alaska, soluble phosphate must be replenished about 200 times annually to match plant uptake, as opposed to about 10 times for mineral N (Bunnell, MacLean & Brown 1975). A substantial amount of P can be immobilized in microbial biomass (e.g. 35% of the soil P content; S. Jonasson *et al.* unpublished data). Thus, periodic declines in soil microbial biomass, up to 75% in 3 weeks at Barrow (Miller & Laursen 1974), could release substantial amounts of nutrients. For example, a 90% decrease in microbial biomass could satisfy the entire annual P requirement of the vascular vegetation at this site (Chapin, Barsdate & Barel 1978).

Compared with N mineralization, P availability is strongly controlled by physicochemical, as well as biological, processes. In moist and wet tundra systems, P input into surface organic soils is small, coming mainly from precipitation (Nadelhoffer *et al.* 1992). As plant roots are effectively isolated from mineral-weathered P in wet tundra soils, recycling of P from organic matter must supply nearly all of the P taken up annually by plants in these ecosystems. In dry tundra ecosystems, where organic horizons are thin, P availability depends more on P release from mineral soil horizons.

Seasonal variations in P availability are high in arctic soils. As with mineral N, soluble and exchangeable phosphate pools in organic horizons are typically greatest at snowmelt and decline as soil temperatures increase through the growing season (Dowding *et al.* 1981). As a result, most P uptake in tundra plants probably occurs just after surface thaw in early summer, just before freezing in autumn, or during other periods when microbial populations decline (Chapin *et al.* 1978). Laboratory incubations of tundra soils also indicate that P availability is highest at temperatures just above freezing, when microbial demands for P are relatively low (Nadelhoffer *et al.* 1991). Increases in soil temperature have been found to increase available P (+9°C, Van Cleve *et al.* 1983), to have no effect (+4°C, Jonasson *et al.* 1993; +0.7°C, Robinson *et al.* 1995) or to both increase and decrease P release (Nadelhoffer *et al.* 1991).

Thus, a large proportion of soil P is immobilized within microbial biomass, and this biological pool is strongly seasonal in its P release. Rock weathering is a significant source of P in arctic soils where the process occurs in close proximity to the soil organic matter. The controls on P availability in arctic soils are less well elucidated than for N or C mineralization.

MICROBIAL COMMUNITY STRUCTURE
AND FUNCTION IN ARCTIC ECOSYSTEMS:
THE FUTURE

The microbial communities are fundamentally very similar in species composition to those at temperate latitudes, although in extreme tundra soils fungal and bacterial biomass may be lower than in temperate ecosystems, and the organisms adapted to low temperatures. The broad microbial processes of C, N and P mineralization which occur are the same as at more southerly latitudes, but they are usually slower, governed by a suite of autocorrelated factors that include low soil temperature, exceptionally high or low soil moisture contents, organic matter of low resource quality, strong seasonality and microclimates unfavourable to decomposition generated by changes in topography.

Since the composition and function of microbial communities in the Arctic do not differ fundamentally from those of more temperate latitudes, the important questions remaining unanswered are, likewise, similar. After 30 years of soils research in arctic environments, the challenge of relating the community structure of microbial decomposers to function still exists, despite some successful attempts from tundra (Flanagan & Scarborough 1974) and temperate (e.g. Boddy 1986; Newsham *et al.* 1992) ecosystems which match organisms and processes together. Flanagan and Scarborough (1974) showed that the structure of the saprotrophic fungal community is closely linked to function at the microscale. More work is necessary at this level of resolution. Frankland (1992) emphasized that many more comparative studies of 'key' fungal species in systems, and their combinations, are necessary to go some way towards understanding fungal community structure and function. Such experiments should bridge entirely artificial media and all the complexities of soil in the field. By applying mainstream ecological theory (e.g. island biogeography (Wildman 1987); strategy theory (Swift 1985)), including modelling (Halley *et al.* 1996), frameworks are given to include microbial diversity in models of decomposition. Since the IBP, little new information on the structure of the microbial community in arctic soils has been gained, and the techniques for fungal isolation, which are still inadequate to reveal completely the microbial community, have not changed. Because microbial biodiversity is inadequately characterized, it is only possible to guess at the community's response to environmental perturbations. It is to be hoped that the functional roles of saprotrophic and mycorrhizal organisms may eventually be revealed with molecular methods currently being developed for use in soils (e.g. Clapp *et al.* 1995).

Scaling up from microbial community structure and function at the individual resource level to the system scale is necessary. At the broad scale, the hypothesis of Swift and Heal (1984), who stated 'the catabolic processes of decomposition ... are non-specific; for every given reaction there is a cohort of organisms with the appropriate capacity ...', should be tested: for example by relating the effects on decomposition and nutrient mineralization of sequential removal of microbial isolates from, or sequential addition of more than two microbial isolates to, a resource. To sum up, the complexity of the decomposer community, the inadequacy of existing techniques, the heterogeneity of arctic soil environments and the effects of extreme climatic conditions lead to considerable gaps in current knowledge in the structure and function of the microbial community in arctic soils. Nevertheless, the function of these communities underpins all arctic ecosystems and a much clearer understanding of their identity, role and responses to perturbation is essential if reliable predictions of ecosystem responses are to be made.

ACKNOWLEDGEMENTS

It is a pleasure to thank our colleagues Dr A.N. Parsons, Ms J.A. Potter, Professor T.V. Callaghan, Professor and Mrs J.A. Lee, Dr M.C. Press, Dr J.M. Welker, Ms O.B. Borisova, Mr J.B. Kirkham and Dr F.R. Livens. Our work in this review was funded by the Natural Environment Research Council (NERC) as part of the Arctic Terrestrial Ecology Special Topic Programme. We thank the staff of the Norsk Polarinstitutt and the Kings Bay Kull Compani at Ny-Ålesund (Svalbard), Professor M. Sonesson and N.-Å. Andersson of the Abisko Naturvetenskapliga Station and Mr N.I. Cox, manager of the NERC Arctic Research Station, Harland Huset, for helping to provide facilities, expertise and logistical support.

REFERENCES

Aber, J.D. & Melillo, J.M. (1980). Litter decomposition: measuring relative contributions of organic matter and nitrogen to forest soils. *Canadian Journal of Botany*, **58**, 416–421.

Alexander, V. (1974). A synthesis of the IBP tundra biome circumpolar study of nitrogen fixation. *Soil Organisms and Decomposition in Tundra* (Ed. by A.J. Holding, O.W. Heal, S.F. MacLean Jr & P.W. Flanagan), pp. 109–121. Tundra Biome Steering Committee, Stockholm.

Anderson, J.M. (1991). The effects of climate change on decomposition processes in grassland and coniferous forests. *Ecological Applications*, **1**, 326–347.

Anderson, J.M., Huish, S.A., Ineson P., Leonard. M.A. & Splatt, P.R. (1985). Interactions of invertebrates, micro-organisms and tree roots in nitrogen and mineral element fluxes in deciduous woodland soils. *Ecological Interactions in Soil* (Ed. by A.H. Fitter, D. Atkinson, D.J. Read & M.B. Usher), pp. 377–392. Blackwell Scientific Publications, Oxford.

Bardgett, R.D. (1991). The use of the membrane filter technique for comparative measurements of hyphal lengths in different grassland sites. *Agriculture, Ecosystems and Environment*, **34**, 115–119.

Berg, B., Ekbohm, G. & McClaugherty, C. (1984). Lignin and holocellulose relations during long-term decomposition of some forest litters. Long-term decomposition in a Scots pine forest IV. *Canadian Journal of Botany*, **62**, 2540–2550.

Billings, W.D., Luken, J.O., Mortensen, D.A. & Petersen, K.M. (1982). Arctic tundra: a sink or source for atmospheric carbon dioxide in a changing environment? *Oecologia*, **53**, 7–11.

Bliss, L.C. (1981). Introduction to tundra biome: past and present. *Tundra Ecosystems: A Comparative Analysis* (Ed. by L.C. Bliss, O.W. Heal & J.J. Moore), p. 3. Cambridge University Press, Cambridge.

Bliss, L.C., Heal, O.W. & Moore, J.J. (Eds) (1981). *Tundra Ecosystems: A Comparative Analysis.* Cambridge University Press, Cambridge.

Boddy, L. (1986). Water and decomposition processes in terrestrial ecosystems. *Water, Fungi and Plants* (Ed. by P.G. Ayres & L. Boddy), pp. 375–398. Cambridge University Press, Cambridge.

de Boois, H.M. (1974). Measurement of seasonal variations in the oxygen uptake of various litter layers of an oak forest. *Plant and Soil,* **40,** 545–555.

Bunnell, F.L., MacLean, S.F. & Brown, J. (1975). Barrow, Alaska, USA. *Structure and Function of Tundra Ecosystems* (Ed. by T. Rosswall & O.W. Heal). *Ecological Bulletins (Stockholm),* **20,** 73–124.

Bunnell, F.L., Tait, D.E.N., Flanagan, P.W. & Van Cleve, K. (1977). Microbial respiration and substrate weight loss I. A general model of abiotic variables. *Soil Biology and Biochemistry,* **9,** 33–40.

Chapin, D. & Bledsoe, L.J. (1992). Nitrogen fixation in Arctic plant communities. *Arctic Ecosystems in a Changing Climate: An Ecophysiological Perspective* (Ed. by F.S. Chapin III, R.L. Jeffries, J.F. Reynolds, G.R. Shaver & J. Svoboda), pp. 301–319. Academic Press, San Diego.

Chapin, D., Bliss, L.C. & Bledsoe L.J. (1991). Environmental regulation of nitrogen fixation in a high arctic lowland ecosystem. *Canadian Journal of Botany,* **69,** 2744–2755.

Chapin, F.S. III, Barsdate, R.J. & Barel, D. (1978). Phosphorus cycling in Alaskan coastal tundra: a hypothesis for the regulation of nutrient cycling. *Oikos,* **31,** 189–199.

Chapin, F.S. III, Jeffries, R.L., Reynolds, J.F., Shaver, G.R. & Svoboda, J. (Eds) (1992). *Arctic Ecosystems in a Changing Climate: An Ecophysiological Perspective.* Academic Press, San Diego.

Chapin, F.S. III & Körner, C. (Eds) (1995). *Arctic and Alpine Biodiversity: Patterns, Causes and Ecosystem Consequences.* Springer, London.

Chapin, F.S. III, Moilanen, L. & Kielland, K. (1993). Preferential use of organic nitrogen by a nonmycorrhizal arctic sedge. *Nature,* **361,** 150–153.

Chapin, F.S. III & Shaver, G.R. (1985). Individualistic growth responses of tundra plant species to environmental manipulations in the field. *Ecology,* **66,** 564–576.

Clapp, J.P., Young, J.P.W., Merryweather, J.W. & Fitter, A.H. (1995). Diversity of fungal symbionts in arbuscular mycorrhizas from a natural community. *New Phytologist,* **130,** 259–265.

Clein, J.S. & Schimel, J.P. (1995). Microbial activity of tundra and taiga soils at sub-zero temperatures. *Soil Biology and Biochemistry,* **27,** 1231–1234.

Coleman, D.C. & Sasson, A. (1980). Decomposer sub-system. *Grasslands, Systems Analysis and Man* (Ed. by A.I. Breymeyer & G.M. Van Dyne), pp. 609–655. Cambridge University Press, Cambridge.

Colwell, P.R., Brayton, B.R., Grimes, D.J., Roszak, D.R., Huq, S.A. & Palmer, L.M. (1985). Viable but non-culturable *Vibrio cholerae* and related pathogens in the environment: implications for the release of genetically engineered microorganisms. *Biotechnology,* **3,** 817–820.

Cragg, J.B. (1981). Preface. *Tundra Ecosystems: A Comparative Analysis* (Ed. by L.C. Bliss, O.W. Heal & J.J. Moore), pp. xxiii–xxv. Cambridge University Press, Cambridge.

Deverall, B.J. (1968). Psychrophiles. *The Fungi: An Advanced Treatise* (Ed. by G.C. Ainsworth & A.S. Sussman), Vol. 3, pp. 129–135. Academic Press, London.

Domsch, K.H., Gams, W. & Anderson, T.-H. (1993). *Compendium of Soil Fungi,* Vols I and II, 2nd edn. Academic Press, London.

Dowding, P., Chapin, F.S. III, Wielgolaski, F.E. & Kilfeather, P. (1981). Nutrients in tundra ecosystems. *Tundra Ecosystems: A Comparative Analysis* (Ed. by L.C. Bliss, O.W. Heal & J.J. Moore), pp. 647–683. Cambridge University Press, Cambridge.

Dowding, P. & Widden, P. (1974). Some relationships between fungi and their environment in tundra regions. *Soil Organisms and Decomposition in Tundra* (Ed. by A.J. Holding, O.W. Heal, S.F. MacLean Jr & P.W. Flanagan), pp. 123–150. Tundra Biome Steering Committee, Stockholm.

Dunican, L.K. & Rosswall, T. (1974). Taxonomy and physiology of tundra bacteria in relation to site characteristics. *Soil Organisms and Decomposition in Tundra* (Ed. by A.J. Holding, O.W. Heal, S.F. MacLean Jr & P.W. Flanagan), pp. 79–92. Tundra Biome Steering Committee, Stockholm.

Dyer, M.L., Meentemeyer, V. & Berg, B. (1990). Apparent controls of mass loss rate of leaf litter on a regional scale. *Scandinavian Journal of Forest Research*, **5**, 1–13.

Farrell, J. & Rose, A. (1967). Temperature effects on microorganisms. *Annual Review of Microbiology*, **21**, 101–120.

Finotti, E., Moretto, D., Marsella, R. & Mercantini, R. (1993). Temperature effects and fatty acid patterns in *Geomyces* species isolated from Antarctic soil. *Polar Biology*, **13**, 127–130.

Flanagan, P.W. & Bunnell, F. (1976). Decomposition models based on climatic variables, substrate variables, microbial respiration and production. *The Role of Terrestrial and Aquatic Organisms in Decomposition Processes* (Ed. by J.M. Anderson & A. Macfadyen), pp. 437–457. Blackwell Scientific Publications, Oxford.

Flanagan, P.W. & Scarborough, A.M. (1974). Physiological groups of decomposer fungi on tundra plant remains. *Soil Organisms and Decomposition in Tundra* (Ed. by A.J. Holding, O.W. Heal, S.F. MacLean Jr & P.W. Flanagan), pp. 159–181. Tundra Biome Steering Committee, Stockholm.

Flanagan, P.W. & Veum, A.K. (1974). Relationships between respiration, weight loss, temperature and moisture in organic residues on tundra. *Soil Organisms and Decomposition in Tundra* (Ed. by A.J. Holding, O.W. Heal, S.F. MacLean Jr & P.W. Flanagan), pp. 249–277. Tundra Biome Steering Committee, Stockholm.

Frankland, J.C. (1992). Mechanisms in fungal succession. *The Fungal Community*, 2nd edn (Ed. by G.C. Carroll & D.T. Wicklow), pp. 383–401. Marcel Dekker, New York.

Frankland, J.C., Dighton, J. & Boddy, L. (1990). Methods for studying fungi in soil and forest litter. *Methods in Microbiology* (Ed. by R. Grigorova & J.R. Norris), **22**, 343–404. Academic Press, London.

French, D.D. (1974). Classification of IBP tundra biome sites based on climate and soil properties. *Soil Organisms and Decomposition in Tundra* (Ed. by A.J. Holding, O.W. Heal, S.F. MacLean Jr & P.W. Flanagan), pp. 3–25. Tundra Biome Steering Committee, Stockholm.

Frostegård, A., Bååth, E. & Tunlid, A. (1993). Shifts in the structure of soil microbial communities in limed forests as revealed by phospholipid fatty acid analysis. *Soil Biology and Biochemistry*, **25**, 723–730.

Funk, D.E., Pullman, E.R., Peterson, K.M., Crill, P. & Billings, W.D. (1994). The influence of water table on carbon dioxide, carbon monoxide and methane fluxes from taiga bog microcosms. *Global Biogeochemical Cycles*, **8**, 271–278.

Gehrke, C., Johanson, U., Callaghan, T.V., Chadwick, D. & Robinson, C.H. (1995). The impact of enhanced UV-B radiation on litter quality, decomposition and fungal growth on *Vaccinium uliginosum* in the subarctic. *Oikos*, **72**, 213–222.

Gersper, P.L., Alexander, V., Barkley, S.A., Barsdate, R.J. & Flint, P.S. (1980). The soils and their nutrients. *An Arctic Ecosystem: The Coastal Tundra at Barrow, Alaska* (Ed. by J. Brown, P.C. Miller, L.L. Tieszen & F.L. Bunnell), pp. 219–254. Dowden, Hutchinson & Ross, Stroudsburg, Penn.

Giblin, A.E., Nadelhoffer, K.J., Shaver, G.R., Laundre, J.A. & McKerrow, A.J. (1991). Biogeochemical diversity along a riverside toposequence in Arctic Alaska. *Ecological Monographs*, **61**, 415–435.

Gill, S., Bellesisles, J., Brown, G., Gagne, S., Lemieux, C., Mercier, J.P. & Dion, P. (1994). Identification of the variability of ribosomal DNA spacer from *Pseudomonas* soil isolates. *Canadian Journal of Microbiology*, **40**, 541–547.

Gorham, E. (1991). Northern peatlands: role in the carbon cycle and probable responses to climatic warming. *Ecological Applications*, **1**, 182–195.

Granhall, U. & Lid-Torsvik, V. (1975). Nitrogen fixation by bacteria and free-living blue-green algae in tundra ecosystems. *Fennoscandian Tundra Ecosystems I. Plants and Microorganisms* (Ed. by F.E. Wielgolaski), pp. 305–315. Springer, Berlin.

Haag, R.W. (**1974**). Nutrient limitations to plant productivity in two tundra communities. *Journal of Botany*, **52**, 103–116.

Halley, J.M., Robinson, C.H., Comins, H.N. & Dighton, J. (**1996**). Predicting straw decomposition by a four-species fungal community: a cellular automaton model. *Journal of Applied Ecology*, **33**, 493–507.

Hart, S.C. & Gunther, A.J. (**1989**). *In situ* estimates of annual net nitrogen mineralisation and nitrification in a subarctic watershed. *Oecologia*, **80**, 284–288.

Heal, O.W. (**1981**). Introduction. *Tundra Ecosystems: A Comparative Analysis* (Ed. by L.C. Bliss, O.W. Heal & J.J. Moore), p. xxvii. Cambridge University Press, Cambridge.

Heal, O.W., Flanagan, P.W., French, D.D. & MacLean, S.F. Jr (**1981**). Decomposition and accumulation of organic matter. *Tundra Ecosystems: A Comparative Analysis* (Ed. by L.C. Bliss, O.W. Heal & J.J. Moore), pp. 587–633. Cambridge University Press, Cambridge.

Heal, O.W. & French, D.D. (**1974**). Decomposition of organic matter in tundra. *Soil Organisms and Decomposition in Tundra* (Ed. by A.J. Holding, O.W. Heal, S.F. MacLean Jr & P.W. Flanagan), pp. 279–309. Tundra Biome Steering Committee, Stockholm.

Heal, O.W., Latter, P.M. & Howson, G. (**1978**). A study of the rates of decomposition of organic matter. *Production Ecology of British Moors and Montane Grasslands* (Ed. by O.W. Heal & D.F. Perkins), pp. 136–159. Springer, New York.

Henry, G.H.R., Freedman, B. & Svoboda, J. (**1986**). Effects of fertilisation on three tundra plant communities at a polar desert oasis. *Canadian Journal of Botany*, **64**, 2502–2507.

Henry, G.H.R. & Svoboda, J. (**1986**). Dinitrogen fixation (acetylene reduction) in high arctic sedge meadow communities. *Arctic and Alpine Research*, **18**, 181–187.

Holding, A.J. (**1981**). The microflora of tundra. *Tundra Ecosystems: A Comparative Analysis* (Ed. by L.C. Bliss, O.W. Heal & J.J. Moore), pp. 561–585. Cambridge University Press, Cambridge.

Holding, A.J., Heal, O.W., MacLean, S.F. Jr & Flanagan, P.W. (eds) (**1974**). *Soil Organisms and Decomposition in Tundra*. Tundra Biome Steering Committee, Stockholm.

Ivarson, K.C. (**1973**). Fungal flora and rate of decomposition of leaf litter at low temperatures. *Canadian Journal of Soil Science*, **53**, 79–84.

Jonasson, S. (**1983**). Nutrient content and dynamics in north Swedish shrub tundra areas. *Holarctic Ecology*, **6**, 295–304.

Jonasson, S., Havström, M., Jensen, M. & Callaghan, T.V. (**1993**). *In situ* mineralisation of nitrogen and phosphorus of arctic soils after perturbations simulating climate change. *Oecologia*, **95**, 179–186.

Kerry, E. (**1990**). Effects of temperature on growth rates of fungi from Subantarctic Macquarie Island and Casey, Antarctica. *Polar Biology*, **10**, 293–299.

Kielland, K. & Chapin, F.S. III (**1992**). Nutrient absorption and accumulation in arctic plants. *Arctic Ecosystems in a Changing Climate: An Ecophysiological Perspective* (Ed. by F.S. Chapin III, R.L. Jeffries, J.F. Reynolds, G.R. Shaver & J. Svoboda), pp. 321–335. Academic Press, San Diego.

Koenig, R.T. & Cochran, V.L. (**1994**). Decomposition and nitrogen mineralisation from legume and non-legume crop residues in a subarctic agricultural soil. *Biology and Fertility of Soils*, **17**, 269–275.

Kohls, S.J., Van Kessel, C., Baker, D.D., Grigal, D.F. & Lawrence, D.B. (**1994**). Assessment of N_2 fixation and N cycling by *Dryas* along a chronosequence within the forelands of the Athabasca Glacier, Canada. *Soil Biology and Biochemistry*, **26**, 623–632.

Kjøller, A. & Struwe, S. (**1982**). Microfungi in ecosystems. *Oikos*, **39**, 389–422.

Latter, P.M. & Heal, O.W. (**1977**). A preliminary study of the growth of fungi and bacteria from temperate and Antarctic soils in relation to temperature. *Soil Biology and Biochemistry*, **3**, 365–379.

Linkens, A.E., Melillo, J.M. & Sinasbaugh, R.L. (**1984**). Factors affecting cellulase activity in terrestrial and aquatic ecosystems. *Current Perspectives in Microbial Ecology* (Ed. by M.J. King & C.A. Reddy), pp. 572–579. American Society for Microbiology, Washington.

Marion, G.M. & Black, C.H. (1987). The effect of time and temperature on nitrogen mineralisation in arctic tundra soils. *Soil Science Society of America Journal*, **51**, 1501–1508.

Meentemeyer, V. (1978). Macroclimate and lignin control of litter decomposition rates. *Ecology*, **59**, 465–472.

Melillo, J.M., Aber, J.D. & Muratore, J.F. (1982). Nitrogen and lignin control of hardwood leaf litter decomposition dynamics. *Ecology*, **63**, 621–626.

Melillo, J.M., Callaghan, T.V., Woodward, F.I., Salati, E. & Sinha, S.K. (1990). Effects on ecosystems. *Climate Change, the IPCC Scientific Assessment* (Ed. by J.T. Houghton, G.T. Jenkins & J.J. Ephraums), pp. 282–310. Cambridge University Press, Cambridge.

Miller, O.K. Jr & Laursen, G.A. (1974). Belowground fungal biomass on US tundra biome sites at Barrow, Alaska. *Soil Organisms and Decomposition in Tundra* (Ed. by A.J. Holding, O.W. Heal, S.F. MacLean Jr & P.W. Flanagan), pp. 151–158. Tundra Biome Steering Committee, Stockholm.

Moore, T.R. (1984). Litter decomposition in a subarctic, spruce–lichen woodland in eastern Canada. *Ecology*, **65**, 299–308.

Nadelhoffer, K.J., Giblin, A.E., Shaver, G.R. & Laundre, J.L. (1991). Effects of temperature and substrate quality on element mineralisation in six arctic soils. *Ecology*, **72**, 242–253.

Nadelhoffer, K.J., Giblin, A.E., Shaver, G.R. & Linkens, A.E. (1992). Microbial processes and nutrient availability in arctic soils. *Arctic Ecosystems in a Changing Climate: An Ecophysiological Perspective* (Ed. by F.S. Chapin III, R.L. Jeffries, J.F. Reynolds, G.R. Shaver & J. Svoboda), pp. 139–168. Academic Press, San Diego.

Newsham, K.K., Boddy, L., Frankland, J.C. & Ineson, P. (1992). Effects of dry-deposited sulphur dioxide on fungal decomposition of angiosperm tree leaf litter. III. Decomposition rates and fungal respiration. *New Phytologist*, **122**, 127–140.

Nömmik, H. (1981). Fixation and biological availability of ammonium on soil clay minerals. *Terrestrial Nitrogen Cycles* (Ed. by F.E. Clark & T. Rosswall), *Ecological Bulletins (Stockholm)*, **33**, 273–280.

Nosko, P., Bliss, L.C. & Cook, F.D. (1994). The association of free-living nitrogen-fixing bacteria with the roots of high arctic graminoids. *Arctic and Alpine Research*, **26**, 180–186.

Oechel, W.C., Hastings, S.J., Vourlitis, G., Jenkins, M., Reichers, G. & Grulke, N. (1993). Recent change of Arctic tundra ecosystems from a net carbon dioxide sink to a source. *Nature*, **361**, 520–523.

O'Lear, H.A. & Seastedt, T.R. (1994). Landscape patterns of litter decomposition in alpine tundra. *Oecologia*, **99**, 95–101.

Parsons, A.N., Press, M.C., Wookey, P.A., Welker, J.M., Robinson, C.H., Callaghan, T.V. & Lee, J.A. (1995). Growth responses of *Calamagrostis lapponica* to simulated environmental change in the sub-arctic. *Oikos*, **72**, 61–66.

Paul, E.A. & Clark, F.E. (1989). *Soil Microbiology and Biochemistry*. Academic Press, New York.

Rayner, A.D.M. & Webber, J.F. (1984). Interspecific mycelial interactions – an overview. *The Ecology and Physiology of the Fungal Mycelium* (Ed. by D.H. Jennings & A.D.M. Rayner), pp. 383–417. Cambridge University Press, Cambridge.

Robinson, C.H., Borisova, O.B., Callaghan, T.V. & Lee, J.A. (1996). Fungal hyphal length in litter of *Dryas octopetala* in a high-Arctic polar semi-desert, Svalbard. *Polar Biology*, **16**, 71–74.

Robinson, C.H., Ineson, P., Piearce, T.G. & Rowland, A.P. (1992). Nitrogen mobilisation by earthworms in limed peat soils under *Picea sitchensis*. *Journal of Applied Ecology*, **29**, 226–237.

Robinson, C.H., Wookey, P.A., Parsons, A.N., Potter, J.A., Callaghan, T.V., Lee, J.A., Press, M.C. & Welker, J.A. (1995). Responses of plant litter decomposition and nitrogen mineralisation to simulated environmental change in a high arctic polar semi desert and subarctic dwarf shrub heath. *Oikos*, **74**, 503–512.

Rosswall, T. & Clarholm, M. (1974). Characteristics of tundra bacterial populations and a comparison with populations from forest and grassland soils. *Soil Organisms and Decomposition in Tundra* (Ed. by A.J. Holding, O.W. Heal, S.F. MacLean Jr & P.W. Flanagan), pp. 93–108. Tundra Biome Steering Committee, Stockholm.

Rosswall, T. & Granhall, U. (1980). Nitrogen cycling in a subarctic ombrotrophic mire. *Ecology of a Subarctic Mire* (Ed. by M. Sonesson). *Ecological Bulletins (Stockholm)*, **30**, 209–234.

Ryan, J.K. (1981). Invertebrate faunas at IBP tundra sites. *Tundra Ecosystems: A Comparative Analysis* (Ed. by L.C. Bliss, O.W. Heal & J.J. Moore), pp. 517–539. Cambridge University Press, Cambridge.

Shaver, G.R. & Chapin, F.S. III (1980). Response to fertilisation by various plant growth forms in an Alaskan tundra: nutrient accumulation and growth. *Ecology*, **61**, 662–675.

Sparrow, S.D., Sparrow, E.B. & Cochran, V.L. (1992). Decomposition in forest and fallow subarctic soils. *Biology and Fertility of Soils*, **14**, 253–259.

Stahl, P.D., Parkin, T.B. & Eash, N.E. (1995). Sources of error in direct microscopic methods for estimation of fungal biomass in soil. *Soil Biology and Biochemistry*, **27**, 1091–1097.

Swift, M.J. (1976). Species diversity and the structure of microbial communities in terrestrial habitats. *The Role of Aquatic and Terrestrial Organisms in Decomposition Processes* (Ed. by J.M. Anderson & A. MacFadyen), pp. 185–222. Blackwell Scientific Publications, Oxford.

Swift, M.J. (1985). Microbial diversity and decomposer niches. *Current Perspectives in Microbial Ecology* (Ed. by M.J. King & C.A. Reddy), pp. 8–16. American Society for Microbiology, Washington.

Swift, M.J. & Heal, O.W. (1984). Theoretical considerations of microbial succession and growth strategies: intellectual exercise or practical necessity? *Microbial Communities in Soil* (Ed. by V. Jensen, A. Kjøller & L.H. Soneson), pp. 115–131. Elsevier, London.

Swift, M.J., Heal, O.W. & Anderson, J.M. (1979). *Decomposition in Terrestrial Ecosystems.* Blackwell Scientific Publications, Oxford.

Syzova, M.V. & Panikov, N.S. (1995). Biomass and composition of microbial communities in soils of Northern Russia. *Global Change and Arctic Terrestrial Ecosystems* (Ed. by T.V. Callaghan, W.C. Oechel, T. Gilmanov, J.I. Holten, B. Maxwell, U. Molau, B. Sveinbjörnsson & M. Tyson), pp. 263–272. Commission of the European Communities Ecosystem Research Report, Brussels.

Tester, M., Smith, S.E. & Smith, F.A. (1987). The phenomenon of 'nonmycorrhizal' plants. *Canadian Journal of Botany*, **65**, 419–431.

Torn, M.S. & Chapin, F.S. III (1993). Environmental and biotic controls over methane flux from Arctic tundra. *Chemosphere*, **26**, 357–358.

Van Cleve, K. (1974). Organic matter quality in relation to decomposition. *Soil Organisms and Decomposition in Tundra* (Ed. by A.J. Holding, O.W. Heal, S.F. MacLean Jr & P.W. Flanagan), pp. 311–324. Tundra Biome Steering Committee, Stockholm.

Van Cleve, K. & Alexander, V. (1981). Nitrogen cycling in tundra and boreal ecosystems. *Terrestrial Nitrogen Cycles* (Ed. by F.E. Clark & T. Rosswall), *Ecological Bulletins (Stockholm)*, **33**, 375–404.

Van Cleve, K., Oliver, L.K., Schlentner, R., Viereck, L.A. & Dyrness, C.T. (1983). Productivity and nutrient cycling in taiga forest ecosystems. *Canadian Journal of Forest Research*, **13**, 747–766.

Van Cleve, K. & Yarie, J. (1986). Interaction of temperature, moisture and soil chemistry in controlling nutrient cycling and ecosystem development in the taiga of Alaska. *Forest Ecosystems in the Alaskan Taiga: A Synthesis of Structure and Function* (Ed. by K. Van Cleve, F.S. Chapin III, P.W. Flanagan, L.A. Viereck & C.T. Dyrness), pp. 160–189. Springer, New York.

Whalen, S.C. & Reeburg, W.S. (1989). A methane flux transect along the trans-Alaskan pipeline haul road. *Tellus Series B*, **42**, 237–249.

Whalen, S.C. & Reeburg, W.S. (1990). Consumption of atmospheric methane to sub-ambient concentrations by tundra soils. *Nature*, **346**, 160.

Wildman, H.G. (1987). Fungal colonisation of resources in soil – an island biogeographical approach. *Transactions of the British Mycological Society*, **88**, 291–297.

Widden, P. & Parkinson, D. (1978). The effects of temperature on growth of four high Arctic fungi in a three-phase system. *Canadian Journal of Botany*, **24**, 415–421.

Widden, P. & Parkinson, D. (1979). Populations of fungi in a high arctic ecosystem. *Canadian Journal of Botany*, **57**, 2408–2417.

Wookey, P.A., Parsons, A.N., Welker, J.M., Potter, J.A., Callaghan, T.V., Lee, J.A. & Press, M.C. (1993). Comparative responses of phenology and reproductive development to simulated environmental change in subarctic and high arctic plants. *Oikos*, **67**, 490–502.

Wookey, P.A., Robinson, C.H., Parsons, A.N., Welker, J.M., Press, M.C., Callaghan, T.V. & Lee, J.A. (1995). Environmental constraints on the growth, photosynthesis and reproductive development of *Dryas octopetala* at a high arctic polar semi-desert, Svalbard. *Oecologia*, **102**, 478–489.

Wookey, P.A., Welker, J.M., Parsons, A.N., Press, M.C., Callaghan, T.V. & Lee, J.A. (1994). Differential growth, allocation and photosynthesis responses of *Polygonum viviparum* L. to simulated environmental change at a high arctic polar semi-desert. *Oikos*, **70**, 131–139.

Wynn-Williams, D.D. (1990). Ecological aspects of Antarctic microbiology. *Advances in Microbial Ecology*, **11**, 71–146.

3. The role of bryophytes and lichens in polar ecosystems

ROYCE E. LONGTON

Department of Botany, The University of Reading, Reading RG6 2AS, UK

INTRODUCTION

This account reviews the role of bryophytes and lichens in arctic and antarctic vegetation, and in energy flow, nutrient cycling and other functional processes in polar ecosystems. Bryophytes and lichens are shown to be important in many tundra communities in terms of cover, production and phytomass. They are generally not freely consumed by herbivores, and many mosses are slow to decompose, resulting in high phytomass to production ratios. The accumulating bryophyte phytomass has high thermal insulating, water-holding and cation exchange capacities. It therefore exerts a powerful influence on soil temperature and water regimes and on nutrient cycling, and it forms a carbon sink of significance in terms of global warming.

Consideration is given to the way in which arctic bryophytes and lichens may be expected to respond to climatic change and other effects of man on the tundra environment. Instructive in this respect is a comparison of Cool-Arctic ecosystems with those on nutrient-rich antarctic islands subject to similar temperatures in summer but to warmer, wetter winters.

POLAR VEGETATION ZONES

Tundra vegetation tends to become progressively lower in stature and simpler in structure with increasing latitude. Vegetation zones may be recognized, and many schemes have been proposed, resulting in an inconsistent terminology (Table 3.1). Three zones are recognized in the Arctic, and four in the Antarctic, in Table 3.1 and Figs 3.1 and 3.2, which present a unified system to facilitate bipolar comparison. Each prefix relates to regions where lowlands support physiognomically similar communities. They can be used specifically, for example Cold-Arctic, or collectively, for example cold-polar regions. The terms sub, high, low and polar desert have been avoided because of inconsistency in past use (Table 3.1). The Mild- and Cool-Arctic correspond with the Low and High Arctic respectively of Bliss and Matveyeva (1992), except that a few northern capes and islands – the arctic polar desert region of Aleksandrova (1988) – are recognized as forming the Cold-Arctic. There is no Frigid-Arctic. The boundaries of antarctic zones are those recognized by recent authors such as Lewis Smith (1984).

TABLE 3.1. Vegetation zones in polar regions. (After Longton 1988.)

Zone	Approximately corresponding zones of previous authors		Highest mean monthly air temperature (°C)	Characteristics of the vegetation
	Arctic	Antarctic		
Mild-polar	Low Arctic (BP) Part of the mid-Arctic (P) Subarctic tundra subregion (A)	Those parts of the southern temperate region lacking arboreal vegetation (GW)	6–10 (–12)	Extensive grass heath, dwarf-shrub heath, mire and other closed phanerogamic vegetation. *Sphagnum* abundant in many mires, though local in the Mild-Antarctic. Fellfields on the drier uplands. Woodland local
Cool-polar	Most of the High Arctic (BP) Part of the mid-Arctic (P) Arctic tundra subregion (A)	Sub-Antarctic zone (G)	3–7	Open fellfields and barrens predominant but mire, dry meadow and other closed angiosperm-dominated communities locally extensive in favourable habitats. Dwarf-shrub heaths of restricted occurrence or absent. *Sphagnum* seldom a major component of mires
Cold-polar	Parts of the High Arctic (B) Arctic polar desert region (A)	Maritime Antarctic (H) Low Antarctic (W)	0–2	Closed stands of bryophytes, lichens or algae extensive where wet or mesic conditions occur, with open cryptogamic vegetation on drier ground. Herbaceous phanerogams subordinate to cryptogams or absent. Liverworts frequent
Frigid-polar	—	Continental Antarctic (H) High Antarctic (W)	<0	Vegetation largely restricted to scattered colonies of mosses, lichens or algae, and to endolithic micro-organisms. Phanerogams absent. Liverworts very rare

A, Aleksandrova 1980; B, Bliss & Matveyeva 1992; G, Greene 1964; H, Holdgate 1964; P, Polunin 1951; W, Wace 1965.

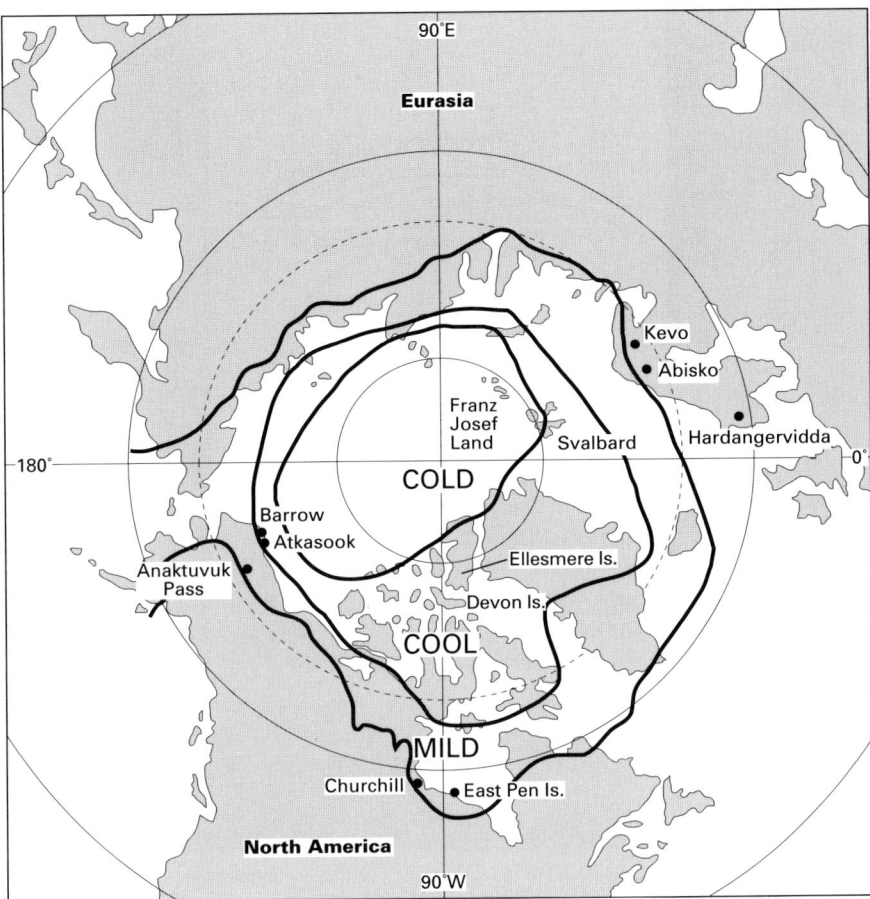

FIG. 3.1. Vegetation zones in Arctic regions. (After Longton 1988.)

The terminology emphasizes that zonation is correlated with mean summer temperature (Table 3.1), although water availability controls the distribution and luxuriance of vegetation within each zone. Other climatic parameters show less correlation, as indicated by the contrast between winter temperature, length of growing season and precipitation in the Cool-Arctic and Cool-Antarctic (Table 3.2). The scheme in Table 3.1, and its limitations, are discussed in Longton (1988), which also documents the important role of bryophytes and lichens in tundra vegetation.

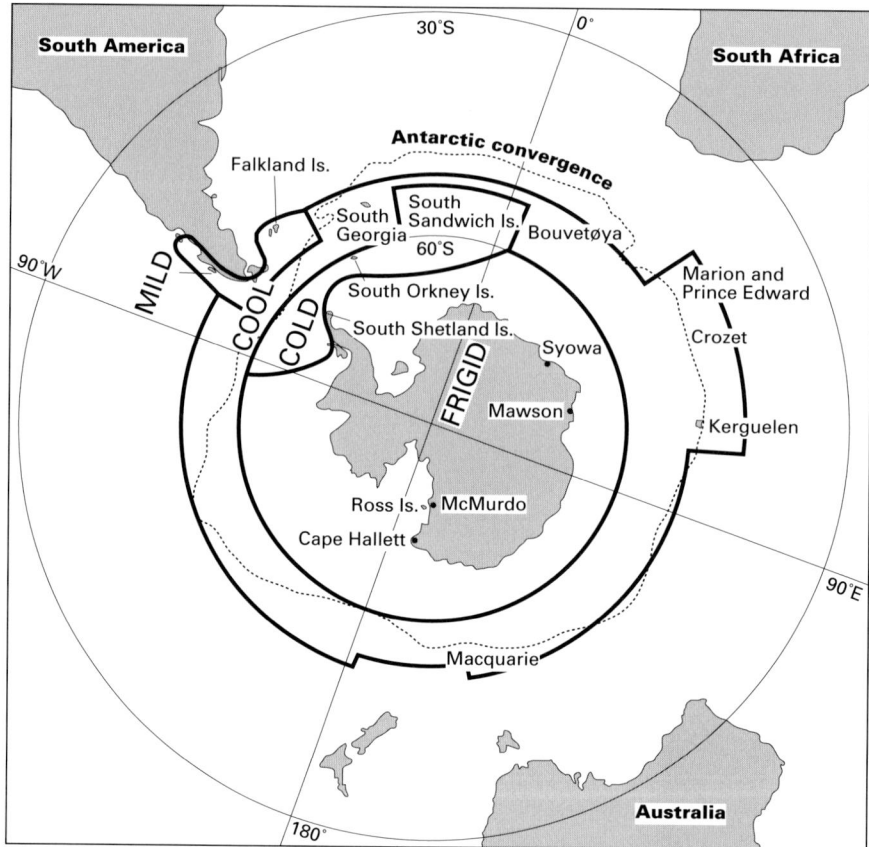

FIG. 3.2. Vegetation zones in Antarctic regions. (After Longton 1988.)

VEGETATION, PRODUCTION AND PHYTOMASS

Frigid-Antarctic

The Frigid-Antarctic (Fig. 3.2) is predominantly glaciated, with ice-free areas confined to coastal lowlands and montane nunataks. Mean monthly air temperatures remain below 0°C throughout the year, and arid conditions prevail. Macrophytic vegetation is restricted to scattered communities of lichens, small cushion- and turf-forming mosses, algae and cyanobacteria. There are no native flowering plants.

Cover is generally less than 5% on dry, stoney ground, where such mosses as *Sarconeurum glaciale* and the cosmopolitan *Bryum argenteum* grow as scattered

TABLE 3.2. Climatic data for representative polar localities. (Data from various sources as indicated in Longton 1988.)

Locality and zone	Mean monthly air temperature (°C)		Number of months with mean air temperature >0°C	Mean annual precipitation (mm rainfall equivalent)	Mean annual relative humidity (%)	Mean cloudiness (tenths)	Mean annual wind speed (m s^{-1})
	Warmest month	Coldest month					
Mild-Arctic Coral Harbour, NWT (64°08'N, 83°10'W)	8.3	−30.4	4	249	—	6.0	5.7
Mild-Antarctic Stanley, Falkland Is. (51°42'S, 57°51'W)	8.8	2.2	12	668	82	7.2	7.0
Cool-Arctic Barrow, Alaska (71°17'N, 156°47'W)	3.9	−27.9	3	110	79	—	5.4
Cool-Antarctic King Edward Point, South Georgia (54°16'S, 36°30'W)	5.3	−1.5	9	1405	75	—	4.3
Cold-Arctic Mys Chelyuskin, USSR (77°43'N, 104°17'E)	0.8	−31.1	2	294	88	—	6.5
Cold-Antarctic Signy Is. (60°43'S, 45°36'N)	1.3	−9.0	4	400	86	8.6	6.9
Frigid-Antarctic McMurdo (77°51'S, 166°40'E)	−3.4	−27.8	0	119	57	—	6.5

short turfs. Where melt water is available in summer, species of *Bryum* and *Schistidium* locally give almost continuous cover, forming elongate, contorted cushions up to 5 cm high and often colonized by crustose lichens and cyanobacteria. Lichen communities on rocks are more widespread and variable, and include *Buellia*, *Caloplaca* and *Xanthoria* spp., as well as larger foliose (e.g. *Umbilicaria* spp.) and fruticose (e.g. *Alectoria*, *Usnea* spp.) taxa. As many as 33 lichen and five moss species may occur at a given site (Kappen 1985). Under the most severely arid conditions, however, vegetation is restricted to crustose lichens on the lower surface of pebbles or to endolithic lichens and micro-organisms.

The moderate size of moss cushions in the more luxuriant communities results from longevity and slow decomposition. The green, photosynthetic layer is only 1–2 mm deep and patterns of branching and rhizoid banding suggest ages of 25–100 years for the older cushions (Matsuda 1968; Seppelt & Ashton 1978). Phytomass may reach $1250\,\mathrm{g\,m^{-2}}$ in closed stands, but annual net production is unlikely to exceed $100\,\mathrm{g\,m^{-2}}$. Production is probably less than $5\,\mathrm{g\,m^{-2}}$, with total phytomass up to $200\,\mathrm{g\,m^{-2}}$, in the more typical, open moss communities (Longton 1974; Ino 1983; Schwarz, Green & Seppelt 1992). Phytomass up to $950\,\mathrm{g\,m^{-2}}$ has been recorded for fruticose lichens (Kappen 1985). Moss and lichen production, although low, represents the major energy input.

Cold-polar regions

Antarctic

Cold-Antarctic (Fig. 3.2) vegetation benefits from frequent rain in summer and an adequate supply of mineral nutrients, including nitrogen, of marine origin. Mean air temperatures are marginally above 0°C in summer. Bryophytes and lichens flourish with little competition from the two native angiosperms.

On coastal cliffs *Verrucaria* spp. are abundant close to the high water mark, giving way to species of *Caloplaca*, *Xanthoria* and other appressed lichens above, while *Buellia*, *Lecanora*, *Lecidia* and *Placopsis* spp. occupy exposed inland rocks. Moister, more sheltered rocks, and upland soil and scree, support communities dominated by foliose (e.g. *Umbilicaria* spp.) or fruticose (e.g. *Himantormia*, *Ramalina*, *Usnea* spp.) lichens.

Bryophytes on soil include open stands of short turf- and cushion-forming acrocarpous mosses (e.g. *Ceratodon*, *Pohlia* spp.) on dry, porous volcanic substrata and cushions of *Andreaea*, *Dicranoweisia* and *Schistidium* spp. on unstable upland soils. Closed communities dominated by tall turf-forming acrocarps (*Chorisodontium*, *Polytrichum* spp.) occur on mesic slopes, giving way to carpet- and hummock-forming pleurocarps (e.g. *Brachythecium*, *Calliergidium*, *Sanionia* spp.) on ground subject to melt-water seepage. Tall turf- and carpet-forming mosses may extend continuously over several hundred square metres, the former in banks overlying peat up to 2 m deep with permafrost below 20–30 cm.

Net annual production and phytomass reach 250 g m^{-2} and 800–1750 g m^{-2} respectively in fruticose lichen communities (Lewis Smith 1984). Production has been estimated as 150–900 g m^{-2} in banks of tall turf-forming mosses and in moss carpets (Longton 1970; Baker 1972; Collins 1977). Lower values may be anticipated for other communities. In the distinctive moss banks, phytomass of green shoots reaches 300–1000 g m^{-2}, with total phytomass 20000–30000 g m^{-2} above permafrost and up to 46000 g m^{-2} to the base of the banks (Lewis Smith 1984). Upward growth rates were estimated as 0.9–1.3 mm year^{-1} over the past 200 years, with annual rates of peat accumulation 90–160 g m^{-2} (Fenton 1980). Radiocarbon dates for the base of *Chorisodontium aciphyllum* banks extend back to about 5500 years BP (Bjorck *et al.* 1991).

Arctic

Cold-Arctic (Fig. 3.1) vegetation has been described in detail by Aleksandrova (1988), who stressed the dominant role of bryophytes and lichens in communities broadly comparable in structure with those in the Cold-Antarctic. The vascular flora is richer than in the geographically isolated Cold-Antarctic. They comprise principally arctic-endemic herbaceous species, occurring in isolated tufts or small cushions and contributing little to cover or living phytomass. The vegetation is best developed on extensively glaciated islands in the Barents Sea sector, where, as in the Cold-Antarctic, melt water, light but frequent precipitation, and prevalent fog, cloud cover and high relative humidity combine to ensure adequate water availability. The most luxuriant growth occurs as a network in channels between soil polygons. Plant distribution is strongly influenced by the duration of winter snow.

Flowering plants are virtually absent where snow is removed by wind. Here, boulders support fruticose lichens, including the bipolar *Neuropogon sulphureus*, and crustose species of *Lecidea*, *Lecanora* and *Rhizocarpon*. Crustose lichens also predominate on the finer sands and gravels. Where snow persists until mid-June, loamy soil on the surface of the polygons may be covered by a colourful mosaic of crustose lichens, including species of *Collema*, *Ochrolechia* and *Pertusaria*, associated with leafy liverworts in a crust about 5 mm thick. Tall turf-forming mosses (e.g. *Ditrichum*, *Polytrichum* spp.) grow in the intervening channels with *Cetraria* spp. and other fruticose lichens. Scattered flowering plants, including *Cerastium*, *Draba* and *Papaver* spp., also grow in the channels, particularly among the mosses. The vegetation is more sparse where snow cover persists longer. The fruticose lichen *Stereocaulon rivularum* is then prominent in mesic habitats, while mat-forming pleurocarpous mosses are abundant in wetter places. A black film of small leafy liverworts (e.g. *Cephaloziella*, *Lophozia* spp.), often associated with cyanobacteria, covers the surface of the polygons.

According to Alexandrova (1970) above-ground living phytomass averages only $10\,g\,m^{-2}$ throughout the east European Cold-Arctic, but is considerably higher locally. Living phytomass was $387\,g\,m^{-2}$ in a complex of communities in Franz Josef Land, with bryophytes contributing 25%, fruticose and foliose lichens 33% and crustose lichens 18%. Angiosperms formed only 4% of the living phytomass above ground and 18% below ground, while dead phytomass, principally roots, was similar to the living ($394\,g\,m^{-2}$).

Cool-polar regions

Antarctic

The Cool-Antarctic comprises isolated oceanic islands (Fig. 3.2) characterized by high precipitation, cool relatively long summers and mild winters. Angiosperm vegetation occurs widely on coastal lowlands giving way to open fellfields, often dominated by mosses and liverworts, on the more arid uplands. The principal vascular plant communities are mires, grass heaths and the characteristically southern hemisphere herbfields and coastal tussock grasslands. Bryophytes and lichens are prominent except in tussock grassland, and even here banks of *Polytrichum* and *Chorisodontium* spp. occur locally.

Mires occupy valleys where graminoids (e.g. the rush, *Rostkovia magellanica*) are rooted in a continuous understorey of *Calliergon* spp., *Drepanocladus* spp. and other, mainly pleurocarpous, mosses. Leafy liverworts are prominent but *Sphagnum* occurs only locally. Grass heaths occupy mesic sites, and are dominated by varying proportions of caespitose grasses, for example *Festuca contracta*, and *Acaena* spp. (Rosaceae). Tall turf-forming mosses, including *Tortula robusta* and the bipolar *Polytrichum alpinum*, occur abundantly in the moister stands with *Peltigera* spp. and other foliose lichens, while *Cladonia* spp. are prominent in drier areas. Herbfields also occur in mesic sites and are dominated by species of *Acaena*, *Crassula* and other angiosperms with perennial, woody stolons giving rise to annual, herbaceous shoots. Lichens are seldom conspicious in herbfields but mosses including *Tortula robusta* and the pleurocarps *Brachythecium rutabulum* and *Sanionia uncinata*, again bipolar taxa, are often abundant.

Estimates of annual production in grass heaths and herbfields range from 850 to $1650\,g\,m^{-2}$ (Table 3.3). Total phytomass may exceed $9500\,g\,m^{-2}$, with roots and rhizomes a major component. The contribution of lichens is low. Mosses give up to 30% of above-ground production ($150–250\,g\,m^{-2}$) and 50% of above-ground phytomass, but low phytomass to production ratios indicate relatively high turnover. Moss production has been recorded at $300–700\,g\,m^{-2}$ in mires, $450–900\,g\,m^{-2}$ in

TABLE 3.3. Representative data for annual net production and phytomass (g m^{-2}) in Cool-Antarctic communities. (Data from Lewis Smith & Walton 1975 for South Georgia, and Jenkin 1975 for Macquarie Is.)

Vegetation type	Locality	Net annual production					Phytomass				
		Vascular plants		Bryophytes	Lichens	Total	Vascular plants		Bryophytes	Lichens	Total
		Above-ground	Below-ground				Above-ground	Below-ground			
Grass heath											
Festuca contracta	South Georgia	340	350	150	2	842	425 (+1598)	1642	500	12	4177
Herbfield											
Acaena magellanica	South Georgia	885	500	250	0	1635	1300 (+517)	7536	221	0	9574
Pleurophyllum hookeri	Macquarie Is.	314	550	146	4	1014	139 (+266)	1920	393	9	2727
Tussock grassland											
Poa foliosa	Macquarie Is.	1890	3670	21	0	5581	912 (+2592)	4800	6	0	8310

Data for above-ground vascular plant phytomass show living (+standing dead). Other phytomass data are totals of living plus attached dead.

moss banks, and 500–1000 g m^{-2} by streams, but only 25–100 g m^{-2} in fellfields and rock faces (Longton 1970; Lewis Smith 1982; Russell 1990). Production and phytomass of *Cladonia rangiferina* in *Festuca* grassland may reach 100 g m^{-2} and 1840 g m^{-2} respectively (Lewis Smith 1993).

Arctic

The Cool-Arctic (Fig. 3.1) experiences short, cool summers and prolonged, severe winters (Table 3.2). Precipitation is typically low and the generally sparse, open nature of the vegetation may be largely attributed to aridity.

Wetland communities occur widely, but seldom extensively, at sites of impeded drainage. They range from homogeneous stands of grasses and sedges (e.g. *Arctagrostis, Carex, Dupontia* spp.) to more varied assemblages showing pronounced patterns of microrelief caused by cryoturbatic disturbances or hummock formation by mosses and graminoids such as *Eriophorum vaginatum*. Homogeneous mires include a closed moss layer of large pleurocarps (e.g. *Calliergon, Drepanocladus, Tomenthypnum* spp.) and tall turf-forming acrocarps (e.g. *Cinclidium, Polytrichum* spp.). The hummocky mires support a wider variety of cryptogams with large mosses in hollows giving way to fruticose lichens (e.g. *Cetraria, Cladonia, Ochrolechia* spp.) and smaller mosses such as *Seligeria polaris* on the hummocks. *Sphagnum*-dominated mires occur only locally, and dwarf-shrub heath is seldom extensive.

Open vegetation such as graminoid heaths, fellfields and barrens dominate the landscape. In graminoid heaths, the tallest plants are tufted species (e.g. *Carex, Kobresia, Luzula* spp.) associated with creeping willows, cushion and rosette plants. However, vascular cover is low, commonly under 20%. Mat- and tall turf-forming mosses (e.g. *Aulacomnium, Ditrichum, Tomenthypnum* spp.) are abundant, with fruticose and foliose lichens often conspicuous but seldom abundant. Both mosses and lichens may be abundant in fellfields, where the angiosperms form a discontinuous cover of cushion and rosette plants (e.g. *Draba, Dryas, Saxifraga* spp.). Mat-forming mosses (e.g. *Hylocomium, Hypnum, Racomitrium* spp.) predominate in relatively moist areas, often associated with turf and cushion forms (e.g. *Andreaea, Ditrichum* spp.) and with leafy liverworts such as *Gymnomitrion corraloides*. Lichens occur abundantly in drier areas, and include crustose (e.g. *Lecidea, Ochrolechia* spp.), fruticose (e.g. *Cladonia, Sphaerophorus* spp.) and foliose (e.g. *Hypogymnia, Parmelia* spp.) taxa. Variation in microrelief associated with frost action again increases species diversity. Barrens occupy extensive tracts of arid scree, gravel and clay and support only scattered angiosperms, with occasional mosses (e.g. *Ditrichum, Hypnum, Racomitrium* spp.) and lichens such as *Thamnolia subobscura*.

Total production ranges from 100 to 300 g m^{-2} in mires and dwarf-shrub heaths, with the below-ground angiosperm component predominating, but is generally

TABLE 3.4. Representative data for annual net production and phytomass (g m^{-2}) in representative Cool-Arctic communities. (Data from Bliss 1977; Bliss & Svoboda 1984.)

Vegetation type	Locality	Annual net production					Phytomass				
		Vascular plants		Bryophytes	Lichens	Total	Vascular plants		Bryophytes	Lichens	Total
		Above-ground	Below-ground				Above-ground	Below-ground			
Wet meadow											
Wet sedge moss meadow	Devon Is.	46	130	103	0	279	78 (+120)	1295	1097	0	2592
Hummocky sedge moss meadow	Devon Is.	45	104	33	0	182	86 (+187)	2023	908	0	3208
Frost-boil sedge moss meadow	Devon Is.	58	119	15	0	193	112 (+202)	1332	1100	0	2748
Grass heath											
Graminoid steppe	Elef Ringness Is., Canada	13	13	32	<1	58	13 (+74)	88	2128	20	2323
Graminoid steppe	Elef Ringness Is., Canada	4	4	1	<1	9	4 (+120)	519	76	9	728
Moss–graminoid meadow	King Christian Is., Canada	5	5	32	<1	42	41*	23	2136	10	2210
Dwarf-shrub heath											
Cassiope tetragona heath	Devon Is.	18	90	20	4	132	159 (+228)	1041	423	48	1899
Fellfield											
Cushion plant–lichen fellfield	Devon Is.	15	3	2	3	23	89 (+298)	57	15	49	508
Cushion plant–moss fellfield	Devon Is.	27	5	20	2	54	126 (+192)	50†	600	23	991
Barren											
Papaver radicata barren	Devon Is.	0.5	1.0	0.1	0	1.5	310 (+82)	0.9	2.4	0	15

Data for above-ground vascular plant phytomass are living (+standing dead). Other phytomass data are living plus attached dead. Total production and phytomass include small algal components in some cases.

* Sum of living plus standing dead.

† Live roots only.

below 60 g m^{-2} in the more xeric communities and is negligible in barrens (Table 3.4). Bryophyte production locally exceeds 100 g m^{-2} in mires, and is generally comparable with, or greater than, the above-ground angiosperm component. Phytomass greatly exceeds production in most communities, and high phytomass to production ratios are characteristic of the mosses, as well as angiosperms, roots and rhizomes. The living and dead components of the bryophyte layer cannot readily be distinguished, as regeneration occurs from brown tissue to a considerable depth, under experimental conditions. The contribution of lichens to both production and phytomass is small in Table 3.4 but may be higher locally, lichens forming up to 44% of the above-ground phytomass in some grass heaths (Webber 1978).

Mild-polar regions

Antarctic

The Mild-Antarctic, comprising the Falkland Islands and the magellanic moorland region of south-west Chile (Fig. 3.2), is marked by cool summers, moderate to high precipitation and mean temperatures above 0°C throughout the year. The vegetation is a complex of coastal tussock grassland, mire, dwarf-shrub heath and grassland dominated by tall-growing species such as *Cortadaria pilosa*. The magellanic moorland region supports occasional stands of woodland with both conifers and broad-leaved evergreens (*Nothofagus* spp.). Pleurocarpous mosses and *Sphagnum* spp. are abundant in most communities, and bryophytes are a striking feature of the woodland where leafy hepatics form thick mats on the ground and on the trees (Moore 1979; Greene, Hassel de Menendez & Matteri 1985).

Arctic

The Mild-Arctic (Fig. 3.1) is warmer in summer than the Cool-Arctic, conditions are less arid, but the winters remain severe. Graminoid mires are extensive in the wetter lowlands and differ from those further north in a greater abundance of *Sphagnum* spp. and the presence of deeper peat. On more mesic terrain, mire commonly intergrades with dwarf-shrub heath dominated by *Betula nana*, *Empetrum* spp. and ericoids (e.g. *Cassiope*, *Ledum*, *Vaccinium* spp.), with a varying admixture of graminoids. Mosses are abundant (e.g. *Aulacomnium*, *Sphagnum*, *Tomenthypnum* spp.), as are both fruticose (e.g. *Cladonia* spp.) and foliose (e.g. *Peltigera* spp.) lichens. Increasingly towards the south of the zone, similar vegetation exists beneath taller shrubs (e.g. *Alnus*, *Salix* spp.).

Fellfields and grass heaths occupy much of the uplands, while in the lowlands the arctic and boreal zones merge in a broad ecotone comprising a mosaic of tundra and open woodland with *Betula*, *Larix* or *Picea* spp. Woodland typically includes a dwarf-shrub stratum over a continuous layer of robust, weft-forming mosses (e.g. *Hylocomium*, *Pleurozium* spp.) or, at drier sites, of fruticose lichens (e.g. *Cetraria*, *Cladonia*, *Stereocaulon* spp.). The trees commonly support festoons of epiphytic lichens (e.g. *Alectoria*, *Usnea* spp.). Also widespread are lichen heaths with only scattered angiosperms and a variety of communities dominated by mosses, notably *Racomitrium* heath on islands such as Iceland and Jan Mayen.

Bliss and Matveyeva (1992) estimate mean standing crop to range from $1470 \, \mathrm{g \, m^{-2}}$ in Mild-Arctic fellfields to $5800 \, \mathrm{g \, m^{-2}}$ in tall-shrub communities, and total organic matter from $4010 \, \mathrm{g \, m^{-2}}$ in fellfields to $43\,310 \, \mathrm{g \, m^{-2}}$ in mires. These authors indicated that annual production probably averages $45 \, \mathrm{g \, m^{-2}}$ in fellfields to $1000 \, \mathrm{g \, m^{-2}}$ in tall-shrub communities.

Annual production of *Sphagnum* spp. in northern mires is generally between $70 \, \mathrm{g \, m^{-2}}$ on hummocks and $400 \, \mathrm{g \, m^{-2}}$ in wet depressions, with that of pleurocarpous mosses in fens in the lower half of this range (Vitt 1990). However, Murray, Tenhunen and Kummerow (1989) reported low *Sphagnum* production ($10 \, \mathrm{g \, m^{-2}}$) in extensive, relatively dry tussock tundra in Alaska, rising locally to $164 \, \mathrm{g \, m^{-2}}$ in wetter, more shaded areas. The difference was attributed to variation between species in response to water availability and high irradiance (Murray, Tenhunen & Nowak 1993). Moss production in mesic spruce woodland is commonly around $100 \, \mathrm{g \, m^{-2}}$, and is roughly comparable with above-ground production of the trees (Weber & Van Cleve 1984).

In lichens, Andreev (1954) showed that annual phytomass increase in stands of *Cladonia* spp. may average $17\text{--}27 \, \mathrm{g \, m^{-2}}$ during the relatively young 'accumulation phase' of the colonies in various Soviet tundra and taiga communities. In Fennoscandia, lichen phytomass reaches $300\text{--}450 \, \mathrm{g \, m^{-2}}$ in alpine lichen heaths (Wielgolaski & Kjelvik 1975) and $300 \, \mathrm{g \, m^{-2}}$ in mature *Cladonia stellaris* woodland (Ahti 1977).

Synthesis and adaptations

Mosses, and to a lesser extent lichens, are dominant in terms of production and phytomass in frigid- and cold-polar regions. Annual production is low in the Frigid-Antarctic, but reaches $1000 \, \mathrm{g \, m^{-2}}$ under humid conditions in the Cold-Antarctic. Here, production of mosses is comparable with total production in some temperate ecosystems, although the annual growth increment may be less than $5 \, \mathrm{mm}$ (Longton 1970). Mosses also contribute significantly to above-ground production and phytomass in mild- and cool-polar communities, and arctic mosses show high phytomass to production ratios indicative of slow turnover. Moss production generally increases with water availability. It is greater in mires than in mesic and xeric communities, and a similar trend is evident in a comparison of different mire communities on Devon Island (Table 3.4). Annual production of mosses, and probably also of lichens, is conspicuously higher in the Cool-Antarctic than over much of the Arctic, a fact probably attributable to the longer growing season, combined with higher precipitation and greater availability of nutrients of marine origin. Lichens are generally less productive than mosses, but they contribute substantial phytomass in lichen heaths and woodlands.

Factors responsible for the success of mosses and lichens in polar regions have been discussed by Kershaw (1985), Longton (1988) and Sveinbjornsson and Oechel (1992). The two groups are remarkably similar in attributes beneficial in severe environments. Both tend to occupy, and indeed to create, relatively favourable microenvironments. Many species exhibit a broad response of net assimilation rate to temperature, with maxima at 10–15°C but with positive net assimilation and dark respiration continuing at or below 0°C. Light compensation and saturation intensities are typically lower in mosses and lichens than in vascular plants, and compensation levels decrease with temperature permitting positive net assimilation under cool, low-light conditions.

Retention within the colonies of CO_2 derived from respiration of the mosses or of micro-organisms may raise the concentration around moss leaves substantially above ambient, although this effect varies with growth form even within a species (Tarnawski et al. 1992). Responses of net assimilation rate to environmental factors commonly show acclimation to changing conditions during the growing season. Photosynthesis in arctic mosses does not appear to be inhibited by increasing concentration of assimilate, nor to decline with senescence late in the growing season. At least in mosses, positive net assimilation can thus begin under snow cover in spring, may be maintained 24 h per day in mid-summer, and shows no significant decline under favourable conditions late in the growing season (Oechel & Sveinbjornsson 1978), despite reduction due to photoinhibition as recorded in both mosses and lichens (Adamson et al. 1988; Kappen, Breuer & Bolter 1991; Murray et al. 1993).

Poikilohydry may also be an adaptive feature in polar bryophytes and lichens. Most species have little access to soil moisture and lack an effective cuticle. This enables them to absorb water through much of their surface but results in rapid water loss under drying conditions. In compensation, the cytoplasm of many mesic and xeric species is relatively desiccation tolerant, the plants becoming inactive when dry, but resuming normal metabolism rapidly on remoistening. The physiological basis for desiccation tolerance in mosses is discussed by Seel, Hendry and Lee (1992). Mosses and lichens are thus adapted to switching rapidly between periods of metabolic activity and rest, utilizing favourable conditions whenever they occur. This may be facilitated by micromorphological features thought to facilitate simultaneous uptake by moss leaves of both water and CO_2 (Noakes & Longton 1989). Poikilohydry may also enhance frost resistance, by conferring tolerance of cytoplasmic dehydration resulting from extracellular ice formation. Freezing-point depression by soluble sugars and, especially in lichens, by polyhydric alcohols may also be important in frost resistance (Melick & Seppelt 1994).

Many polar bryophyte and lichen species also occur widely in boreal and temperate regions, and the features discussed above cannot be regarded as specific adaptations to polar conditions. However, apparently adaptive morphological and physiological variation between polar and temperate populations has been reported within moss species (Sveinbjornsson & Oechel 1983; Longton 1994a).

ENERGY FLOW AND CARBON BALANCE

Herbivory

Tundra ecosystems are marked by limited herbivory, both in the Cold-Antarctic, where the cryptogamic vegetation supports an entirely invertebrate fauna (Davis 1981), and in the Arctic with its wider range of plants and animals (Bliss 1986). Lichens and moss gametophytes are less freely consumed than angiosperms in the Arctic (Chernov 1985), as elsewhere (Lawrey 1986; Davidson, Harborne & Longton 1990).

Data reviewed in Longton (1988) suggest that the energy content of bryophytes and lichens, and the concentrations of organic nutrients, are only slightly lower than in associated angiosperms. Low digestibility related to a high crude fibre content, resulting in part from lignin-like phenolic compounds in the cell walls (Hebant 1977; Davidson, Harborne & Longton 1989), is thought to be a major cause of the low consumption of mosses. This is consistent with observations that arthropods with sucking mouthparts, such as mites, appear to be most commonly implicated in bryophyte herbivory. However, consumption of mosses by tardigrades, dipteran and lepidopteran larvae, and by weevils in the Cool-Antarctic, has also been reported. Green capsules are more often consumed than gametophyte shoots (Smith 1977; Gerson 1987; Davidson *et al.* 1990), but some arctic animals eat moss gametophytes and lichens on a scale seldom reported elsewhere. Prins (1982) speculated that animals such as geese may benefit from arachidonic acid which is present in mosses. This highly unsaturated fatty acid could increase limb mobility at low temperature and protect cell membranes against cold.

Lichens are well known to be crucial for caribou and reindeer (*Rangifer tarandus*), and therefore to indigenous human populations reliant on these animals. Graminoids, willow leaves and forbs are preferred in summer, but lichens often represent some 60–70% of winter food (Boertje 1984; Chernov 1985). *Cladonia* and *Cetraria* spp., dug from beneath snow in woodland and lichen heath, appear to be eaten in preference to the nitrogen-fixing *Stereocaulon* spp. (Kallio 1975). Arboreal species of *Alectoria*, *Evernia* and *Usnea* are also taken. There are some

700000 semi-domesticated reindeer in Fennoscandia alone, each requiring up to 5 kg dry wt of lichen daily in winter and grazing about 2000 m^{-2} in 6 months (Bliss 1975; Moser, Nash & Thomson 1979). Management to prevent overgrazing is complicated because the relationship between net assimilation rate and mat thickness in fruticose lichens reflects a balance between the positive effect of increasing mat thickness on water content and the negative effect of respiration in lower, non-photosynthetic parts of the deeper colonies (Andreev 1954; Sveinbjornsson 1987).

Lichens are rich in carbohydrate, are readily digested by caribou though not by other mammals, and form an effective and available energy source in winter when a high basal metabolism is required to generate body heat (Scotter 1972; White & Trudell 1980). They are deficient in protein, lipids and several essential mineral elements and captive animals lose weight on a lichen diet without a nitrogen supplement. In nature the low-quality diet is accommodated by utilizing fat reserves and breaking down muscle, with replenishment through feeding on nutritious young angiosperm leaves in spring (Richardson & Young 1977; Klein 1982; Skogland 1984; Staaland, Jacobsen & White 1984). Batzli, White and Bunnell (1981) argue that *R. tarandus* recycles nitrogen from urea to the rumen, has specialized kidneys which concentrate urea when dietary protein is low, and that low protein intake may be beneficial in winter by decreasing urinary and faecal water loss and in turn the energy required to heat ingested frozen water. Klein (1982) suggested that *R. tarandus* evolved in response to a lichen-based food niche unoccupied by other animals.

Consumption of mosses by *R. tarandus* is low, except on Cool-Arctic islands where preferred lichens are scarce and the animals are unable to migrate southwards in autumn. Here, rumen samples in winter contained 30–60%, and occasionally up to 86% moss (Reimers 1977; Thomas & Edmonds 1983). Digestibility of moss by *R. tarandus* appears to be low (Thomas & Kroeger 1981), and Svalbard reindeer have an unusually large caecum–colon complex, possibly as an adaptation to a low-quality diet (Staaland, Jacobsen & White 1979).

Mosses form a minor part of the diet of several arctic rodents (Batzli & Sobaski 1980; West 1982), and are more freely consumed by *Lemmus* spp. In peak years consumption by lemmings (*L. sibericus*) in Alaska may reach 25% of above-ground primary production, with mosses forming 5–20% of the diet in summer and 30–40% in winter. Grazing can then damage bryophyte and lichen vegetation through both consumption and disturbance as the animals seek graminoid shoot bases beneath snow (Bunnell, MacLean & Brown 1975). Mosses are not freely digested by lemmings. Their value could lie in availability during winter, or as a source of unsaturated fatty acids (Prins 1982) or of mineral nutrients such as calcium, magnesium and iron (Batzli *et al.* 1981).

Decomposition

Slow decomposition is characteristic of *Sphagnum*, particularly the hummock-forming species (Johnson & Damman 1993), and of many other mosses. Decomposition of *Dicranum* and *Drepanocladus* spp., as well as *Sphagnum* spp., at Fennoscandian tundra sites was only 4–7% within 1 year (Rosswall, Veum & Karenlampi 1975). Similar values were reported for woody tissue, whereas loss of herbaceous tissues ranged from 18% (*Pinus sylvestris* leaves) to 37% (*Eriophorum vaginatum*). In laboratory tests on boreal forest mosses and tree foliage, decomposition rate in *Pleurozium schreberi* was conspicuously low, with that in *Hylocomium splendens* at the lower end of the range for leaves (Fyles & McGill 1987). In the Cold-Antarctic, annual decomposition was less than 2% of production in mesic, semi-ombrogenous moss turfs and 4–25% in waterlogged soligenous moss carpets (Baker 1972; Yarrington & Wynne-Williams 1985).

Slow decomposition, as reflected in the high bryophyte phytomass in wetland and mesic communities (Table 3.4), implies that mosses have contributed significantly to the Arctic carbon sink discussed by Oechel *et al.* (see Chapter 12). Based on conservative estimates in Clymo and Hayward (1982), northern peatlands contain 120000×10^6 t of carbon, equivalent to 24 years' emission from fossil fuels at the present rate, and more than 50% of emissions since 1860, with half derived from *Sphagnum* and a further component from other mosses. Annual carbon fixation by slowly decomposing mosses in peatlands and boreal forests was estimated as 6.5% of current emissions from fossil fuels (Longton 1992). Gorham (1991) believes peatlands to contain substantially more carbon than estimated above, with the effect on climate change complicated by CH_4 release.

Not all mosses decompose slowly – as exemplified by *Tortula robusta*, an abundant moss in Cool-Antarctic *Acaena* herbfield. In one set of observations, phytomass of *T. rubusta* fell from $500 \mathrm{g\,m}^{-2}$ in early spring to $125 \mathrm{g\,m}^{-2}$ in mid-summer, as increasing leaf area in *Acaena* reduced moss photosynthesis and the lower parts of the shoots decomposed. Leaf fall in *Acaena* allowed renewed moss growth later in the summer, and moss phytomass recovered to $425 \mathrm{g\,m}^{-2}$ in autumn (Walton 1973). Similar seasonal change in moss phytomass has been reported in a Cool-Antarctic mire (Russell 1990). On present evidence, bryophyte production is higher, but phytomass lower in mesic communities in the Cool-Antarctic than in the Cool-Arctic (Tables 3.3 & 3.4).

Rates of moss decomposition tend to increase along a gradient of increasing tissue mineral content, N:C ratios and soil moisture (Russell 1990), but rates are reduced under the acid, waterlogged, anaerobic conditions in *Sphagnum* mires. There are related, negative correlations between decomposition and crude fibre content of the tissues, estimated on South Georgia as 15–23% in *Tortula robusta* compared with 36% in the more slowly decomposing *Polytrichum alpestre* (Pratt

& Lewis Smith 1982). A similar mechanism may thus inhibit both digestion (p. 83) and decomposition.

NUTRIENT CYCLING

Arctic ecosystems receive a higher proportion of nutrient input from precipitation and nitrogen fixation than do temperate systems, because chemical weathering is inhibited by low temperature and permafrost. Mosses and lichens have a major influence on nutrient cycling in tundra and other northern ecosystems through their role in nitrogen fixation, and the ability of mosses to accumulate and retain elements from precipitation (see Chapter 10; Longton 1984, 1988; Oechel & Van Cleve 1986). Retention of precipitation by bryophytes is also likely to reduce losses by leaching of nutrients already present in the soil.

Nitrogen fixation

Biological fixation is of major significance in nitrogen-deficient northern eco-systems, although fixation rates are lower than at temperate latitudes (Dowding *et al.* 1981). Fixation by *Stereocaulon paschale* and *Nephroma arcticum*, lichens with cyanobacterial photobionts, at rates of $380 \, mg \, N \, m^{-2} \, year^{-1}$ and locally up to $3 \, g \, N \, m^{-2} \, year^{-1}$, has been reported in Fennoscandian lichen woodland (Kallio 1975). However, Gunther (1989) indicated fixation by lichens to be of minor significance in Alaskan forests.

Fixation by *Peltigera* and *Stereocaulon* spp. also occurs in the more xeric tundra communities (Alexander, Billington & Schell 1978), but free-living cyanobacteria and heterotrophic bacteria appear to be generally more important. These organisms occur only locally in polar soils (Van Cleve & Alexander 1981), but are abundant in bryophyte colonies in wetland and mesic habitats. Nitrogen-fixing micro-organisms, associated with *Calliergon* and *Drepanocladus* spp. in wet tundra, with *Sphagnum* spp. in northern mires, with *Hylocomium* and *Pleurozium* spp. in wood-lands and with *Funaria hygrometrica* during primary succession, are considered to represent major contributors of fixed nitrogen (Rodgers & Henriksson 1976; Alexander *et al.* 1978; Billington & Alexander 1978; Rosswall & Granhall 1980). Studies with ^{15}N have confirmed the release of NH_4–N from cyanobacteria growing on tundra mosses and its transfer to the moss and to vascular plants (Alexander *et al.* 1978).

Nutrient uptake and retention

Bryophytes act as efficient filters of nutrients arriving in precipitation, throughfall or litter by absorbing them directly into their tissues, or retaining them externally in solution in capillary spaces. Many species also take up nutrients from the soil. The annual growth increment of the moss layer at an Alaskan taiga site was found to contain nutrients in excess of inputs from throughfall. Interception by mosses of nitrogen in precipitation and throughfall has been shown to result in an increase in concentration in the soil, and uptake of nitrogen from precipitation into lichens also occurs (Crittenden 1983; Oechel & Van Cleve 1986; Bowden 1991).

Elements such as phosphorus and potassium that occur primarily in solution in the cytoplasm are likely to be recycled from older to younger tissue in both mosses and lichens, possibly permitting higher growth rates than could be supported by the external nutrient supply alone. However, some may be released by leaching, particularly following membrane damage by desiccation or freezing (Brown 1984; Crittenden 1991). Mosses accounted for 75% of the annual accumulation of phosphorus in an Alaskan black spruce (*Picea mariana*) forest, and Chapin *et al.* (1987) suspected that mycorrhizal fungi may be involved in transferring it from the moss to tree roots.

Other elements are strongly retained in moss phytomass even after death of the plants. This applies particularly to divalent cations (e.g. Ca^{2+} and Mg^{2+}) associated with anionic exchange sites in the cell wall (Brown 1982). Also, some nitrogen is often retained in association with phenolic compounds in the cell wall (Berg 1984). These elements are made available to other organisms on decomposition of the moss. Weetman (1968) noted that tree roots in black spruce (*Picea mariana*) woodland are concentrated near the base of the moss layer, and suggested that decomposing mosses provide a collecting point for elements absorbed from throughfall. Weber and Van Cleve (1984) demonstrated ^{15}N uptake by boreal forest feather mosses and reached a similar conclusion.

Nutrient immobilization in slowly decomposing bryophyte phytomass may thus have a major influence in restricting recycling, and therefore in controlling ecosystem development and productivity. Dowding *et al.* (1981) estimated that the brown layer of mosses contained 50% of the calcium present in dead phytomass in mesic tundra meadows on Devon Island, with a turnover time of 22 years. Similarly in mires, absorption of nitrogen and other elements by *Sphagnum* reduces availability to other plants (Pakarinen 1978; Woodin, Press & Lee 1985). In boreal forests cover and phytomass of mosses tend to increase as succession proceeds from deciduous woodland (*Betula*, *Populus* spp.) through a white spruce (*Picea glauca*) stage to the black spruce climax. Increasing nutrient immobilization in the mosses is thought to be responsible for reducing vascular plant production as succession proceeds (Oechel & Van Cleve 1986). Bryophytes may therefore increase the pools of nutrients in northern ecosystems, but reduce availability to other organisms.

OTHER EFFECTS OF
BRYOPHYTES AND LICHENS

Mosses and lichens also affect tundra ecosystems in other ways. They are particularly prominent in seral communities, and may be influential in driving processes of primary, secondary and cyclic succession (Longton 1988, 1992). They provide microenvironments of vital importance for invertebrates, and in some communities for the establishment of vascular plants although the relationships may be complex (Bliss & Svoboda 1984; Sohlberg & Bliss 1987). Lichens release compounds capable of suppressing the growth of associated vascular plants and bryophytes (Lawrey 1986). *Sphagnum* spp. control the environment of mires by lowering pH, probably by releasing H^+ ions in exchange for other cations (Gagnon & Glime 1992), and creating waterlogged, anaerobic conditions to which only a characteristic range of other organisms is adapted.

A layer of moss or lichen acts as an effective mulch, retaining moisture in the upper layers of the soil. Mosses and their undecomposed remains are particularly efficient in thermal insulation when dry, thus restricting heat penetration into arctic soils in summer. When wet and frozen in winter, their effect in reducing heat flux away from the soil is reduced. The net effect of mosses or peat in decreasing soil temperatures in summer is generally greater than the converse effect in winter, and over much of the Arctic the distribution of permafrost is positively correlated with that of mire vegetation underlain by moss and peat (Ives 1974). Thermocarst resulting from destriction of the vegetation by the summer use of tracked vehicles during early stages of arctic oil exploration (Bliss 1970) demonstrated vividly the importance of the moss layer in maintaining permafrost. This phenomenon could occur more widely if warmer, possibly drier conditions led to extensive melting of permafrost, both directly and by accelerating the decomposition of organic matter.

THE FUTURE

It is believed that arctic environments will become subject to enhanced CO_2, higher temperatures particularly in winter, and increased winter precipitation. Predictions differ regarding changes in soil moisture in summer. Some models indicate increased summer precipitation in certain areas, while others suggest that drying of tundra soils, already in evidence locally, will increase (Maxwell 1992). However, enhanced winter precipitation would result in greater topographically related variation in soil moisture, and thus in vegetation. Acid deposition will add nutrients, notably nitrogen, and UV-B radiation may rise. There have been many contrasting predictions of the associated changes in arctic ecosystems, including the likelihood that accelerated decomposition of existing phytomass, much of it derived from mosses, will have a significant feedback effect on global CO_2 levels (Johnson & Damman 1993; Gignac & Vitt 1994; see Chapter 12).

The cover of vascular plants is likely to increase, reducing the degree to which cryptogams dominate the vegetation, particularly in cold-polar regions. Already, a rapid expansion in populations of the two native vascular plants in the Cold-Antarctic has been noted at some sites and correlated with rising temperature (Fowbert & Lewis Smith 1994). Bryophytes and lichens may nevertheless remain abundant as an understorey, as in many existing communities. Encroachment of tall shrubs into mires could increase production of some *Sphagnum* species, whose growth Tenhunen *et al.* (1992) believe currently to be restricted by high irradiance. However, Chapin *et al.* (Chapter 4) report that nutrient addition and artificial increase in temperature led to a decrease in the relative contribution of mosses to phytomass in an Alaskan mire over a period of years.

The abundance and productivity of bryophytes will undoubtedly be influenced by any changes in precipitation. A warming trend combined with reduced summer precipitation would result in a shift towards more xeric species, which may have inherently low growth rates (Noakes & Longton 1989), resulting in reduced productivity. However, mesic and hydric communities would probably persist in topographically favourable sites fed by melt water from winter snow.

Conversely, bryophyte production would probably be enhanced where warming is accompanied by increased precipitation in summer by analogy with the taiga region. Cool-Antarctic islands provide a further analogue, having cool summers but warmer winters and higher precipitation than much of the Arctic (Table 3.2). These oceanic islands are also less nutrient deficient than much of the Arctic, and current data suggest that total productivity, and bryophyte productivity, are both higher. Overall phytomass to production ratios are also high, but bryophyte turnover appears to be more rapid (Tables 3.3 & 3.4), and the influence of mosses in sequestering nutrients is thus probably less than in the Arctic.

Here, it must be emphasized that increases in nitrogen deposition, and possible release of nutrients from existing phytomass, must be considered when predicting the responses of arctic vegetation to climatic change, for nutrient availability is currently a major limiting factor (Chapin 1983). Again the responses are difficult to predict. Vascular plant growth is likely to be enhanced by the moderate increases in nitrogen likely at high latitudes. For bryophytes, there is evidence of variation between species in response to enhanced nutrient availability (Li & Vitt 1994). A deterioration in *Racomitrium* heath at British montane sites has been attributed to nitrogen deposition (Thompson & Baddeley 1991). Growth of some *Sphagnum* species may also be inhibited by nitrogen deposition (Lee, Parsons & Baxter 1993), but growth of *S. fallax*, currently an invasive species in European mires, is promoted by treatment with nitrate and ammonia (Twenhoven 1992).

Such differences in response could result in significant changes in the floristic composition of tundra bryophyte vegetation. The new conditions might favour competitive species with seasonally high rates of growth and decomposition, and with growth enhanced by nutrient addition. Such taxa are well adapted to growing under a herbaceous canopy, which reduces light availability during much of the growing season but also provides an annual nutrient flux from litter deposition. They could tend to replace conservative species showing lower growth rates, sustained over longer periods and not significantly increased by nutrient addition, and lower rates of decomposition (Rincon 1988). The analogy of the Cool-Antarctic islands (page 89) is particularly relevant to this scenario.

There is some evidence that increasing atmospheric CO_2 concentrations may enhance nitrogen fixation in lichens (Norby & Sigal 1989), thus further raising availability. While this may be damaging to some bryophytes, increasing CO_2 concentrations could have a greater long-term stimulating effect on photosynthesis and productivity in bryophytes than in vascular plants because of the lack of end-product inhibition (Sveinbjornsson & Oechel 1992). The results of Jauhiainen, Vasander and Silvola (1994) with *Sphagnum* suggest, however, that the response may be complex and interact with the effects of increasing nutrient availability.

Finally, it must be emphasized that while bryophytes and lichens are particularly susceptible to atmospheric pollutants because of direct uptake into photosynthetic cells, they are, above all, opportunistic organisms, and are likely to be relatively resilient to climatic change. The majority of tundra species range widely in other biomes. In bryophytes, wide phenotypic plasticity permitted by individual genotypes may be important in this respect, as well as genetic differentiation within species (Longton 1994a,b), while environmental relationships of CO_2 exchange show acclimation to changes in ambient conditions of temperature and other factors in many mosses and lichens (Kershaw 1985; Longton 1988). Genetic variability within species, and thus capacity for adaptation, could be enhanced by increasing UV radiation as suggested by Fahselt's (1993) studies on lichens, which appear to be less prone to damage by UV-B than bryophytes and other plants.

Poikilohydry, allowing the plants to switch rapidly between states of metabolic activity and rest as dictated by environmental conditions, contributes to the ecological flexibility of both mosses and lichens. Moreover, bryophytes can survive as small populations in microhabitats where conditions are far different from ambient. Some arctic endemics have their closest relatives among tropical floras, and Steere (1978) believed them to have survived in the Arctic since a period of temperate or subtropical conditions in the early Tertiary. While some rare taxa may be at risk, polar bryophytes in general appear to be less in danger of extinction through climatic change than is the case with the substantial number of arctic endemic angiosperms.

REFERENCES

Adamson, H., Wilson, M., Selkirk, P. & Seppelt, R. (**1988**). Photoinhibition in Antarctic mosses. *Polarforschung*, **58**, 103–111.

Ahti, T. (**1977**). Lichens of the boreal coniferous zone. *Lichen Ecology* (Ed. by M.R.D. Seaward), pp. 145–181. Academic Press, London.

Aleksandrova, V.D. (**1980**). *The Arctic and Antarctic; their Division into Geobotanical Areas.* Cambridge University Press, Cambridge.

Aleksandrova, V.D. (**1988**). *Vegetation of the Soviet Polar Deserts.* Cambridge University Press, Cambridge.

Alexander, V., Billington, M. & Schell, D.M. (**1978**). Nitrogen fixation in arctic and alpine tundra. *Vegetation and Production Ecology of an Alaskan Arctic Tundra* (Ed. by L.L. Teiszen), pp. 539–558. Springer, New York.

Alexandrova, V.D. (**1970**). The vegetation of the tundra zones in the USSR and data about its productivity. *Productivity and Conservation in Northern Circumpolar Lands* (Ed. by W.A. Fuller & P.G. Kevan), pp. 93–114. Morges, Switzerland.

Andreev, V.N. (**1954**). Pirost kormovykh lishainikov i priemy ego regulirovaniia. *Geobotanica*, **9**, 11–74.

Baker, J.H. (**1972**). The rate of production and decomposition in *Chorisodontium aciphyllum* (Hook. f. et Wils.) Broth. *British Antarctic Survey Bulletin*, **27**, 123–129.

Batzli, G.O. & Sobaski, S.T. (**1980**). Distribution, abundance and foraging patterns of ground squirrels near Atkasook, Alaska. *Arctic and Alpine Research*, **12**, 501–510.

Batzli, G.O., White, R.G. & Bunnell, F.L. (**1981**). Herbivory; a strategy of tundra consumers. *Tundra Ecosystems: A Comparative Analysis* (Ed. by L.C. Bliss, O.W. Heal & J.J. Moore), pp. 359–375. Cambridge University Press, Cambridge.

Berg, A. (**1984**). Decomposition of moss litter in a mature Scots pine forest. *Pedobiologia*, **26**, 301–308.

Billington, M. & Alexander, V. (**1978**). Nitrogen fixation in a black spruce (*Picea mariana* (Mill) B.S.P.) forest in Alaska. *Environmental Role of Nitrogen-fixing Blue-green Algae and Asymbiotic Bacteria* (Ed. by U. Granhall), pp. 209–215. *Ecological Bulletins (Stockholm)*, **26**.

Bjorck, S., Malmer, N., Hjort, C., Sandgren, P., Ingolfsson, O., Wallen, B. Lewis Smith, R.I. & Jonsson, B.L. (**1991**). Stratigraphic and paleoclimatic studies of a 5500-year-old moss bank on Elephant Island, Antarctica. *Arctic and Alpine Research*, **23**, 361–374.

Bliss, L.C. (**1970**). Oil and the ecology of the Arctic. *Transactions of the Royal Society of Canada*, Series IV, **8**, 361–372.

Bliss, L.C. (**1975**). Tundra grasslands, herblands and shrublands and the role of herbivores. *Geoscience and Man*, **10**, 51–79.

Bliss, L.C. (**1977**). *Truelove Lowland, Devon Island, Canada; a High Arctic Ecosystem.* University of Alberta Press, Edmonton.

Bliss, L.C. (**1986**). Arctic ecosystems; their structure, function and herbivore carrying capacity. *Grazing Research at Northern Latitudes* (Ed. by O. Gudmundsson), pp. 5–25. Plenum Press, New York.

Bliss, L.C. & Matveyeva, N.V. (**1992**). Circumpolar arctic vegetation. *Arctic Ecosystems in a Changing Climate* (Ed. by F.S. Chapin, R.L. Jeffries, J.F. Reynolds, J. Svoboda & E.W. Chu), pp. 59–90. Academic Press, San Diego.

Bliss, L.C. & Svoboda, J. (**1984**). Plant communities and plant production in the western Queen Elizabeth Islands. *Holarctic Ecology*, **7**, 325–344.

Boertje, R. (**1984**). Seasonal diets of the Denali caribou herd, Alaska. *Arctic*, **37**, 161–165.

Bowden, R.D. (**1991**). Inputs, outputs, and accumulation of nitrogen in an early successional moss (*Polytrichum*) ecosystem. *Ecological Monographs*, **61**, 207–223.

Brown, D.H. (1982). Mineral nutrition. *Bryophyte Ecology* (Ed. by A.J.E. Smith), pp. 383–444. Chapman & Hall, London.

Brown, D.H. (1984). Uptake of mineral elements and their use in pollution monitoring. *The Experimental Biology of Bryophytes* (Ed. by A.F. Dyer & J.G. Duckett), pp. 229–255. Academic Press, London.

Bunnell, F.L., MacLean, S.F. & Brown, J. (1975). Barrow, Alaska, USA. *Structure and Function of Tundra Ecosystems* (Ed. by T. Rosswall & O.W. Heal), pp. 73–124. *Ecological Bulletins (Stockholm),* **20**.

Chapin, F.S. (1983). Direct and indirect effects of temperature on Arctic plants. *Polar Biology,* **2**, 47–52.

Chapin, F.S., Oechel, W.C., Van Cleve, K. & Lawrence, W. (1987). The role of mosses in the phosphorous cycling of an Alaskan black spruce forest. *Oecologia,* **74**, 310–315.

Chernov, Y. (1985). *The Living Tundra.* Cambridge University Press, Cambridge.

Clymo, R.S. & Hayward, P.M. (1982). The ecology of *Sphagnum. Bryophyte Ecology* (Ed. by A.J.E. Smith), pp. 229–289. Chapman & Hall, London.

Collins, N.G. (1977). The growth of mosses in two contrasting communities in the maritime Antarctic; measurement and prediction of net annual production. *Adaptations within Antarctic Ecosystems* (Ed. by G.A. Llano), pp. 921–933. Smithsonian Institution, Washington.

Crittenden, P.D. (1983). The role of lichens in the nitrogen economy of subarctic woodlands: nitrogen loss from the nitrogen-fixing lichen *Stereocaulon paschale* during rainfall. *Nitrogen as an Ecological Factor* (Ed. by J.A. Lee, S. McNeill & I.H. Rorison), pp. 43–68. Blackwell Scientific Publications, Oxford.

Crittenden, P.D. (1991). Ecological significance of necromass production in mat-forming lichens. *Lichenologist,* **23**, 323–331.

Davidson, A.J., Harborne, J.B. & Longton, R.E. (1989). Identification of hydroxycinnamic and phenolic acids in *Mnium hornum* and *Brachythecium rutabulum* and their possible role in protection against herbivory. *Journal of the Hattori Botanical Laboratory,* **67**, 415–422.

Davidson, A.J., Harborne, J.B. & Longton, R.E. (1990). The acceptability of mosses as food for generalist herbivores, slugs in the Arionidae. *Botanical Journal of the Linnean Society,* **104**, 99–113.

Davis, R.C. (1981). Structure and function of two Antarctic terrestrial moss communities. *Ecological Monographs,* **51**, 125–143.

Dowding, P., Chapin, F.S., Wielgolaski, F.E. & Kilfeather, P. (1981). Nutrients in tundra ecosystems. *Tundra Ecosystems: A Comparative Analysis* (Ed. by L.C. Bliss, O.W. Heal & J.J. Moore), pp. 647–683. Cambridge University Press, Cambridge.

Fahselt, D. (1993). UV absorbance by thallus extracts of umbilicate lichens. *Lichenologist,* **25**, 415–422.

Fenton, J.H.C. (1980). The rate of peat accumulation in Antarctic moss banks. *Journal of Ecology,* **68**, 211–228.

Fowbert, J.A. & Lewis Smith, R.I. (1994). Rapid population increases in native vascular plants in the Argentine Islands, Antarctic Peninsula. *Arctic and Alpine Research,* **26**, 290–296.

Fyles, J.W. & McGill, W.B. (1987). Decomposition of boreal forest litters from central Alberta under laboratory conditions. *Canadian Journal of Forestry Research,* **17**, 109–114.

Gagnon, Z.E. & Glime, J.M. (1992). The pH-lowering ability of *Sphagnum magellanicum* Brid. *Journal of Bryology,* **17**, 47–57.

Gerson, U. (1987). Mites which feed on mosses. *Symposia Biologica Hungarica,* **35**, 721–724.

Gignac, L.D. & Vitt, D.H. (1994). Responses of northern peatlands to climatic change; effects of bryophytes. *Journal of the Hattori Botanical Laboratory,* **75**, 119–132.

Gorham, E. (1991). Northern peatlands; role in the carbon cycle and probable responses to climatic warming. *Ecological Applications,* **1**, 182–195.

Greene, S.W. (1964). Plants of the land. *Antarctic Research* (Ed. by R.J. Adie & G. de Q. Robin), pp. 240–253. Butterworth, London.

Greene, S.W., Hassel de Menendez, G.G. & Matteri, C.M. (1985). La contribucion de les briofitas en la vegetacion de la transecta. *Transecta Botanica de la Patagonia Austral* (Ed. by O. Boelcke, D.M. Moore & F.A. Roig), pp. 557–591. Consego Nacional de Investigaciones Cientificas y Tecnicas, Buenos Aires.

Gunther, A.J. (1989). Nitrogen fixation by lichens in a subarctic watershed. *The Bryologist*, **92**, 202–208.

Hebant, C. (1977). *The Conducting Tissues of Bryophytes*. Cramer, Vaduz.

Holdgate, M.W. (1964). Terrestrial ecology in the maritime Antarctic. *Biologie Antarctique* (Ed. by R. Carrick, M.W. Holdgate & J. Prevost), pp. 181–194. Hermann, Paris.

Ino, Y. (1983). Estimation of primary production in moss community on East Ongul Island, Antarctica. *Antarctic Record*, **80**, 30–38.

Ives, J.D. (1974). Permafrost. *Arctic and Alpine Environments* (Ed. by J.D. Ives & R.G. Barry), pp. 159–194. Methuen, London.

Jauhaianen, J., Vasander, H. & Silvola, J. (1994). Response of *Sphagnum fuscum* to N deposition and increased CO_2. *Journal of Bryology*, **18**, 83–96.

Jenkin, J.F. (1975). Macquarie Island, Subantarctic. *Structure and Function of Tundra Ecosystems* (Ed. by T. Rosswall & O.W. Heal), pp. 375–397. *Ecological Bulletins (Stockholm)*, **20**.

Johnson, L.C. & Damman, A.W.H. (1993). Decay and its regulation in *Sphagnum* peatlands. *Advances in Bryology*, **5**, 249–296.

Kallio, P. (1975). Kevo, Finland. *Structure and Function of Polar Ecosystems* (Ed. by T. Rosswall & O.W. Heal), pp. 193–223. *Ecological Bulletins (Stockholm)*, **20**.

Kappen, L. (1985). Vegetation and ecology of ice-free areas of northern Victoria Land, Antarctica. I. The lichen vegetation of Birthday Ridge and an inland mountain. *Polar Biology*, **4**, 213–235.

Kappen, L., Breuer, M. & Bolter, M. (1991). Ecological and physiological investigations in continental Antarctic cryptogams III. Photosynthetic production of *Usnea sphacelata*; diurnal courses, models and the effects of photoinhibition. *Polar Biology*, **11**, 393–401.

Kershaw, K.A. (1985). *The Physiological Ecology of Lichens*. Cambridge University Press, Cambridge.

Klein, D.R. (1982). Fire, lichens and caribou. *Journal of Range Management*, **35**, 390–395.

Lawrey, J.D. (1986). Ecological role of lichen substances. *Bryologist* **89**, 111–122.

Lee, J.A., Parsons, A.N. & Baxter, R. (1993). *Sphagnum* species and polluted environments, past and future. *Advances in Bryology*, **5**, 297–314.

Lewis Smith, R.I. (1982). Growth and production in South Georgian Bryophytes. *Comite National Francais des Recherches Antarctiques*, **51**, 229–239.

Lewis Smith, R.I. (1984). Terrestrial plant biology of the sub-Antarctic and Antarctic. *Antarctic Ecology*, Vol. 1 (Ed. by R.M. Laws), pp. 61–162. Academic Press, London.

Lewis Smith, R.I. (1993). Dry coastal ecosystems on sub-Antarctic islands. *Dry Coastal Ecosystems; Polar Regions and Europe* (Ed. by E. van der Maarl), pp. 73–93. Elsevier, Amsterdam.

Lewis Smith, R.I. & Walton, D.H.W. (1975). South Georgia, Subantarctic. *Structure and Function of Tundra Ecosystems* (Ed. by T. Rosswall & O.W. Heal), pp. 399–423. *Ecological Bulletins (Stockholm)*, **20**.

Li, Y. & Vitt, D.H. (1994). The dynamics of moss establishment; temporal responses to nutrient gradients. *The Bryologist*, **97**, 357–364.

Longton, R.E. (1970). Growth and productivity of the moss *Polytrichum alpestre* Hoppe in Antarctic regions. *Antarctic Ecology*, Vol. 2 (Ed. by M.W. Holdgate), pp. 818–837. Academic Press, London.

Longton, R.E. (1974). Microclimate and biomass in communities of the *Bryum* association on Ross Island, continental Antarctica. *The Bryologist*, **77**, 109–127.

Longton, R.E. (1984). The role of bryophytes in terrestrial ecosystems. *Journal of the Hattori Botanical Laboratory*, **55**, 147–163.

Longton, R.E. (1988). *Biology of Polar Bryophytes and Lichens*. Cambridge University Press, Cambridge.

Longton, R.E. (1992). The role of bryophytes and lichens in terrestrial ecosystems. *Bryophytes and Lichens in a Changing Environment* (Ed. by J.W. Bates & A.M. Farmer), pp. 32–76. Oxford University Press, Oxford.

Longton, R.E. (1994a). Genetic differentiation within the moss *Polytrichum alpestre* Hoppe. *Journal of the Hattori Botanical Laboratory*, **75**, 1–13.

Longton, R.E. (1994b). Reproductive biology in bryophytes; the challenge and the opportunities. *Journal of the Hattori Botanical Laboratory*, **76**, 159–172.

Matsuda, T. (1968). Ecological study of the moss community and microorganisms in the vicinity of Syowa Station, Antarctica. *Japanese Antarctic Research Expedition Scientific Reports*, **E29**, 1–58.

Maxwell, B. (1992). Arctic climate; potential for change under global warming. *Arctic Ecosystems in a Changing Climate* (Ed. by F.S. Chapin, R.L. Jeffries, J.F. Reynolds, G.R. Shaver, J. Svoboda & E.W. Chu), pp. 11–34. Academic Press, San Diego.

Melick, D.R. & Seppelt, R.D. (1994). Seasonal investigation of soluble carbohydrates and pigment levels in Antarctic bryophytes and lichens. *The Bryologist*, **97**, 13–19.

Moore, D.M. (1979). Southern oceanic wet heathlands (including magellanic moorland). *Heathlands and Related Shrublands of the World: A Descriptive Study* (Ed. by R.L. Specht), pp. 489–497. Elsevier, Amsterdam.

Moser, T.J., Nash, T.H. & Thomson, J.W. (1979). Lichens of Anaktuvuk Pass, Alaska, with emphasis on the impact of caribou grazing. *The Bryologist*, **82**, 393–408.

Murray, K.J., Tenhunen, J.D. & Kummerow, J. (1989). Limitations on *Sphagnum* growth and net primary production in the foothills of the Philip Smith Mountains, Alaska. *Oecologia*, **80**, 256–262.

Murray, K.J., Tenhunen, J.D. & Nowak, R.S. (1993). Photoinhibition as a control on photosynthesis and production in *Sphagnum* mosses. *Oecologia*, **96**, 200–207.

Noakes, T.D. & Longton, R.E. (1989). Studies on water relations in mosses from the cold-Antarctic. *University Research in Antarctica* (Ed. by R.B. Heywood), pp. 103–116. British Antarctic Survey, Cambridge.

Norby, R.J. & Sigal, L.L. (1989). Nitrogen fixation in the lichen *Lobaria pulmonaria* at elevated atmospheric carbon dioxide. *Oecologia*, **79**, 566–568.

Oechel, W.C. & Sveinbjornsson, B. (1978). Primary production processes in arctic bryophytes at Barrow, Alaska. *Vegetation and Production Ecology of an Alaskan Arctic Tundra* (Ed. by L.L. Tieszen), pp. 269–298. Springer, New York.

Oechel, W.C. & Van Cleve, K. (1986). The role of bryophytes in nutrient cycling in the taiga. *Forest Ecosystems in the Alaskan Taiga* (Ed. by K. Van Cleve, F.S. Chapin, P.W. Flanagan, L.A. Viereck & C.T. Dyrness), pp. 121–137. Springer, New York.

Pakarinen, P. (1978). Production and nutrient ecology of three *Sphagnum* species in southern Finnish raised bogs. *Annales Botanici Fennici*, **15**, 15–26.

Polunin, N. (1951). The real Arctic; suggestions for its delimitation, subdivision and characterization. *Journal of Ecology*, **39**, 308–315.

Pratt, R.M. & Lewis Smith, R.I. (1982). Seasonal trends in chemical composition of reindeer forage plants on South Georgia. *Polar Biology*, **1**, 13–22.

Prins, H.H.T. (1982). Why are mosses eaten in cold environments only? *Oikos*, **38**, 374–380.

Reimers, E. (1977). Population dynamics of two populations of reindeer in Svalbard. *Arctic and Alpine Research*, **9**, 369–381.

Richardson, D.H.S. & Young, C.M. (1977). Lichens and vertebrates. *Lichen Ecology* (Ed. by M.R.D. Seaward), pp. 121–144. Academic Press, London.

Rincon, E. (1988). The effects of herbaceous litter on bryophyte growth. *Journal of Bryology*, **15**, 209–218.

Rodgers, G.A. & Henriksson, E. (1976). Association between the blue-green algae *Anabaena variabilis* and *Nostoc muscorum* and the moss *Funaria hygrometrica* with reference to the colonisation of Surtsey. *Acta Botanica Islandica*, **4**, 10–15.

Rosswall, T. & Granhall, U. (1980). Nitrogen cycling in a subarctic ombrotrophic mire. *Ecology of a Subarctic Mire* (Ed. by M. Sonesson), pp. 209–234. *Ecological Bulletins (Stockholm)*, **30**.

Rosswall, T., Veum, A.K. & Karenlampi, L. (1975). Plant litter decomposition at Fennoscandian tundra sites. *Fennoscandian Tundra Ecosystems. I. Plants and Microorganisms* (Ed. by F.E. Wielgolaski), pp. 268–278. Springer, New York.

Russell, S. (1990). Bryophyte production and decomposition in tundra ecosystems. *Botanical Journal of the Linnean Society*, **104**, 3–22.

Schwarz, A.M.J., Green, T.G.A. & Seppelt, R.D. (1992). Terrestrial vegetation at Canada Glacier, Southern Victoria Land, Antarctica. *Polar Biology*, **12**, 397–404.

Scotter, G.W. (1972). Chemical composition of forage plants from the Reindeer Preserve, North West Territories. *Arctic*, **25**, 21–27.

Seel, W.E., Hendry, G.A.F. & Lee, J.A. (1992). Effects of desiccation on some activated oxygen processing enzymes and anti-oxidants in mosses. *Journal of Experimental Botany*, **43**, 1031–1037.

Seppelt, R.D. & Ashton, D.H. (1978). Studies on the ecology of the vegetation of Mawson Station, Antarctica. *Australian Journal of Ecology*, **3**, 373–388.

Skogland, T. (1984). Wild reindeer foraging – niche organisation. *Holarctic Ecology*, **7**, 345–379.

Smith, V.R. (1977). Notes on the feeding of *Ectomnorrhinus similis* Waterhouse (Curculionidae) adults on Marion Island. *Oecologia*, **29**, 269–273.

Sohlberg, E.H. & Bliss, L.C. (1987). Responses of *Ranunculus sabinei* and *Papaver radicatum* to removal of the moss layer in a high-Arctic meadow. *Canadian Journal of Botany*, **65**, 1224–1228.

Staaland, H., Jacobsen, E. & White, R.G. (1979). Comparison of the digestive tract in Svalbard and Norwegian reindeer. *Arctic and Alpine Research*, **11**, 457–466.

Staaland, H., Jacobsen, E. & White, R.G. (1984). The effect of mineral supplements on nutrient concentrations and pool sizes in the alimentary tract of reindeer fed on lichens and concentrates during winter. *Canadian Journal of Zoology*, **62**, 1232–1241.

Steere, W.C. (1978). Floristics, phytogeography and ecology of Arctic Alaskan bryophytes. *Vegetation and Production Ecology of an Alaskan Arctic Tundra* (Ed. by L.L. Tieszen), pp. 141–167. Springer, New York.

Sveinbjornsson, B. (1987). Reindeer lichen productivity as a function of mat thickness. *Arctic and Alpine Research*, **19**, 437–441.

Sveinbjornsson, B. & Oechel, W.C. (1983). The effect of temperature preconditioning on the temperature sensitivity of CO_2 flux in geographically diverse populations of the moss *Polytrichum commune* Hedw. *Ecology*, **64**, 1100–1108.

Sveinbjornsson, B. & Oechel, W.C. (1992). Controls of growth and productivity of bryophytes; environmental limitations under current and anticipated conditions. *Bryophytes and Lichens in a Changing Environment* (Ed. by J.W. Bates & A.M. Farmer), pp. 77–102. Oxford University Press, Oxford.

Tarnawski, M., Melick, D., Roser, D., Adamson, E., Adamson, H. & Seppelt, R. (1992). *In situ* carbon dioxide levels in cushion and turf forms of *Grimmia antarctici* at Casey Station, East Antarctica. *Journal of Bryology*, **17**, 241–249.

Tenhunen, J.D., Lange, O.L., Hahn, R., Siegwolf, R. & Oberbauer, S.F. (1992). The ecosystem role of poikilohydric tundra plants. *Arctic Ecosystems in a Changing Climate* (Ed. by F.S. Chapin, R.L. Jeffries, J.F. Reynolds, G.R. Shaver, J. Svoboda & E.W. Chu), pp. 213–238. Academic Press, San Diego.

Thomas, D.C. & Edmonds, J. (1983). Rumen content and habitat selection of Peary caribou in winter, Canadian Arctic Archipelago. *Arctic and Alpine Research*, **15**, 97–105.

Thomas, D.C. & Kroeger, P. (1981). Digestibility of plants in ruminal fluids of barren-ground caribou. *Arctic*, **34**, 321–324.

Thompson, D.B.A. & Baddeley, J. (1991). Some effects of acid deposition on montane *Racomitrium* *lanuginosum* heaths. *The Effects of Acid Deposition on Nature Conservation in Great Britain* (Ed. by S.J. Woodin & A.M. Farmer), pp. 17–28. Nature Conservancy Council, Peterborough.

Twenhoven, F.L. (1992). Competition between two *Sphagnum* species under different deposition levels. *Journal of Bryology*, **17**, 71–80.

Van Cleve, K. & Alexander, V. (1981). Nitrogen cycling in tundra and boreal ecosystems. *Terrestrial Nitrogen Cycles* (Ed. by F.F. Clark & T. Rosswall), pp. 375–404. *Ecological Bulletins (Stockholm)*, **33.**

Vitt, D.H. (1990). Growth and production dynamics of boreal mosses over climatic, chemical and topographic gradients. *Botanical Journal of the Linnean Society*, **104,** 35–59.

Wace, N.M. (1965). Vascular plants. *Biogeography and Ecology in Antarctica* (Ed. by P. van Oye & J. van Mieghem), pp. 201–266. Junk, The Hague.

Walton, D.W.H. (1973). Changes in standing crop and dry matter production in an *Acaena* community on South Georgia. *Proceedings of the Conference on Primary Production and Production Processes, Tundra Biome* (Ed. by L.C. Bliss & F.E. Wielgolaski), pp. 185–190. IBP Tundra Biome Steering Committee, Edmonton.

Webber, P.J. (1978). Spatial and temporal variation of the vegetation and its production, Barrow, Alaska. *Vegetation and Production Ecology of an Alaskan Arctic Tundra* (Ed. by L.L. Tieszen), pp. 37–112. Springer, New York.

Weber, M.G. & Van Cleve, K. (1984). Nitrogen transformations in feather moss and forest floor layers of interior Alaska black spruce ecosystems. *Canadian Journal of Forestry Research*, **14,** 278–290.

Weetman, G. (1968). The relationship between feather moss growth and the nutrition of black spruce. *Proceedings of the Third International Peat Congress* (Ed. by C. Lafleur & J. Butler), pp. 366–370. International Peat Society, Quebec City.

West, S.D. (1982). Dynamics of colonisation and abundance in central Alaskan populations of the northern red-backed vole, *Clethrionomys rutilans. Journal of Mammalogy*, **63,** 128–143.

White, R.G. & Trudell, J. (1980). Habitat preference and forage consumption by reindeer and caribou near Atkasook, Alaska. *Arctic and Alpine Research*, **12,** 511–529.

Wielgolaski, F.E. & Kjelvik, S. (1975). Production of plants (vascular plants and cryptogams) in alpine tundra, Hardangervidda. *Primary Production and Production Processes, Tundra Biome* (Ed. by L.C. Bliss & F.E. Wielgolaski), pp. 75–86. IBP Tundra Biome Steering Committee, Edmonton.

Woodin, S., Press, M.C. & Lee, J.A. (1985). Nitrate reductase activity in *Sphagnum fuscum* in relation to wet deposition of nitrate from the atmosphere. *New Phytologist*, **99,** 381–388.

Yarrington, M.R. & Wynne-Williams, D.D. (1985). Methanogenesis and the anaerobic micro-biology of a wet moss community at Signy Island. *Antarctic Nutrient Cycling and Food Webs* (Ed. by W.R. Siegfried, P.R. Condy & R.M. Laws), pp. 329–333. Springer, Berlin.

4. The role of arctic vegetation in ecosystem and global processes

F. STUART CHAPIN III, JOSEPH P. McFADDEN
AND SARAH E. HOBBIE
Department of Integrative Biology, University of California, Berkeley,
CA 94720-3140, USA

INTRODUCTION

As a result of the increasing atmospheric concentrations of radiatively active greenhouse gases, global temperatures are expected to rise (Houghton, Jenkins & Ephraums 1990), particularly at high latitudes. Current temperature trends are consistent with these predictions (Chapman & Walsh 1993), suggesting that this global warming may already have begun. In the Arctic, climatic warming will be accompanied by a longer snow-free season, increased summer cloudiness as sea ice melts and allows greater evaporation from northern oceans (Maxwell 1992), greater soil temperature and depth of soil thaw (Kane *et al.* 1992), altered (probably reduced) soil moisture (Rind 1987; Kane *et al.* 1992), and increased nutrient availability (Nadelhoffer *et al.* 1991; Chapin *et al.* 1992), although the exact nature and magnitude of these changes are less certain than those of temperature change. Other, more direct human impacts on arctic ecosystems in the coming decades include air pollution (see Chapter 10), disturbance associated with resource extraction, and altered grazing regime due to changing patterns of reindeer husbandry and hunting. Together these environmental changes are certain to affect arctic plant communities. The challenge is to predict the effects of these changes on community composition and ecosystem processes and on the role of arctic ecosystems in arctic and global processes. In this chapter we present a framework for predicting how arctic plant communities will respond to such change and discuss the implications for ecosystem and global processes.

RESPONSE TO GLOBAL CHANGE

Growth-form response

All arctic plant species differ to some degree in their physiological response to the environment and, therefore, in their distribution along environmental gradients (Whittaker 1953; Webber 1978), their response to environmental manipulations (Chapin & Shaver 1985; Havström, Callaghan & Jonasson 1993; Wookey *et al.* 1993; Parsons *et al.* 1994; Chapin *et al.* 1995) and their rates and patterns of migration in response to climatic change (Brubaker, Anderson & Hu 1995). Species responses to environment are often consistent within a growth form (e.g. deciduous shrub vs. evergreen shrub) and can, therefore, be predicted a priori, as described below.

Most arctic plant species show an initial positive response to increased nutrient supply (Shaver & Chapin 1980; Jonasson 1992) because slow rates of nutrient mineralization in cold waterlogged soils cause nutrient limitation (Gersper *et al.* 1980; Nadelhoffer *et al.* 1991). The nutrient response is most pronounced in forbs, graminoids and deciduous shrubs, and least pronounced in evergreen shrubs, mosses and lichens, leading in the long term to dominance by deciduous shrubs and graminoids and decreased abundance of evergreen shrubs, mosses and lichens (Shaver & Chapin 1986; Jonasson 1992; Chapin *et al.* 1995). Thus, enhanced soil fertility predictably increases biomass and production of most species within a few years but changes species composition in the longer term.

Tundra plants are well adapted to low temperature and generally show relatively small direct growth responses to increased temperature in the Low Arctic (Chapin & Shaver 1985; Havström *et al.* 1993; Wookey *et al.* 1993). Some species respond positively to increased temperature, whereas others respond negatively, in ways that currently are difficult to predict a priori (Chapin & Shaver 1985). However, in the High Arctic, growth of all plants increases substantially in response to increased air temperature caused by microtopography (Warren Wilson 1966) or temperature manipulation (Romer, Cummins & Svoboda 1983; Havström *et al.* 1993; Wookey *et al.* 1993). Thus, we expect a stimulation of community biomass and production in response to higher temperatures to occur primarily in the High Arctic.

Decreases in light intensity below current ambient levels (as might occur with increased summer cloudiness) generally have relatively small initial effects on most arctic species (Chapin & Shaver 1985), as might be expected from their high carbohydrate concentrations (Billings & Mooney 1968). However, production of all but the most shade-tolerant species declines with reduced light intensity over the long term (Chapin *et al.* 1995), particularly in the High Arctic (Havström *et al.* 1993).

Graminoids and mosses both respond positively to increased soil moisture in the short term, although presumably for different reasons. Graminoids respond more to the effects of water movement on lateral nutrient transport (Chapin *et al.* 1988) than to the direct effects of moisture on plant water relations (Oberbauer & Miller 1982). As with nutrients, long-term increases in soil moisture change community composition, leading to dominance by aerenchymatous graminoids and decreases in shrubs and lichens (Webber 1978). Both the short-term and long-term community responses to changes in soil moisture lead us to expect changes in functioning of ecosystems.

Increased CO_2 concentration has little effect on photosynthesis by the dominant species of tussock tundra (Tissue & Oechel 1987) or on net ecosystem CO_2 flux

(Grulke *et al.* 1990; Oechel *et al.* 1994), so we assume that the trend of increasing atmospheric CO_2 will have minor direct effects on arctic ecosystems, at least in the short term.

In summary, changes in soil resources (nutrients and water) and decreases in light intensity have predictable, contrasting effects on different growth forms of plants and may, therefore, alter ecosystem processes, whereas the direct effects of CO_2 and temperature in the Low Arctic are more subtle and less predictable. If growth forms differ in their long-term response to environment, what is the net effect on the ecosystem as a whole?

Community and ecosystem responses

Although most arctic species initially respond strongly to changes in environment, these responses are substantially buffered at the ecosystem level. For example, in tussock tundra of arctic Alaska, where light, temperature and nutrients were manipulated for a decade to simulate global changes in these variables, there was no significant effect of nutrient addition on total community biomass (Fig. 4.1)

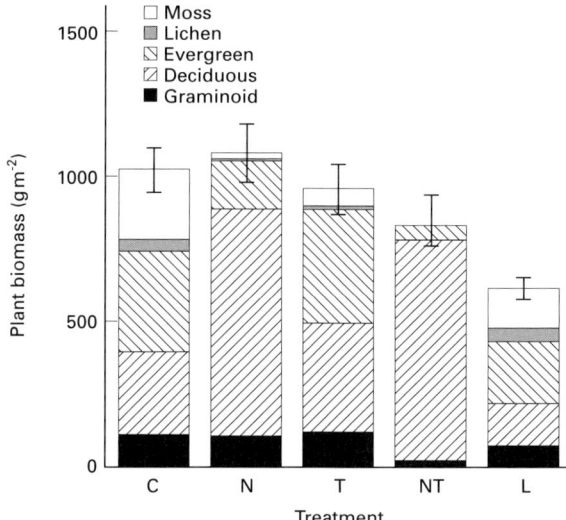

Fig. 4.1. Total peak-season biomass (excluding roots) of Alaskan tussock tundra vegetation by growth form in response to environmental manipulations after 9 years of treatments. Treatments are control (C), nutrient addition (N), temperature increase (T), nutrient and temperature increases (NT), and light attenuation (L). The nutrient treatment received $10\,g\,m^{-2}$ of nitrogen and $5\,g\,m^{-2}$ of phosphorus annually. In the temperature treatment, summer air temperature was raised by 3°C with a plastic greenhouse. In the light attenuation treatment, light was reduced by 50% by optically neutral shade cloth. Data are means ± SE, $n = 4$ plots. (After Chapin *et al.* 1996.)

despite the strong initial positive effect of nutrients on most species. The negligible effect at the ecosystem level occurred because the strong positive response of all species of deciduous shrubs was compensated by decreased abundance of all species of evergreen shrubs, lichens and mosses (Chapin *et al.* 1995). The small decline in total biomass with increased temperature was also the result of large compensatory changes by individual species. However, in contrast to the nutrient response, each species of a given growth form tended to have a different response to temperature. All except the most shade-tolerant understorey species responded negatively to reduced light intensity, resulting in a strong decline in biomass at the ecosystem level.

The stability of ecosystem biomass and net primary production (NPP) follows logically from physiological differences among species and resulting shifts in competitive balance. Many of these competitive interactions are asymmetric; for example, increased biomass of woody shrubs (which are limited more strongly by nutrients than by light) in response to nutrient addition decreases light availability to light-limited understorey shrubs and mosses. Conversely, increased water availability stimulates mosses more than shrubs, allowing the mosses to immobilize nutrients and reduce shrub growth. The small changes in biomass and NPP at the ecosystem level (relative to changes in individual species) are observed in response to manipulations (Chapin *et al.* 1995), geographic variation in environment (Webber 1978), annual variation in weather (Lauenroth, Dodd & Simms 1978; Chapin & Shaver 1985), or longer-term changes in climate (COHMAP 1988; Brubaker *et al.* 1995). Because of this compensation at the ecosystem level, we expect ecosystem properties such as NPP or rates of biogeochemical cycling in closed communities to be less sensitive to climatic change than would be predicted from extrapolations of the environmental responses of single species. The eventual dominance of fertilized plots by deciduous shrubs and the decline in evergreen shrubs is consistent with descriptive ordinations of undisturbed tundra vegetation, where high-fertility sites are dominated by deciduous species and low-fertility sites by evergreens (Webber 1978; Shaver & Chapin 1991).

In contrast to the stabilization of biomass and NPP at the ecosystem level, species diversity was strongly reduced by manipulations simulating climatic change. After 9 years, the strong dominance by *Betula* in plots simulating climatic change was associated with loss of four to eight species in these treatments (Fig. 4.2) – a decline of 30–50% in vascular-plant species richness. Most concern about loss of species diversity with climatic change has focused on the tropics, where the largest number of species is being lost (Solbrig 1991). However, because the tundra has so few species, any species loss has a larger *proportional* impact on tundra diversity than in more species-rich communities. Presumably, over longer time scales the species diversity of arctic tundra will recover in response to climatic warming as new species invade from warmer climates.

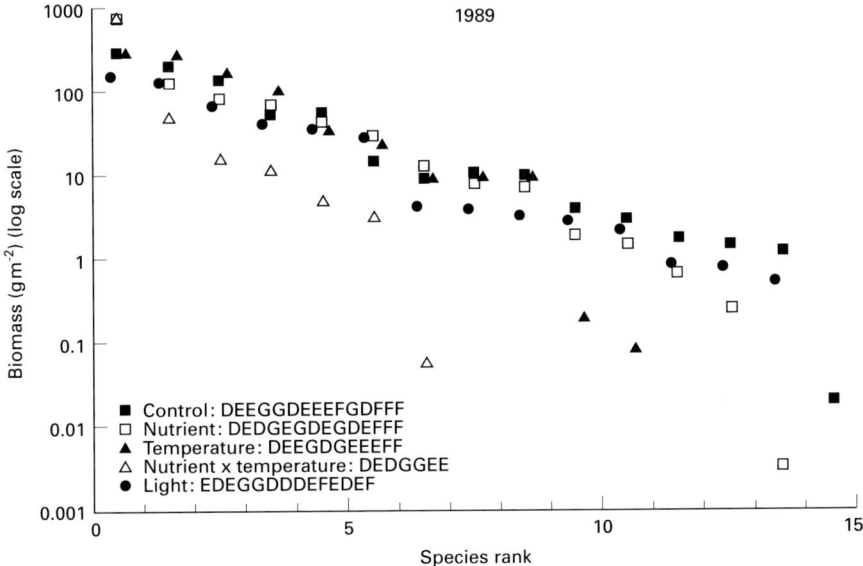

Fig. 4.2. Vascular-plant dominance–diversity curves based on total biomass (excluding roots) of individual species sampled 9 years after initiation of the treatments described in Fig. 4.1. Also shown is the sequence in abundance of growth forms represented by each species in each treatment: evergreen shrub (E), deciduous shrub (D), graminoid (G), and forb (F). (After Chapin *et al.* 1995.)

The loss of species from tundra in response to manipulations simulating global change represented loss of entire functional groups of plants (lichens, mosses and forbs). This shift in functional diversity could have important ecosystem consequences. For example, inflorescences of forb species, which disappeared or were strongly reduced in treatments simulating climatic warming, are nutritionally important and selectively grazed by caribou during lactation (White & Trudell 1980) and are the major plant species used by bumble bees and other pollinators (Williams & Batzli 1982). Lichens are critical to the overwinter nutrition of caribou (White & Trudell 1980) and mosses strongly influence soil thermal regime (Tenhunen *et al.* 1992). Thus, the loss of even a few species under climatic warming could have disproportionate effects on animal-mediated processes and on soil thermal regime.

Regional climatic warming may already be altering vegetation composition of Alaskan arctic tundra. Total biomass of the dominant sedge, *Eriophorum vaginatum*, declined to 30% of its initial value in unmanipulated control plots during our 11-year study (Chapin *et al.* 1995), conducted during the warmest decade on record (Trenberth 1990; Oechel *et al.* 1993), as predicted by our manipulations simulating a warmer, more fertile environment (Fig. 4.3).

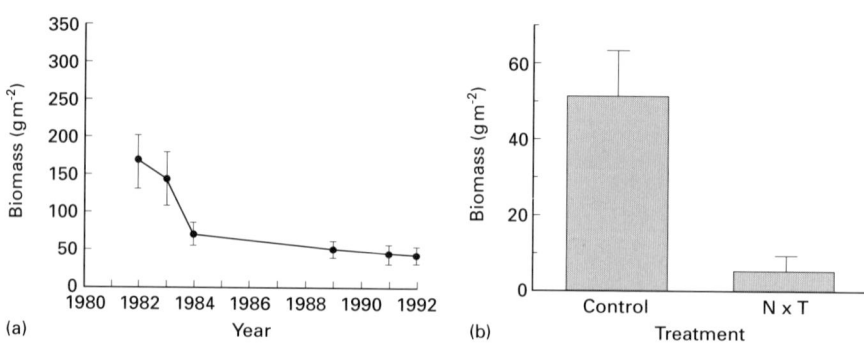

FIG. 4.3. (a) The time course of mid-summer total biomass (mean ± SE) of *Eriophorum vaginatum* in Alaska in unmanipulated control plots from 1982 to 1992 and (b) the biomass response after 9 years of simulated climatic warming with increased mineralization (nutrient × temperature, N × T). (After Chapin *et al.* 1995.)

COMMUNITY EFFECTS ON ECOSYSTEM PROCESSES

Although our experimental manipulations simulating climate change had relatively modest effects on ecosystem traits such as total biomass, we suggest that global environmental change in the long term will strongly affect ecosystem processes through changes in species composition because species differ strongly in their effect on ecosystem processes.

Biogeochemical processes are strongly and predictably altered by those changes in community composition that alter: (i) plant uptake, internal recycling and loss of nutrients (Chapin 1980; Shaver 1981; Shaver & Melillo 1984); or (ii) litter quantity, quality and location (Hobbie 1992). Deciduous shrubs and graminoids have higher rates of growth and nutrient cycling per unit biomass than do evergreen shrubs or lichens (Berendse & Aerts 1987; Shaver & Chapin 1991). From these observations, we predict that increases in nutrient supply and subsurface water flow will enhance rates of nutrient cycling as a result of the associated increased abundance of deciduous shrubs and graminoids.

The quality and location of litter inputs differ strongly among plant growth forms and could profoundly affect nutrient cycling (Fig. 4.4). Mosses have low concentrations of nitrogen and high concentrations of lignin-like polymers (Chapin, McKendrick & Johnson 1986), explaining their generally low rates of decomposition (Johnson & Damman 1991) and their important role in peat formation (Clymo & Hayward 1982). Deciduous shrubs have leaves with high nitrogen and low lignin concentrations (Chapin *et al.* 1986; Chapin & Shaver 1988) but a large stem allocation (Shaver 1986) and, therefore, high overall lignin and low overall nitrogen concentrations (Hobbie 1995). Evergreen shrubs have low leaf-litter quality and a substantial stem allocation, suggesting low rates of decomposition and mineralization. Graminoids have a low stem allocation but a high allocation to

R *Rubus* S *Salix* B *Betula* E *Eriophorum*
L *Ledum* A *Aulacomnium* C *Cetraria*

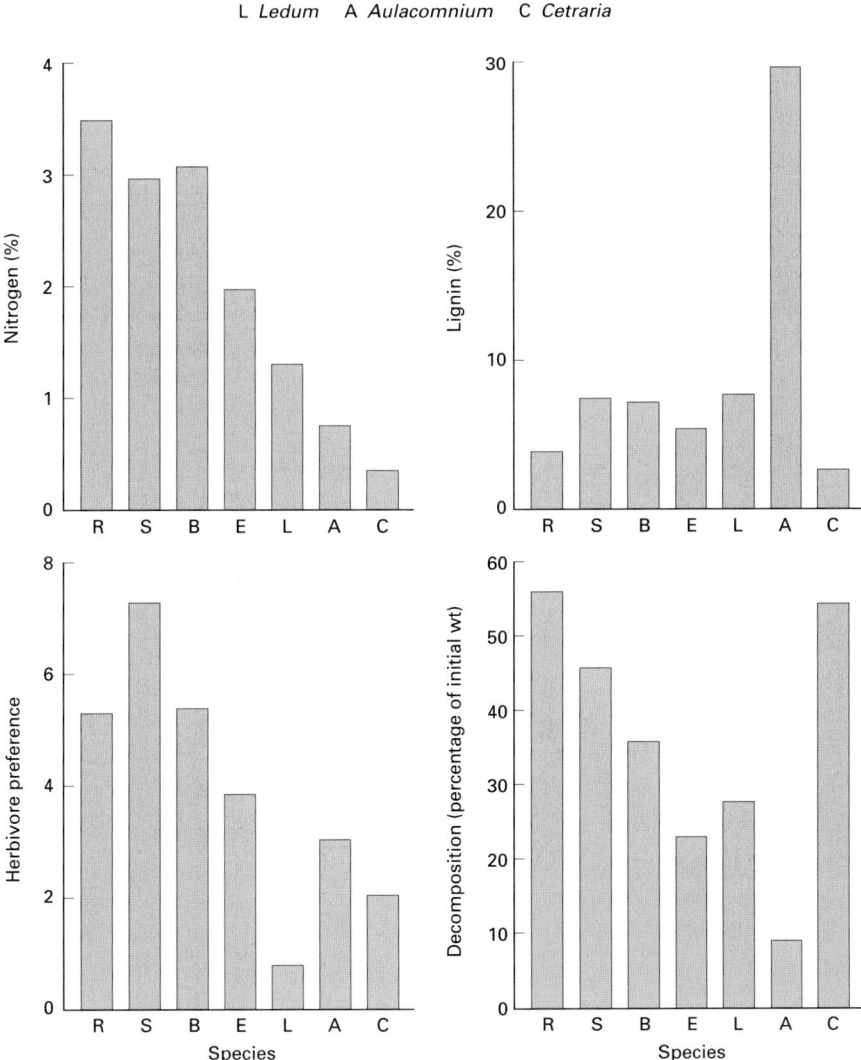

FIG. 4.4. Comparison of major species from Alaskan tussock tundra with respect to mid-season leaf nitrogen and lignin concentrations, average preference by seven species of vertebrate and invertebrate herbivores, and decomposition. (After Chapin & Körner 1996. Data from Chapin *et al.* 1986 and Shaver *et al.* 1995.)

roots (Shaver & Billings 1975), where low temperature, low oxygen concentration and high lignin content retard root decomposition *in situ* (Chapin *et al.* 1988). The net consequence of these multiple species' effects on decomposition requires further study. However, we predict that, under field conditions, mosses most strongly reduce decomposition and promote carbon storage through their effective thermal insulation and low litter quality, followed by deciduous shrubs with their high leaf but low stem litter quality and high stem allocation, followed by evergreen shrubs with their low leaf and stem litter quality, followed by graminoids with their relatively high decomposibility (Hobbie 1995).

The response of herbivores to variation in tissue quality is similar to that of decomposition (see Fig. 4.4), so we expect that the same changes in species composition that stimulate nutrient cycling will promote energy transfer through food chains and perhaps increase the length of food chains (Oksanen 1990).

EFFECTS OF ARCTIC VEGETATION ON ECOSYSTEM AND GLOBAL PROCESSES

In this section we suggest that arctic vegetation change in response to either environmental change or human disturbance has the potential to affect global climate by changing fluxes of radiatively active trace gases and to affect regional climate by altering water and energy exchange with the atmosphere.

Global coupling

Arctic ecosystems directly affect *global* climate primarily through fluxes of the radiatively active trace gases CO_2 and methane (CH_4) (Fig. 4.5). Together, global increases in CO_2 and CH_4 account for 70% of the increase in the warming potential

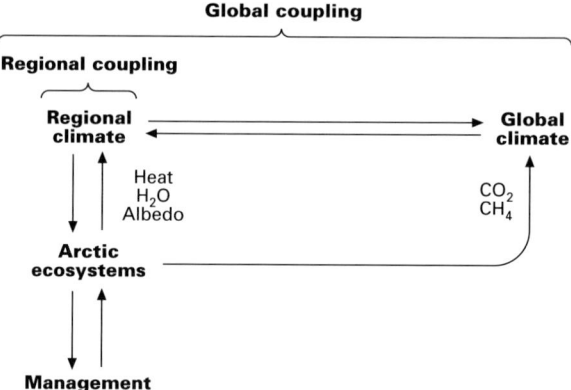

FIG. 4.5. The processes that couple arctic ecosystems to regional climate (through albedo and water/ energy flux) and to global climate (through CO_2 and CH_4 flux). Management influences this coupling through effects on ecosystems.

of the atmosphere (Houghton *et al.* 1990). Atmospheric CO_2 and CH_4 are present at highest concentrations and exhibit greatest seasonal amplitude above 53°N (Zimov *et al.* 1993), suggesting that high latitudes could be important sources of these two gases. Indeed, high latitudes may have switched from being a net sink to a net source of CO_2 in recent years (Oechel *et al.* 1993; Zimov *et al.* 1993), perhaps in response to regional warming and drying. Furthermore, high-latitude wetlands are a globally important source of atmospheric CH_4 (Reeburgh & Whalen 1992). How might vegetation influence the fluxes of these greenhouse gases?

Mosses (especially *Sphagnum* spp.) are universally associated with high-latitude peat accumulation (Clymo & Hayward 1982; Gorham 1991). Lowlands with high soil moisture favour moss growth and peat accumulation (Gorham 1991). Similarly, periods of wet climate show greater peat accumulation than do periods of dry climate. Although the low soil oxygen concentration associated with wet soils strongly inhibits decomposition, mosses exert an important additional effect because their high concentrations of lignin-like polymers (Chapin *et al.* 1986) inhibit decomposition (Johnson & Damman 1991), and their low thermal conductivity (when dry) reduces soil temperature and soil thaw. Mosses tend to maintain the water table close to the soil surface; when the water table is high, mosses have a high rate of evapotranspiration because they lack stomatal control of water loss, but when the water table descends below the soil surface, their lack of roots minimizes evapotranspiration. By contrast, deciduous shrubs, which exploit wet soils that are oxygenated from lateral water movement, and graminoids, which exploit wet anoxic soils, have higher rates of evapotranspiration than mosses when the water table is deep. Graminoids and deciduous shrubs have higher rates of litter decomposition than mosses (see Fig. 4.4). Thus, we expect that the relative abundance of mosses and vascular plants strongly influences the balance between carbon gain and decomposition in wet environments. Experiments in upland tussock tundra (Chapin *et al.* 1995) suggest that climatic warming will favour shrub over moss growth, which should promote carbon loss from the system. The relative magnitude of direct environmental effects vs. vegetation effects on net carbon storage remains to be determined.

Aerenchymatous vascular plants are the major avenue of CH_4 efflux from wetland ecosystems. In Alaskan tundra, sedge removal greatly reduced CH_4 efflux from wet meadow tundra (Fig. 4.6; Torn & Chapin 1993). By contrast, in upland tundra, which can be a modest source or sink of CH_4, vegetation composition has a minimal effect on CH_4 flux. Thus, changes in sedge biomass could have a substantial influence on CH_4 flux in wet-meadow tundra.

In summary, vegetation could have a substantial influence on fluxes of CO_2 and CH_4 from tundra ecosystems and, therefore, the coupling of the Arctic to global climate (see Fig. 4.5). However, the magnitude and nature of controls by vegetation on these fluxes are understood only in broad outline.

F.S.CHAPIN III *et al.*

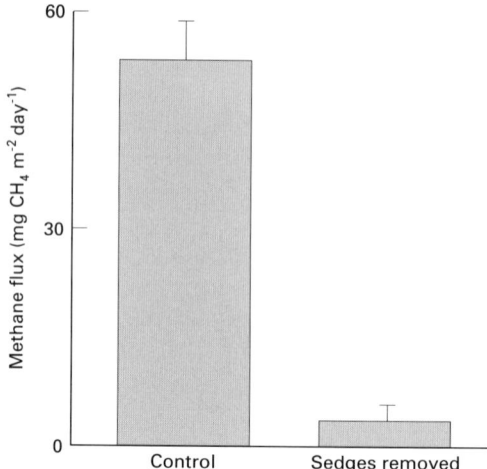

FIG. 4.6. Methane (CH_4) flux from Alaskan wet-meadow tundra before (control) and after (sedges removed) all sedges were pulled from the soil. Data are means ± SE, $n = 5$ plots. (After Torn & Chapin 1993.)

Regional coupling

Arctic ecosystems are coupled to regional climate (see Fig. 4.5) through surface energy exchange processes that are strongly influenced by vegetation. Snow has a much higher albedo (reflectance) than vegetation or bare ground and, therefore, reduces radiation absorption. Any increase in canopy height relative to snow depth, due to increased height growth of shrubs or invasion of tundra by tall shrubs or trees, will mask the snow and reduce the albedo, thus raising the temperature of the overlying air (Bonan, Pollard & Thompson 1992).

The quantity of absorbed radiation in turn influences air temperature. Modelling simulations suggest that conversion of the boreal forest to snow-covered tundra would reduce annual average air temperature in the boreal zone by 6°C and that this temperature effect would be large enough to extend into the tropics (Bonan *et al.* 1992; Bonan, Chapin & Thompson 1995). Similarly, when temperature rose at the end of the Pleistocene, the treeline moved northward, reducing the regional albedo and increasing energy absorption. Approximately half of the climatic warming that occurred at the end of the Pleistocene is estimated to be due to the northward movement of the treeline, with the remaining warming due to changes in solar input (Foley *et al.* 1994; Fig. 4.7). The warmer regional climate, in turn, favours tree reproduction and establishment at the treeline (Payette & Filion 1985), providing a positive feedback to regional warming. Thus, large changes in vegetation height, relative to snow depth, could strongly influence regional

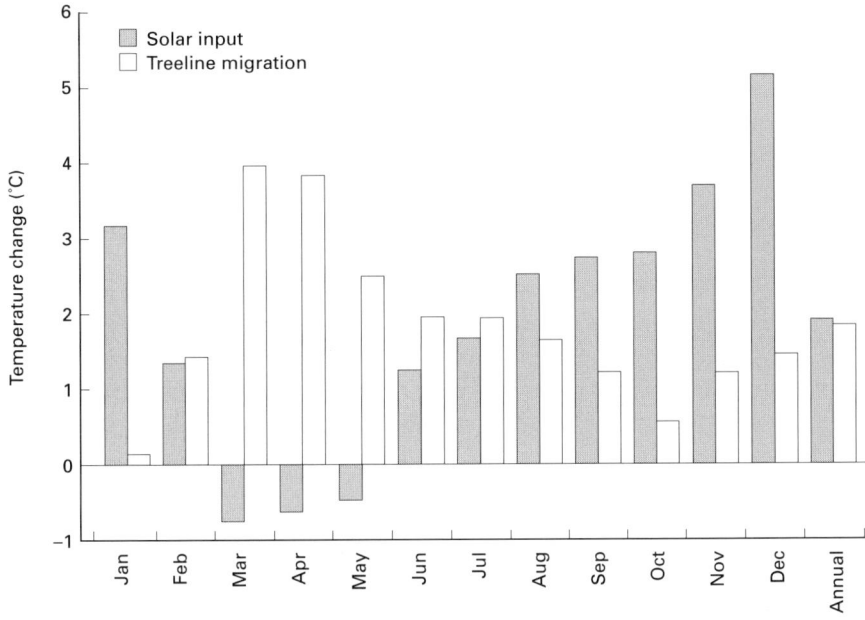

FIG. 4.7. The monthly pattern of change in air temperature (°C) of the Arctic from the Pleistocene to the Holocene caused either directly by increased solar input or indirectly by reduced albedo associated with northward movement of the treeline. Results were estimated by simulations with a general circulation model. (After Foley *et al.* 1994.)

climate. Moreover, summer differences in albedo between forest tundra may be large enough to contribute to the positioning of the arctic front at the forest–tundra border, creating a strong temperature gradient at this point and stabilizing the position of treeline (Pielke & Vidale, 1996). This difference between forest and tundra in biophysical impacts on regional climate is probably quantitatively the largest effect that northern ecosystems exert on climate at the global scale – and it occurs through regional coupling between vegetation processes and regional climate (see Fig. 4.5).

In the snow-free season, vegetation strongly influences the partitioning of net radiation into sensible and latent heat fluxes. Vegetated surfaces all have a similar albedo during the growing season, and, therefore, must dissipate a similar amount of energy. However, the manner in which this energy is dissipated depends strongly on both vegetation and topography. As expected, lowlands with standing water have greater evapotranspiration in proportion to sensible heat loss compared to dry upland heath sites (J.P. McFadden & F.S. Chapin unpublished data). However, on moderately drained upland slopes, vegetation exerts a strong effect, with greater sensible heat flux and less evapotranspiration in areas dominated by shrubs relative to moss-dominated tundra (J.P. McFadden & F.S. Chapin unpublished data). The ratio of sensible to latent heat fluxes (Bowen ratio) can vary two-fold among these vegetation types. These measurements suggest that vegetation exerts a strong effect on the manner in which energy is dissipated at the surface. Is this effect strong enough to be important?

Summer air masses that move from the Arctic Ocean into arctic Canada carry only enough moisture to account for 25% of the precipitation that occurs on land (Walsh *et al.* 1994). Thus, the remaining 75% of precipitation must originate from evapotranspiration over land; that is, recycling of moisture between the land surface and the atmosphere accounts for most of the precipitation in this region of the Arctic. The greater the magnitude of this recycling, the greater the cloudiness (and reflectance of incoming solar radiation) and the cooler the regional climate. Although we do not yet know the relative importance of topography and vegetation in controlling regional coupling, it is clear that land–atmosphere interactions within the Arctic are crucial in determining summer climate.

Human influence on climatic coupling

Human activities can influence both global and regional coupling between arctic ecosystems and climate (see Fig. 4.5). The pollutants advected into the Arctic from lower latitudes create an arctic haze that deposits substantial amounts of heavy metals and acidity on arctic ecosystems (Jaffe *et al.* 1991; see Chapter 10). Mosses and lichens are quite sensitive to these pollutants. During the past 40 years there has been a regional warming of permafrost that cannot be explained by regional climatic warming (Lachenbruch & Marshall 1986) and, therefore, presumably reflects changes in the insulative properties of the tundra surface. Zimov *et al.* (1993) suggest that decreased moss cover as a result of pollution effects may be responsible for the recent warming of permafrost and release of CO_2 from arctic ecosystems. Thus, human impacts may affect the coupling between the Arctic and global climate via their effects on vegetation.

Human activities could strongly influence regional coupling of high-latitude ecosystems with climate through changes in tree cover in the boreal zone. The economic difficulties of Russia and the depletion of forest resources in south-east Asia make the Russian boreal forest a likely target for exploitation in the coming decades. It remains to be seen whether changes in logging intensity and associated changes in fire frequency would be large enough to significantly alter regional albedo and regional temperature regime.

CONCLUSIONS

Recent experiments demonstrate that arctic ecosystems are quite sensitive to manipulations simulating climatic change. Some characteristics of arctic ecosystems, such as biomass and productivity, are relatively insensitive to altered climate because increased biomass of deciduous shrubs is compensated by declines in non-vascular plants and evergreen shrubs. Other community traits, such as species richness, are strongly affected by experimental manipulations simulating short-term (10-year) climatic change. Growth forms of arctic plants differ strongly in their effects on ecosystem processes. The changes in community composition induced by climatic change could strongly affect animal food chains and soil thermal regime as well as alter litter quality and rates of nutrient cycling. We suggest that these secondary effects of climatic change mediated by vegetation change may be just as important as direct climatic effects on ecosystem processes. Arctic vegetation also affects global climate (via effects on CO_2 and CH_4 fluxes) and regional climate (via effects on albedo and water/energy exchange). Human activities could serve as a positive feedback to global warming by enhancing CO_2 efflux and as a negative feedback to regional warming through boreal deforestation and increased albedo.

ACKNOWLEDGEMENTS

We thank Gus Shaver for participation in the research and discussion of the ideas presented in this chapter. This work was funded by the US National Science Foundation programs in Arctic System Science (DPP-9214906 and OPP-9318532), the Department of Energy (92-SC-DOE-1003) and the National Aeronautics and Space Administration (NAGW-3769).

REFERENCES

Berendse, F. & Aerts, R. (1987). Nitrogen-use efficiency: A biologically meaningful definition? *Functional Ecology*, **1**, 293–296.

Billings, W.D. & Mooney, H.A. (1968). The ecology of arctic and alpine plants. *Biological Review*, **43**, 481–529.

Bonan, G.B., Chapin, F.S. III & Thompson, S.L. (1995). Boreal forest and tundra ecosystems as components of the climate system. *Climatic Change*, **29**, 145–167.

Bonan, G.B., Pollard, D. & Thompson, S.L. (1992). Effects of boreal forest vegetation on global climate. *Nature*, **359**, 716–718.

Brubaker, L.B., Anderson, P.M. & Hu, F.S. (1995). Arctic tundra biodiversity: a temporal perspective from late Quaternary pollen records. *Arctic and Alpine Biodiversity: Patterns, Causes and Ecosystem Consequences* (Ed. by F.S. Chapin III & Ch. Körner), pp. 111–125. Springer, Berlin.

Chapin, F.S. III (1980). The mineral nutrition of wild plants. *Annual Review of Ecology and Systematics*, **11**, 233–260.

Chapin, F.S. III, Bret-Harte, M.S., Hobbie, S.E. & Zhong, H. (1996). Plant functional types as predictors of the transient response of arctic vegetation to global change. *Journal of Vegetation Science*, **7**, 347–358.

Chapin, F.S. III, Fetcher, N., Kielland, K., Everett, K.R. & Linkins, A.E. (1988). Productivity and nutrient cycling of Alaskan tundra: enhancement by flowing soil water. *Ecology*, **69**, 693–702.

Chapin, F.S. III, Jefferies, R.L., Reynolds, J.F., Shaver, G.R. & Svoboda, J. (Eds) (1992). *Arctic Ecosystems in a Changing Climate: An Ecophysiological Perspective.* Academic Press, San Diego.

Chapin, F.S. III & Körner, Ch. (1996). Arctic and alpine biodiversity: its patterns, causes, and ecosystem consequences. *Functional Roles of Biodiversity: Global Perspectives,* SCOPE 0263 (Ed. by H.A. Mooney, J.H. Cushman, E. Medina, O.E. Sala & E.-D. Schulze). John Wiley & Sons, Chichester.

Chapin, F.S. III, McKendrick, J.D. & Johnson, D.A. (1986). Seasonal changes in carbon fractions in Alaskan tundra plants of differing growth form: implications for herbivores. *Journal of Ecology,* 74, 707–731.

Chapin, F.S. III & Shaver, G.R. (1985). Individualistic growth response of tundra plant species to environmental manipulations in the field. *Ecology,* 66, 564–576.

Chapin, F.S. III & Shaver, G.R. (1988). Differences in carbon and nutrient fractions among arctic growth forms. *Oecologia,* 77, 506–514.

Chapin, F.S. III, Shaver, G.R., Giblin, A.E., Nadelhoffer, K.G. & Laundre, J.A. (1995). Response of arctic tundra to experimental and observed changes in climate. *Ecology,* 76, 694–711.

Chapman, W.L. & Walsh, J.E. (1993). Recent variations of sea ice and air temperature in high latitudes. *Bulletin of the American Meteorological Society,* 74, 33–47.

Clymo, R.S. & Hayward, P.M. (1982). The ecology of *Sphagnum. Bryophyte Ecology* (Ed. by A.J.E. Smith), pp. 229–289. Chapman & Hall, London.

COHMAP (1988). Climatic changes of the last 18 000 years: observations and model simulations. *Science,* 241, 1043–1052.

Foley, J.A., Kutzbach, J.E., Coe, M.T. & Levis, S. (1994). Feedbacks between climate and boreal forests during the Holocene epoch. *Nature,* 371, 52–54.

Gersper, P.L., Alexander, V., Barkley, S.A., Barsdate, R.J. & Flint, P.S. (1980). The soils and their nutrients. *An Arctic Ecosystem: The Coastal Tundra at Barrow, Alaska* (Ed. by J. Brown, P.C. Miller, L.L. Tieszen & F.L. Bunnell), pp. 219–254. Dowden, Hutchinson & Ross, Stroudsburg, Penn.

Gorham, E. (1991). Northern peatlands: role in the carbon cycle and probable responses to climatic warming. *Ecological Applications,* 1, 182–195.

Grulke, N.E., Reichers, G.H., Oechel, W.C., Hjelm, U. & Jaeger, C. (1990). Carbon balance in tussock tundra under ambient and elevated CO_2. *Oecologia,* 83, 485–494.

Havström, M., Callaghan, T.V. & Jonasson, S. (1993). Differential growth responses of *Cassiope tetragona,* an arctic dwarf-shrub, to environmental perturbations among three contrasting high- and sub-arctic sites. *Oikos,* 66, 389–402.

Hobbie, S.E. (1992). Effects of plant species on nutrient cycling. *Trends in Ecology and Evolution,* 7, 336–339.

Hobbie, S.E. (1995). Direct and indirect effects of plant species on biogeochemical processes in arctic ecosystems. *Arctic and Alpine Biodiversity: Patterns, Causes and Ecosystem Consequences* (Ed. by F.S. Chapin III & Ch. Körner), pp. 213–224. Springer, Berlin.

Houghton, J.T., Jenkins, G.J. & Ephraums, J.J. (Eds) (1990). *Climate Change: The IPCC Scientific Assessment.* Cambridge University Press, Cambridge.

Jaffe, D.A., Honrath, R.E., Herring, J.A., Li, S.M. & Kahl, R.D. (1991). Measurements of nitrogen oxides at Barrow, Alaska during spring: evidence for regional and northern hemispheric sources of pollution. *Journal of Geophysical Research,* 96, 7395–7405.

Johnson, L.C. & Damman, A.W.H. (1991). Species controlled *Sphagnum* decay on a south Swedish raised bog. *Oikos,* 61, 234–242.

Jonasson, S. (1992). Plant responses to fertilization and species removal in tundra related to community structure and clonality. *Oikos,* 63, 420–429.

Kane, D.L., Hinzman, L.D., Woo, M. & Everett, K.R. (1992). Arctic hydrology and climate change. *Arctic Ecosystems in a Changing Climate: An Ecophysiological Perspective* (Ed. by F.S. Chapin III, R.L. Jefferies, J.F. Reynolds, G.R. Shaver & J. Svoboda), pp. 35–57. Academic Press, San Diego.

Lachenbruch, A.H. & Marshall, B.V. (1986). Climate change: geothermal evidence from permafrost in the Alaskan arctic. *Science*, **34**, 689–696.

Lauenroth, W.K., Dodd, J.L. & Simms, P.L. (1978). The effects of water- and nitrogen-induced stresses on plant community structure in a semiarid grassland. *Oecologia*, **36**, 211–222.

Maxwell, B. (1992). Arctic climate: potential for change under global warming. *Arctic Ecosystems in a Changing Climate: An Ecophysiological Perspective* (Ed. by F.S. Chapin III, R.L. Jefferies, J.F. Reynolds, G.R. Shaver & J. Svoboda), pp. 11–34. Academic Press, San Diego.

Nadelhoffer, K.J., Giblin, A.E., Shaver, G.R. & Laundre, J.A. (1991). Effects of temperature and substrate quality on element mineralization in six arctic soils. *Ecology*, **72**, 242–253.

Oberbauer, S. & Miller, P.C. (1982). Growth of Alaskan tundra plants in relation to water potential. *Holarctic Ecology*, **5**, 194–199.

Oechel, W.C., Cowles, S., Grulke, N., Hastings, S.J., Lawrence, W., Prudhomme, T., Riechers, G., Strain, B., Tissue, D. & Vourlitis, G. (1994). Transient nature of CO_2 fertilization in arctic tundra. *Nature*, **371**, 500–503.

Oechel, W.C., Hastings, S.J., Vourlitis, G., Jenkins, M., Riechers, G. & Grulke, N. (1993). Recent change of Arctic tundra ecosystems from a net carbon dioxide sink to a source. *Nature*, **361**, 520–523.

Oksanen, L. (1990). Predation, herbivory, and plant strategies along gradients of primary productivity. *Perspectives on Plant Competition* (Ed. by J.B. Grace & D. Tilman), pp. 445–474. Academic Press, San Diego.

Parsons, A.N., Welker, J.M., Wookey, P.A., Press, M.C., Callaghan, T.V. & Lee, J.A. (1994). Growth responses of four sub-arctic dwarf shrubs to simulated environmental change. *Journal of Ecology*, **82**, 307–318.

Payette, S. & Filion, L. (1985). White spruce expansion at the tree line and recent climatic change. *Canadian Journal of Forest Research*, **15**, 241–251.

Pielke, R.A. & Vidale, P.L. (1996). The boreal forest and the polar front. *Journal of Geophysical Research—Atmospheres*, **100D**, 25755–25758.

Reeburgh, W.S. & Whalen, S.C. (1992). High latitude ecosystems as CH_4 sources. *Ecological Bulletin (Copenhagen)*, **42**, 62–70.

Rind, D. (1987). Components of the ice age circulation. *Journal of Geophysical Research*, **92D**, 4241–4281.

Romer, M.J., Cummins, W.R. & Svoboda, J. (1983). Productivity of native and temperate 'crop' plants in the Keewatin District, N.W.T. *Naturaliste Canadienne*, **110**, 85–93.

Shaver, G.R. (1981). Mineral nutrition and leaf longevity in an evergreen shrub, *Ledum palustre* ssp. *decumbens*. *Oecologia*, **49**, 362–365.

Shaver, G.R. (1986). Woody stem production in Alaskan tundra shrubs. *Ecology*, **56**, 401–410.

Shaver, G.R. & Billings, W.D. (1975). Root production and root turnover in a wet tundra ecosystem, Barrow, Alaska. *Ecology*, **56**, 401–410.

Shaver, G.R. & Chapin, F.S. III (1980). Response to fertilization by various plant growth forms in an Alaskan tundra: nutrient accumulation and growth. *Ecology*, **61**, 662–675.

Shaver, G.R. & Chapin, F.S. III (1986). Effect of fertilizer on production and biomass of tussock tundra, Alaska, U.S.A. *Arctic and Alpine Research*, **18**, 261–268.

Shaver, G.R. & Chapin, F.S. III (1991). Production : biomass relationships and element cycling in contrasting arctic vegetation types. *Ecological Monographs*, **61**, 1–31.

Shaver, G.R., Giblin, A.E., Nadelhoffer, K.J. & Rastetter, E.B. (1996). Plant functional types and ecosystem change in arctic tundras. *Plant Functional Types* (Ed. by T. Smith, H.H. Shugart & F.I. Woodward). Cambridge University Press, Cambridge, in press.

Shaver, G.R. & Melillo, J.M. (1984). Nutrient budgets of marsh plants: efficiency concepts and relation to availability. *Ecology*, **65**, 1491–1510.

Solbrig, O.T. (Ed.) (1991). *From Genes to Ecosystems: A Research Agenda for Biodiversity*. International Union of Biological Sciences, Cambridge, Mass.

Tenhunen, J.D., Lange, O.L., Hahn, S., Siegwolf, R. & Oberbauer, S.F. (1992). The ecosystem role of poikilohydric tundra plants. *Arctic Ecosystems in a Changing Climate: An Ecophysiological Perspective* (Ed. by F.S. Chapin III, R.L. Jefferies, J.F. Reynolds, G.R. Shaver & J. Svoboda), pp. 213–237. Academic Press, San Diego.

Tissue, D.T. & Oechel, W.C. (1987). Response of *Eriophorum vaginatum* to elevated CO_2 and temperature in the Alaskan tussock tundra. *Ecology*, **68**, 401–410.

Torn, M.S. & Chapin, F.S. III (1993). Environmental and biotic controls over methane flux from arctic tundra. *Chemosphere*, **26**, 357–368.

Trenberth, K.E. (1990). Recent observed interdecadal climate changes in the northern hemisphere. *Bulletin of the American Meteorological Society*, **71**, 988–993.

Walsh, J.E., Zhou, X., Portis, D. & Serreze, M. (1994). Atmospheric contribution to hydrologic variations in the arctic. *Atmosphere-Ocean*, **32**, 733–755.

Warren Wilson, J. (1966). An analysis of plant growth and its control in arctic environments. *Annals of Botany*, **30**, 383–402.

Webber, P.J. (1978). Spatial and temporal variation of the vegetation and its productivity, Barrow, Alaska. *Vegetation and Production Ecology of an Alaskan Arctic Tundra* (Ed. by L.L. Tieszen), pp. 37–112. Springer, New York.

White, R.G. & Trudell, J. (1980). Habitat preference and forage consumption by reindeer and caribou near Atkasook, Alaska. *Arctic and Alpine Research*, **12**, 511–529.

Whittaker, R.H. (1953). A consideration of climax theory: The climax as a population and pattern. *Ecological Monographs*, **23**, 41–78.

Williams, J.B. & Batzli, G.O. (1982). Pollination and dispersion of five species of lousewort (*Pedicularis*) near Atkasook, Alaska, U.S.A. *Arctic and Alpine Research*, **14**, 59–74.

Wookey, P.A., Parsons, A.N., Welker, J.M., Potter, J.A., Callaghan, T.V., Lee, J.A. & Press, M.C. (1993). Comparative responses of phenology and reproductive development to simulated environmental change in sub-arctic and high arctic plants. *Oikos*, **67**, 490–502.

Zimov, S.A., Zimova, G.M., Daviodov, S.P., Daviodova, A.I., Voropaev, Y.V., Voropaeva, Z.V., Prosiannikov, S.F., Prosiannikova, O.V., Semiletova, I.V. & Semiletov, I.P. (1993). Winter biotic activity and production of CO_2 in Siberian soils: a factor in the greenhouse effect. *Journal of Geophysical Research*, **98D**, 5017–5023.

5. Habitat fragility as an aid to long-term survival in arctic vegetation

ROBERT M.M.CRAWFORD

Plant Science Laboratory, Sir Harold Mitchell Building, St Andrews University, St Andrews, Fife KY16 9AL, UK

INTRODUCTION

The factors that create the arctic habitat are usually described in terms of short growing seasons, low resource availability and extreme conditions of cold and exposure. Apart from these conventional stress factors, the Arctic is also an area of marked climatic oscillations (Dowdeswell 1995; Dowdeswell & White 1995), physical disturbance and fluctuations in herbivore populations (Oksanen 1983). The combination of environmental stress and disturbance from habitat instability and the possibility of periods of intense grazing imposes a particularly testing combination of adverse conditions for plant survival. Seasonal variations in snowfall create considerable changes in growing season length and even mild winters can have subsequent adverse effects when winter rain falls on frozen soil causing severe ice-encasement (Gudleifsson 1994). In addition, the physical nature of the terrain with constant movement through cryoperturbation and solifluction contributes to the fragility of arctic habitats. The diminutive stature of the vegetation, the low number of species and lack of continuous plant cover, leaves a landscape that is prone to movement and physical disturbance from wind, water, ice and gravity. Disturbance of arctic soils from constant physical movement and erosion creates the impression of a region occupied tenuously by fragile communities that might be irreversibly destroyed by additional disturbance from climatic warming and human interference. This chapter presents the case that the physical fragility of the habitat, coupled with the variability of growing season length, instead of presenting risks to the stability of the high arctic flora, actively improves the long-term fitness of some plant species and preadapts them to the consequences of climatic change.

Preadaptation can be defined as 'an adaptation evolved in one adaptive zone which, quite by chance, proves especially advantageous in an adjacent zone and so allows the organisms to radiate into it' (Allaby 1994). An alternative definition is 'the possession by an organism of characters or traits that would favour its survival in a new or changed environment' (Lincoln, Boxshall & Clark 1982). Both these definitions consider the definition at the level of the individual and are couched in terms of examples such as a bird possessing feathers which then preadapts the individual for flight. Plants, however, provide examples of localized ecotypic variation that differ from those in animals, where interfertile variants rarely live in close proximity to one another. In the Arctic the mesic condition, where community development buffers environmental variation, is strikingly absent. Consequently, change in habitat from dry ridge to snow patch or from drought-prone to flooded is abrupt and plant populations adapted to opposing conditions live in close proximity to one another. It is argued that the polymorphic and interfertile nature of many plant populations provides a means of preserving character traits which can be readily selected should conditions alter and that arctic plant populations are thus preadapted to environmental change due to the variability of the terrain that they inhabit.

HABITAT FRAGILITY IN THE ARCTIC

Arctic ecosystems are frequently described as fragile with low species diversity being cited as the underlying cause which renders them susceptible to disturbance (Elton 1927). Many official conservation policy documents emphasize the vulnerability of the Arctic in terms of its small numbers of plant and animal species and slow biological, biochemical and chemical processes (Larsen 1985). These somewhat global statements do not distinguish adequately what parts of arctic ecosystems are fragile. It can be argued that it is not the plant and animal populations that are fragile but the terrain, and that physical fragility of the habitat should not be assumed to imply that the animal and plant populations are equally delicate. Animal populations regularly run the gamut from superabundance to near extinction (Batzli *et al.* 1980) and this capacity should perhaps be regarded as resilience not fragility. Similarly, in plants there can be prolonged adverse climatic periods when seed production becomes impossible (Barnes 1966) and yet the species persists ready to reproduce sexually whenever conditions permit.

A recent review of extensive Russian records of historical fluctuations in the arctic environment and their effects on plant and animal populations points out that they have a remarkable ability to recover after drastic diminution from environmental disruptions (Krupnik 1993). In animal species, two features aid the capacity for recovery from near extinction. The first is a loose attachment to any particular habitat, and the second is a high regeneration capacity. The ability to change habitats periodically allows animal populations the strategic use of the large spatial dimensions that are available at high latitudes (Dunbar 1973). The movement of animal populations on a north–south axis has been frequently noted, as warmth-demanding boreal species such as elk, red fox, bear, wolf, pine marten and otter move north in warm years and more cold-tolerant species, polar bear, arctic fox, narwhal, reindeer, beluga and arctic cod, move south in cold spells (Krupnik 1993). However, with plant populations migrations do not take place in such short time intervals, either northwards or southwards. This difference is demonstrated in the response of plants as compared with insects to climatic warming at the beginning of the Holocene when beetle assemblages in Europe had already shown significant changes before 13 kaBP (Atkinson, Briffa & Coope 1987), while the principal vegetation changes took place somewhat later. Insects can expand their species range rapidly when changing climatic conditions favour their potential for sexual reproduction. When conditions are suitable, they can either exploit a new area with great rapidity or expand residual populations in just a few seasons. Similarly, they disappear with equal celerity when reproduction is curtailed by climatic deterioration. By contrast, plant survival is much less dependent on sexual reproduction. Plant persistence in an area with a deteriorating climate can be maintained by resorting to long periods of purely vegetation re-production probably over thousands of years as demonstrated in the case of the American aspen in Utah (Barnes 1966). During shorter periods of adversity, plant survival and variation may be maintained by the natural preservation in the soil of extensive seed banks (Fox 1983; McGraw & Vavrek 1989). In addition to these properties, which affect survival and fitness within populations, plants also have the capacity for rapid genetic adjustment of their populations as conditions change. The extensive capacity for plant species to possess many ecotypes provides a unique ability to exploit micrometeorological differences in terrain. This extension of species range as a result of population selection for different habitats also provides a degree of polymorphism that can adapt to abrupt climatic change (see below for examples).

The static nature of plant populations in an environment that is constantly exposed to disturbance and change requires a fresh approach to the study of physiological adaptation in the Arctic. Instead of the conventional topics of cold and drought tolerance and the acquisition of nutrients in limited supply, a more dynamic view is needed which takes account of the changing spatial and temporal window that is available for plant growth as a result of fluctuations in growing season length and the ever-present threat of physical disturbance. Current life-strategy theory suggests that extremes of environmental stress are not compatible with a high risk of disturbance (Grime 1993). According to Grime's three-strategy model, plants can survive loss of biomass from stress factors (stress-tolerators) or from disturbance (ruderals) but the combination of extreme stress and extreme disturbance produces a habitat where there is no viable plant growth strategy. The term *ruderal* comes from the Latin *rudera* (plural of *rudus*), 'broken stones', a term which conjures up an image of many arctic landscapes with screes, shore ridges and gravel sites that are constantly at risk of disturbance from wind, water and ice. The Arctic is the epitomy of what is generally recognized as a stressed habitat for plant life and the disturbance and fragility of the terrain is evident from ice-scoured shores to the unstable screes and slopes of montane habitats (Figs 5.1 & 5.2). The ability of plants to live in these areas may therefore not lie in tolerance of the

FIG. 5.1. View from west side of Mesters Vig Fjord (north-east Greenland, 72°15′N) illustrating disturbance-prone arctic terrain. Note unstable scree in right foreground and erosion-risk, flood-prone plain in the middle distance.

conventional stress factors such as cold or drought (the Arctic is neither the coldest nor the most drought-prone region of the Earth), but rather in their ability to adjust to growing season fluctuations and the physical uncertainty of the habitat. A re-examination is therefore needed of the properties that allow arctic plants to survive these dual adversities of stress and disturbance.

The approach of most plant physiological studies on arctic species can be grouped into two categories: (i) adaptations to the extremes of climatic severity in terms of cold and lack of resources (Billings 1987; Sonesson & Callaghan 1991); and (ii) adaptations that are unique to arctic plants and distinguish them from other cold-adapted species or populations of similar life form that live in montane habitats at lower latitudes (Bliss 1956; Mooney & Billings 1960, 1961; Billings *et al.* 1971; Billings 1974). Although tolerance of cold and minimal resource requirements are basic requirements for arctic vegetation, more positive adaptations for survival in low temperature regimes are needed than just being able to withstand adverse conditions.

FIG. 5.2. Mid-summer (14 July) in the High Arctic at Biskayerhuken, Spitsbergen (79°40′N). The late-lying snow-banks demonstrate the ever-present risk of plants living in their proximity of missing one or more growing seasons. This particular shore has extensive stands of the polar willow (*Salix polaris*).

The following sections examine some ecophysiological case studies of long-term fitness in arctic species, not in terms of tolerance of extreme conditions, but in the potential role of phenological and physiological diversity enabling populations to respond rapidly to the fragility of the habitats and uncertainty of the climate.

PHENOLOGICAL VARIATION AT HIGH LATITUDES

The brevity of the high arctic growing season might suggest that the need for rapid completion of the annual growth cycle of flowering and seed production allows little scope for phenological variation. The compaction of the flowering period has been cited as a possible cause for the degree of polyploidy that is found at high latitudes (Stebbins 1971, 1984). The existence of such rarities as *Saxifraga narthorstii* has been attributed to the simultaneous flowering in the ultra-short growing season of shore habitats in north-east Greenland of *S. oppositifolia* and *S. aizoides*, leading to the hybrid production of *S. narthorstii* (Böcher 1983). The loss of dominance of the grasses as opposed to the sedges (*Carex* spp.) in the High Arctic is attributed to their differing phenologies with the slower sequential production of leaves in grasses being maladaptive in short growing seasons in comparison with the prompt production of sedge leaves.

Despite the shortness of the high arctic growing season, there exist nevertheless comparatively wide variations in the length of time available to different populations of the same species for completing the annual growth cycle. The brevity of the arctic growing season makes a delay of resumption of growth of just 1 week result in a substantial percentage reduction in length of growing season (Table 5.1). Sites where winter snow and ice first disappear can become limiting for plant growth due to water shortage as the growing season advances to its warmest period. Plants that are able to resume growth only later in the season, due to late snow melt, are usually in habitats that are provided with a greater supply of water and nutrients. The plant populations of these wetter areas have therefore the possibility of compensating for the shorter growth period with increased photosynthetic activity and higher growth rates (Crawford & Abbott 1994; Crawford, Chapman & Smith 1995). The extent of this variation alters from year to year depending on the amount of snowfall or ice-encasement that takes place over winter and the time taken for melting in summer.

The response of plants to the position of the summer snow-line differs in the Arctic as compared with alpine and montane habitats at lower latitudes. In alpine areas the summer snow-line is usually more predictable and is marked by an abrupt disappearance of vegetation cover. However, in the Arctic, annual variation in snow-lie can be very variable (Table 5.1) reflecting in coastal sites the large interannual variations in arctic sea ice (Wadhams 1995). The variability of the duration of snow cover can result in plants spending an entire growing season under an unmelted snow bank. Plants that grow in areas prone to late snow-lie are always exposed to the risk of missing an entire growing season. Some species can survive under snow cover for 1 year or possibly more, but only a few species (e.g. *Oxyria digyna*) are able to survive a series of non-emergent years in succession. Populations that live

TABLE 5.1. Comparison of expected growing season length on beach-ridge and low-shore sites as observed at Kongsfjord 1991–95.

Site	Expected date of emergence from snow	Expected date of end of growing season	Length of growing season (days)
Beach-ridge	20–30 June	12–20 August	44–62
Low-shore	7–19 July	12–20 August	24–44

in areas where snow can sometimes lie for more than 12 months usually consist of diminutive specimens of certain widespread species such as *Polygonum viviparum*, *Salix polaris*, *Ranunculus pygmaeus*, *Sibbaldia procumbens*, *Potentilla hyparctica* and *Saxifraga oppositifolia* (Pielou 1994). How much phenotypic plasticity as opposed to genetic differentiation contributes to snow-patch survival is not yet fully understood (see below).

Differentiation in the extent of summer warming also leads to distinct phenological divisions between plants that can inhabit drought-prone slopes and moraines as compared with those that live in wet valley bottoms where peat and permafrost impede free drainage. One much studied case in relation to phenology is *Dryas octopetala*, which throughout its wide polar range shows much visible ecotypic variation (Elkington 1971) and where differences in lateness of snow-lie have resulted in the evolution of two mutually exclusive ecotypes, namely fellfield and snow-bed forms (McGraw & Antonovics 1983) which although interfertile and separated by only 2 weeks in flowering time, remain distinct due to the extremely short growing seasons of the High Arctic (Fig. 5.3a). These ecotypes have been described as varying in shade tolerance and thus having specific specializations for survival in either fellfield or snow-bed communities (McGraw & Antonovics 1983; McGraw 1985, 1987). However, differentiation of populations in relation to these two contrasting habitats, apart from the immediate role in increasing the habitat range available to this species, has a more long-term significance in maintaining separate populations that are adapted respectively for survival in long and short growing seasons.

This example of ecotypic diversity in high arctic regions is therefore not just an aid to exploitation of existing habitats, but is an important property which allows species to respond quickly to changing climatic conditions through site selection of the better-adapted ecotype. Theoretically, it could be possible within one generation for a melting snow-bed population to be replaced by the fellfield form, through site selection from a mixed seedling population.

An example similar to *Dryas octopetala* in having fellfield and snow-bed ecotypes is found in high arctic populations of *Saxifraga oppositifolia*, one of the hardiest plants of polar regions. On dry, exposed ridges in Spitsbergen with warmer temperatures and a longer growing season than the adjacent low shore, the purple saxifrage has a semi-erect form that starts growing before the snow melts in adjacent low-lying shore habitats. By contrast, on the late, cold shore sites this species is found usually as a trailing prostrate plant (Fig. 5.3b) that does not give an impression of being particularly robust. However, appearances are misleading and this frail-looking prostrate saxifrage has a facility for *metabolic rate compensation* (Hochachka & Somero 1973), in which the organism increases its ability to function at low temperature by augmenting the capacity of an existing process, rather than evolving a more efficient alternative. This property of metabolic rate compensation

R.M.M.CRAWFORD

Dryas octopetala

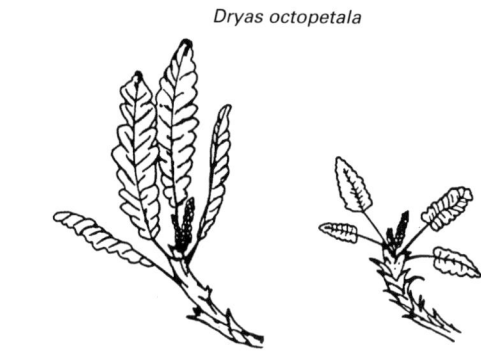

(a)

Saxifraga oppositifolia

(b)

FIG. 5.3. Snow-patch (left) and exposed ridge (right) forms of (a) *Dryas octopetala* and (b) *Saxifraga oppositifolia*. (Reproduced from McGraw & Antonovics 1983 and Lid & Lid 1994, respectively; (b) drawn by Dagny Tande Lid)

(Figs 5.4 & 5.5) allows the cold-wet shore ecotype to outperform the more robust type from the beach ridge in gross photosynthetic capacity, respiration and shoot growth (Crawford *et al.* 1993, 1995). Physiologically, the semi-erect form on the beach ridge is much more drought tolerant, and conserves carbohydrate for periods of stress. The plants of the shore habitat are by contrast less able to conserve carbon resources and use a much greater portion of their energy gains immediately for rapid growth. The two forms have developed opposing strategies which aid their survival in their particular microhabitat, but which would disadvantage them if conditions were to change. Plants of the semi-erect and creeping ecotypes of purple saxifrage can also be recognized in the former having overlapping petals while the latter have non-overlapping petals. The differences in petal insertion are maintained in cultivation and, although interfertile and with intermediate ecotypes, the two extreme forms are recognized as separate subspecies (Lid & Lid 1994) and are

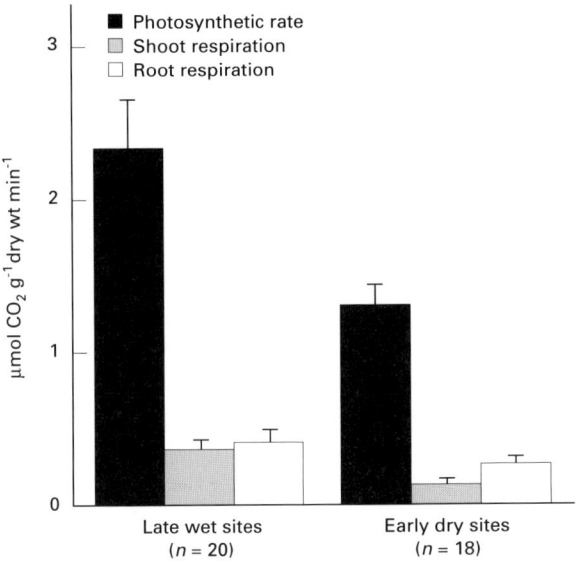

FIG. 5.4. Comparison of photosynthesis, shoot and root respiration in populations of *Saxifraga oppositifolia* as measured in the prostrate form growing in late wet sites (see text) and the semi-erect form found in early dry sites at Ny-Ålesund, Spitsbergen. (Vertical bars = standard deviations.)

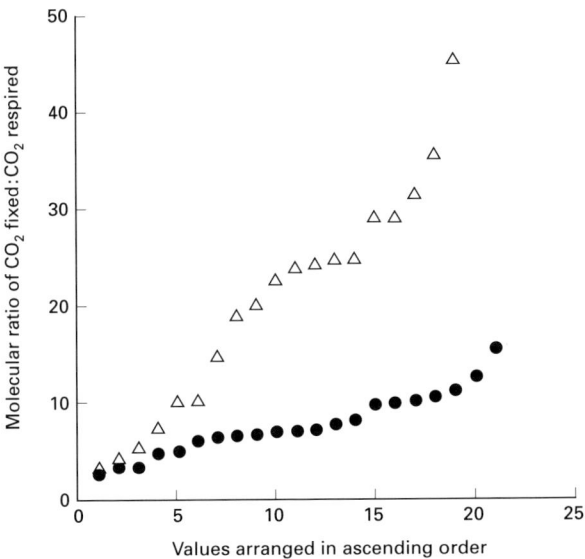

FIG. 5.5. Molecular ratio of CO_2 fixed (gross respiration) to respiration in shoots of *Saxifraga oppositifolia*. Prostrate plants taken from late wet sites (●) compared with semi-erect plants taken from early dry sites (△).

genetically distinct. It would appear that these two forms and intermediates aid the continued long-term presence of the species as they provide a mutual support mechanism for survival in an environment with variable and uncertain growing seasons. In unfavourable years the plants in the low-lying shore habitat are unable to set seed, but nevertheless produce pollen which fertilizes some of the adjacent plants on the beach ridge and thus ensures a production of seedlings at the next generation. A genetically mixed rain of seeds then washes down to the shore where habitat selection results in the preferential survival of the creeping ecotype (Teeri 1973). Although these ecotypes differ in their carbon metabolism, and morphology, the ultimate basis of their differentiation is in their ability to inhabit early and late sites and therefore the ultimate distinction between them is one of phenology.

The genus *Draba* in the Nordic area alone (including Spitsbergen) has at least 16 recognizable species, three diploid and 13 polyploid (Brochmann & Elven 1992). There is much interbreeding particularly among the polyploid species and within this subgroup there can be found marked differentiation in preferences for late and early sites. Late snow-bed polyploid species of *Draba* include *D. crassifolia*, *D. alpina* and *D. norvegica*, while *D. oxycarpa*, *D. corymbosa*, *D. adamsii* and *D. arctica* are not common in late snow-beds and are more frequent on the earlier exposed sites (Brochmann & Elven 1992). Although these taxa are recognized as species they represent in essence the same ecological situation as found in the two subspecies of *Saxifraga oppositifolia* and are capable of genetic exchange which will presumably allow these species to adapt to climatic change from the existing populations without the need for immigration of plants from other climates. Hybridization within related Nordic and arctic species of *Draba* has been suggested as a means whereby this closely related group, which has a high incidence of polyploidy, can achieve rapid rates of evolution (Brochmann, Soltis & Soltis 1992c).

PHYSIOLOGICAL VARIATION AT HIGH LATITUDES

The phenological variation described above for semi-erect populations of *Saxifraga oppositifolia* from dune ridges as compared with prostrate plants from shore habitats is matched by physiological variation (Crawford *et al.* 1993). The shore plants are able to compensate for a late start to the growing season by higher metabolic and growth rates. Both photosynthesis and respiration rates are greater as is also the production of green shoots once the growing season has started. In a detailed study of ecotypic variation in *S. oppositifolia* in a high arctic site at Truelove Lowland, Devon Island, NWT, Canada, plants from the beach ridges were found to be adapted to a dry, moisture-limited environment, controlled transpiration more efficiently and maintained photosynthetic activity at lower leaf water potential values than did plants from the adjacent meadow or snow-bank sites. The latter, however, were found to withstand surface and substrate flooding during the growing season which beach ridge plants could not tolerate (Teeri 1972). Given the distinct morphological differentiation between the shore and beach ridge ecotypes of *S. oppositifolia* with their different flower forms that are maintained in cultivation, it would appear that these are genetically distinct populations that are maintained by habitat conditions.

Other cases exist where plant form varies between populations but is entirely phenotypic and imposed by the habitat. Such phenotypic ecoforms rather than genetically distinct ecotypes can be found in *Polygonum viviparum* (Wookey *et al.* 1994), *Saxifraga cernua* and *Salix polaris* (personal observation). A common perception of diminutive growth in arctic plants is as a passive response due to a lack of resources for growth. However, cases have been noted where plants of different stature differ in respiratory activity. A field study of respiration rate in leaves of arctic plants (McNulty, Cummins & Pellizari 1988) showed that the cyanide-resistant alternative-pathway respiration (APR) had higher activities in plants of short stature, living close to the ground in an environment with fluctuating temperatures, than in taller plants living in more equitable habitats (Table 5.2). Similarly, in a study of the extent of the use of APR in a number of temperate plants (Collier & Cummins 1989) it was found that there was an inverse relationship between plant height and APR. A comparison was made of five understorey species with five ruderal species and the latter were shown to have a greater capacity for APR than understorey species. This was interpreted as conferring on the ruderal species a physiological flexibility that aided survival in fluctuating environments. The ruderal environment is less buffered against environmental fluctuations than that of the understorey species. APR allows plants to 'burn off' excess carbohydrate. Respiration by APR in plants exposed to ample light, but with limited nutrient resources, resulted in over 65% of the available energy being lost as heat. The disposal of excess carbohydrate has been suggested as a means whereby plants can adjust to environmental fluctuations. To be able to 'burn off' carbohydrate and not use it for growth during untypical warm periods would prevent plants that

TABLE 5.2. Rates of total dark respiration (TDR), alternative-pathway respiration (APR) in pmol mm^{-2} min^{-1} and APR/TDR. (Numbers in parentheses = observations.) (Data from McNulty & Cummins 1987.)

Species	TDR		APR		APR/TDR
Cotton	36	(2)	5	(2)	14
Sunflower	66	(3)	5	(3)	8
Peas	60	(3)	1	(1)	2
Spinach	72	(5)	17	(4)	24
Saxifraga cernua	142	(10)	61	(5)	43

Fig. 5.6. *Saxifraga cernua* growing in a sheltered bird-cliff ledge near Kongsfjord, Spitsbergen (78°53′N). This species when growing in long photoperiods 'burns off' a high percentage of its carbohydrate supply through the alternative respiration pathway (see text and McNulty & Cummins 1987).

live in exposed habitats from growing to a size where they would suffer from wind and exposure. In the arctic *Saxifraga cernua* (Fig. 5.6) there can be observed a high level of APR activity (McNulty & Cummins 1987), which was again interpreted as the result of selective pressure from fluctuating temperatures producing a means whereby this species could alter dark respiration rates which increased during acclimation to low growth temperatures (Table 5.2). A subsequent study of this species (McNulty & Cummins 1989) showed that 70–80% of the total measurable respiratory activity of the leaves could be attributed to APR. An active APR activity will result in less adenosine triphosphate (ATP) energy being available for plant growth, and plants that grow less and remain near the surface of the soil will suffer less wind abrasion and physical damage and are also in a microclimate that suffers less from cooling by water loss.

The above examples provide a glimpse of a long known, but relatively understudied and unappreciated aspect of survival in stressed habitats, namely the Montgomery Effect (Montgomery 1912) which states that 'in areas of low environmental potential plants with low growth rates have a selective advantage'. The present case history is particularly illuminating in that it demonstrates that low growth rate in these arctic habitats is not just a passive response due to low resource availability, but an active process which 'burns-off' carbohydrate, as presumably a low growth rate has greater survival value than an overstimulation of growth through active photosynthesis in the long days experienced by plants at high latitudes.

ANOXIA TOLERANCE
IN PLANTS LIABLE TO ICE-ENCASEMENT

The snow-bank, with its provision of winter shelter and wet soils that counteract the lack of summer rainfall, provides a habitat that is favoured by many arctic species. Dependence on this habitat is, however, not without its dangers due to the uncertainty as to how readily the snow will melt in the brief arctic summer. Snow melting followed again by freezing can encase plants in ice and thus deprive their tissues for long periods of access to oxygen (Gudleifsson & Larsen 1993). Most plants if deprived of oxygen die rapidly. Some flood-tolerant species possess perennating organs, such as rhizomes or tubers, that are capable of regenerating new shoots once the anoxic stress has passed (Crawford 1992). A number of arctic species of

Fig. 5.7. Portable anaerobe jar as used for testing anoxia tolerance of high arctic plants in Spitsbergen. Intact plants with roots are placed in Petri dishes in the jar. Water added to a sachet evolves hydrogen which with the aid of a catalyst removes all oxygen. The successful removal of oxygen is monitored by observing the bleaching of a methylene blue strip visible in the upper half of the jar.

areas that are liable to ice-encasement (Fig. 5.7) exhibit a very high tolerance of anoxia (Table 5.3) which can be tested readily *in situ* by placing intact plants in anaerobe jars (Fig. 5.8). This high tolerance of oxygen deprivation is all the more remarkable, as both perennating organs and the foliage survive anoxia in some species for up to 3 weeks at temperatures above 0°C (Crawford, Chapman & Hodge 1994). A parallel situation is seen in some antarctic mites where adults and nymphs of *Alaskozetes antarcticus* show a surprising tolerance of 28 days of anoxia with less than 40% mortality (Block & Sømme 1982). Examination of length of tolerance of oxygen deprivation in some arctic grasses (Fig. 5.9) shows that in the most tolerant species mortality suddenly increases after 50 days of anoxia, suggesting that this is the time necessary for the depletion of some essential reserves, most probably carbohydrates (Crawford & Braendle 1996). North American cranberry (*Vaccinium macrocarpon*) growers have learnt to risk flooding the cranberry bogs to protect the vines from frost only when they are well supplied with carbohydrate reserves, otherwise oxygen-deficiency syndrome symptoms appear (Eck 1990).

Comparison of arctic populations of *Saxifraga oppositifolia* from Spitsbergen with populations from Scotland showed that only the high arctic population

TABLE 5.3. List of species tested in Spitsbergen for anoxia tolerance. Intact plants were kept under total anoxia in anaerobe jars for 7 days at 5°C in the dark. Plants were judged to be anoxia tolerant if after 48 h of re-exposure to air they showed no wilting or discoloration of their shoots.

Anoxia tolerant (including leaves)	Anoxia intolerant
Saxifraga caespitosa	*Equisetum arvense*
S. oppositifolia	*Saxifraga hieracifolia*
S. foliosa	*S. cernua*
Ranunculus sulphureus	*Ranunculus pygmaeus*
Cardamine nymani	*Oxyria digyna*
Huperzia selago	*Pedicularis hirsuta*
Eriophorum scheuchzeri	*Cochlearia groenlandica*
Juncus biglumis	*Polygonum viviparum*
Carex misandra	*Poa alpina* (mature plant—
Luzula arctica	plantlets only tolerant)
L. arcuata ssp. *confusa*	*Draba oxycarpa*
Dryas octopetala	
Puccinellia vahliana	
Alopecurus borealis	
Poa alpina (pseudo-viviparous plantlets only)	
Deschampsia alpina	
Salix polaris (leaves not tolerant)	

FIG. 5.8. A cold north-facing shore which is prone to late snow cover and ice-encasement at Kongsfjord, Spitsbergen with a colony of the anoxia-tolerant sedge *Carex misandra* in the foreground and middle distance.

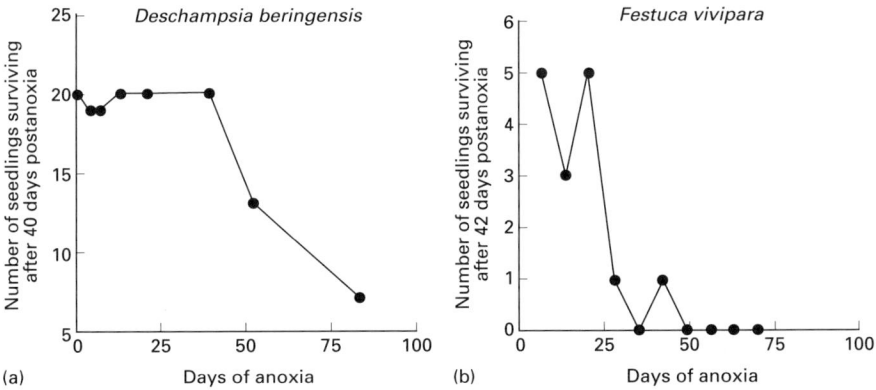

FIG. 5.9. Comparison of anoxia tolerance in grass seedlings of (a) an arctic population of *Deschampsia beringensis* Hult. from Alaska and (b) a subarctic population of *Festuca vivipara* L. from Iceland. The seedlings were grown in a cold room at 10°C using seed for *D. beringensis* and clonal pseudo-viviparous plantlets for *F. vivipara*. The *y*-axis records the number of seedlings surviving after allowing over a month of postanoxic recovery following different lengths of dark anoxic incubation at 5°C.

exhibited a tolerance of anoxia (Crawford *et al.* 1994). Thus, not only does this species show population differences within the Arctic but there are also physiological distinctions between arctic and non-arctic populations which permit the high-latitude plants to survive the uncertainty of their habitat in relation to the risk of prolonged encasement in ice. Snow-patches in more southerly montane habitats, as in Scotland, retreat in a predictable manner most summers and prolonged ice-encasement is therefore less likely.

LIFE-STRATEGY THEORY IN RELATION TO SURVIVAL IN THE ARCTIC

The life-strategy theory of Grime, in which plants are classified as competitors, stress tolerators or ruderals – the three-way CSR theory – postulates that plant productivity can be reduced by stress or by disturbance but that the extreme condition of high stress and high disturbance does not permit a viable strategy (Grime 1979). Thus, where such areas exist, as in arid deserts with shifting sands, there is no viable plant strategy for survival. The Arctic may, however, be an exception to the CSR plant life-strategy theory. The stress factors of the High Arctic present an impressive array of hostile environments for plant survival (Fig. 5.10). The long winter in which plants may be encased in ice and deprived of oxygen for up to 8 or even 9 months is a stress not met with elsewhere. In exposed sites, wind and frost desiccation often combined with abrasion are risks that persist throughout the year with only a very short growing season to repair the damage. To these physical stresses there is the ever present movement and disturbance that comes from cryoperturbation, solifluction and the general susceptibility to movement that a landscape suffers when plant cover is limiting.

The world's most northern plant communities, the diminutive plants of the polar deserts, are exceptional in that they are able to profit even from physical disturbance as it fractionates their clones and spreads portions of rhizomes and stolons through moving soils and gravel screes. Disturbance, which is normally considered as a negative factor in formulating life-strategy theories, has perhaps to be reconsidered as not an entirely negative factor in the Arctic. In classical life-strategy theory (Grime, Hodgson & Hunt 1988) disturbance is considered as likely to favour genotypes in which rapid growth and early reproduction increase the probability that sufficient offspring will be produced to allow survival and re-establishment of the population. In the Arctic, growth is not rapid and reproduction both sexually and vegetatively is not usually achieved at an early stage of establishment. However, once established, disturbance aids dispersal through the fractionation of clones.

Grazing is a specific form of disturbance which is also viewed usually as a negative factor at lower latitudes. However, grazing is essential in the Arctic both

FIG. 5.10. A colony of *Luzula confusa* (Hartm.) Lindeb. growing on an exposed mountain ridge above Isfjord, Spitsbergen (79°N). This colony survives in an area that is prone to severe environmental stress and physical disturbance of the terrain.

for nutrient recycling and to prevent accumulation of old foliage. When there is no grazing the insulation of the soil from the brief period of summer warming can prevent the thawing of the rooting zone and thus kill the plant. Nevertheless, in unproductive habitats removal of shoot material will cause significant loss in productivity and, if continued unremittingly, plant survival would be threatened. However, recent studies in relation to microtine grazing in fellfield communities have highlighted the cyclical influence of grazers in controlling vegetation development even in barren landscapes at high latitudes, where fluctuating rodent populations can cause marked changes in above-ground phytomass even in unproductive communities (Oksanen 1983, 1990). The ability of these arctic plant species to endure the loss of one or more seasons' productive growth through grazing can be linked to their capacity to survive one or more climatically adverse growing seasons with no net gain in carbohydrate reserves. Thus, the same physiological endurance property that allows such plants to recover from missed growing seasons also enables them to survive cyclical grazing pressure and subsequently profit from the increased nutrient availability and removal of insulating litter from the soil surface. Tolerance by these unproductive plants of climatically variable regions even with high grazing pressure is yet another example of how adaptation to one aspect of habitat variability, for example uncertain growing season length, can preadapt the species to other potential stresses such as herbivory.

As with the other physical stresses discussed above, population differences can add to the survival prospects of the species as a whole. The dioecious arctic willows can be viewed as two separate populations, the males and the females (Fig. 5.11). Throughout the Arctic, the sex ratio of these populations is biased in favour of the females (Crawford & Balfour 1983, 1990). One cause of this bias may be from selective grazing of the bark. In *Salix myrsinifolia-phylicifolia*, the bark of the male plant is preferred to that of the female by winter grazing voles (Fig. 5.12; Danell *et al.* 1985). Destruction of male plants rather than females could allow the population to recover more rapidly after the vole population's crash.

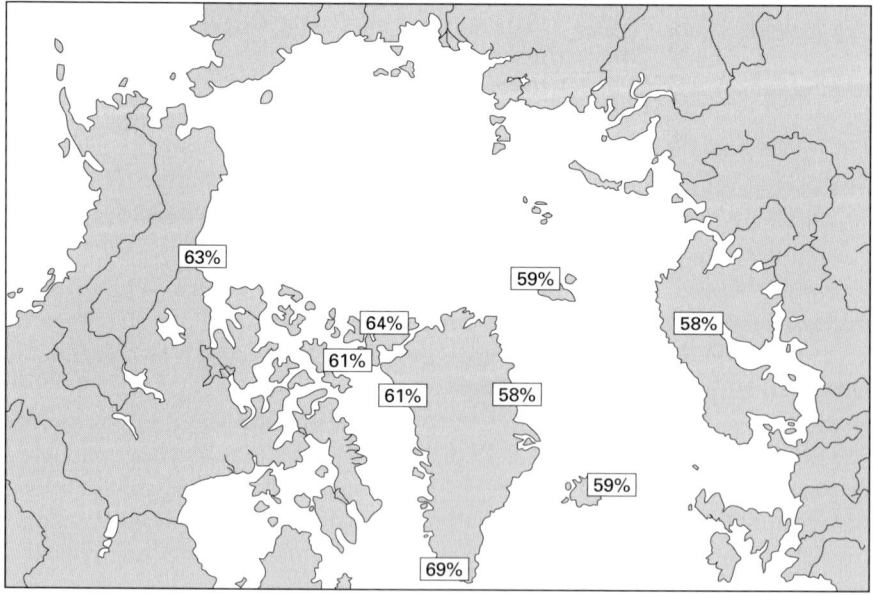

FIG. 5.11. Distribution of some female-biased sex ratios as observed in a number of arctic willow species. (Data from Crawford & Balfour 1990.)

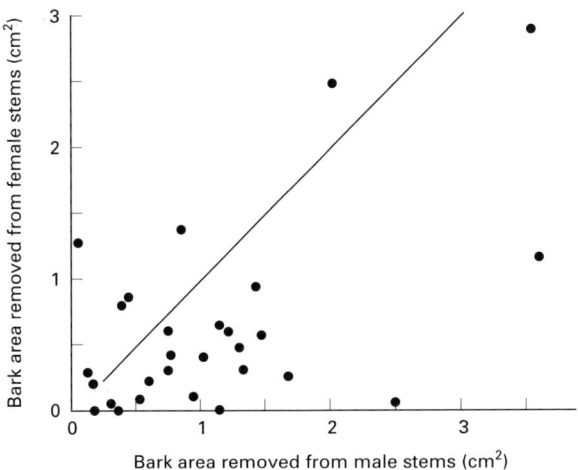

FIG. 5.12. Bark area of stem parts removed from willow (*Salix myrsinifolia-phylicifolia*) by five male vs. five female subadult field voles. (After Danell *et al.* 1985.)

The cyclical nature of grazing pressure in the Arctic allows plants with low growth rates to recover during periods when their particular herbivores suffer a population crash. Why there should be these great surges in grazing populations, followed by equally dramatic reductions, has long been a subject to debate (Batzli *et al.* 1980). Overgrazing alone has not received universal acceptance as an explanation of why rodent populations should show such simultaneous declines over wide areas after a population surge. Recent research on the role of a wound-induced proteinase inhibitor by northern populations of *Carex bigelowii* and *Eriophorum angustifolium* in response to lemming population cycles suggests a means whereby variation in the physiology of the plants ensures that grazing pressure fluctuates, allowing plants a period of recovery in which they can profit from the benefits of grazing in accelerating nutrient cycling (Seldal, Andersen & Högstedt 1994). The most important activity of proteinase inhibitors is to inhibit the activity of trypsin, an enzyme activating other proteolytic enzymes in the intestinal lumen which are secreted as proenzymes from the pancreas thereby reducing the digestive activity of the gut. The wound-induced activation of proteinase inhibitors appears to be followed by drastic reductions of cyclic populations. Reduction in body size, loss in body weight and hypertrophy of the pancreas in lemmings and other cyclic herbivore populations have been suggested as consequences of consuming vegetation with high proteinase activity induced as a result of wounding by grazing (Seldal *et al.* 1994).

The high degree of adaptation in arctic plant populations to cyclical adversity, whether it be climatic, edaphic or biotic, provides a population structure that can recover from the simultaneous or serial imposition of stress and disturbance. Grime's CSR life strategy, although providing a conceptual framework for viewing adaptation in the more steady-state conditions of temperate vegetation, cannot be extended to the fluctuating conditions of the Arctic where even activities such as grazing and physical disturbance can have beneficial as well as damaging effects. However, although the conventional life-strategy theory may not be adequate to describe plant survival strategies at high latitudes, it is instructive as it serves to demonstrate the distinctive nature of plant adaptations in the Arctic. As outlined in the introduction to this chapter, plant survival at these high latitudes is not a question of stress tolerance in the conventional physical sense but of tolerance of physical and temporal uncertainty.

PREADAPTATION TO CLIMATIC CHANGE FROM ECOTYPIC VARIATION

Preadaptation (defined above), whether physiological or morphological, arises essentially as a property which by chance increases fitness in a new situation. Thus, in the Arctic, populations able to survive under snow-banks or other short-growing season habitats are preadapted to any climatic deterioration that reduces growing season length. Conversely, plants able to withstand drought on exposed ridges or gravel banks and screes will be preadapted to survive the effects of early snow melt and greater incidence of drought that may arise from climatic warming at high latitudes. The examples cited above for *Dryas octopetala*, *Saxifraga oppositifolia* and *Draba* spp. show how subspeciation or ecotypic variation provides a facility for expanding the species or genus into a wider habitat range. At the same time, this ability to survive in contrasting habitats provides a source of genetic variability which allows these populations to accommodate substantial changes in climate either warming or cooling. This phenomenon exists in other regions outside the Arctic and is noticeable, for example, in the diversity of form in some boreal tree species which allows the trees to accommodate varying degrees of winter snow load on their branches (Kinloch, Westfall & Forrest 1986). However, in the Arctic the contrast between adjacent habitats is very marked and is accentuated both by the brevity of the growing season and the lack of environmental buffering capacity in the plant community structure. Therefore, it is suggested that the variation that is so noticeable in arctic populations of many plant species is a significant factor in making them preadapted to climatic change.

Speed of evolutionary change will also be facilitated in sympatric plant species that owe their identity to the environmentally labile limits of ecological separation rather than to breeding barriers (Crawford 1989). As these limits are altered or removed by environmental change, new possibilities are immediately available for the recombination of genotypes. Hybridization within related nordic and arctic species of *Draba* has been demonstrated as a means whereby the arctic species of this genus can achieve rapid rates of evolution (Brochmann *et al.* 1992c). Application of molecular techniques to the nordic species of *Draba* has indicated the possibility of independent origins on at least three separate occasions of the octoploid *D. cacuminum* (Brochmann, Soltis & Soltis 1992b). The facility for hybridization between such closely related species is genetically equivalent to gene exchange between ecotypes as it provides a reservoir of readily transferable variation. These studies of *Draba* show how closely related species can achieve rapid evolution from an existing gene pool (Brochmann, Soltis & Soltis 1992a,b) without recourse to immigration from distant sites.

CONCLUSIONS

Geomorphologically, the Arctic is prone to disturbance and the landscape easily scarred, as seen through human eyes (Forbes 1992). Nevertheless, this does not necessarily imply that it is automatically rendered inhospitable to the native flora. There is a case for considering the arctic flora as being well adapted to physical instability. Reproductive plasticity, gene exchange between adjacent populations, and certain morphological and physiological adaptations, provide striking and visible evidence that many species of flowering plants in the Arctic are able to respond and even profit from the physical fragility and uncertainty of their environment. Clonal reproduction by ramet fractionation and movement is actually facilitated as a result of disturbance and cryoperturbation. The high proportion of clonal species, combined with great longevity of individual plants, provides a biological answer to perturbation and produces a system which benefits from disturbance as it creates new opportunities for dispersal and colonization. As animal populations fluctuate, so are plant populations subjected to periods of intense grazing followed by periods of respite, when the animal populations eventually crash. The capacity of plant populations to recover, often with great rapidity, is another testimony to the resilience of most arctic species to periods of adversity. Periods of overgrazing can be followed by renewed vigorous growth, as nutrients are recycled and the shade of old vegetation is removed and the summer sun is not impeded from melting the frozen soil.

The cyclical nature of disturbance and stress at high latitudes maintains a high degree of polymorphism in arctic plant populations. Such balanced polymorphisms not only confer immediate increased species fitness by increasing ecological tolerance, but also enhance long-term fitness which will preadapt many arctic species to long-term climatic oscillations.

ACKNOWLEDGEMENT

This research was supported by a grant from the Natural Environment Research Council which is gratefully acknowledged.

REFERENCES

Allaby, M. (1994). *The Concise Oxford Dictionary of Ecology.* Oxford University Press, Oxford.

Atkinson, T.C., Briffa, K.R. & Coope, G.R. (1987). Seasonal temperatures in Britain during the past 22 000 years, reconstructed using beetle remains. *Nature,* **325,** 587–592.

Barnes, B.V. (1966). The clonal growth of the American aspen. *Ecology,* **47,** 439–447.

Batzli, G.O., White, R.G., Maclean Jr, S.F., Pitelka, F.A. & Collier, B.D. (1980). The herbivore-based trophic system. *An Arctic Ecosystem* (Ed. by J. Brown, P.C. Miller, L.L. Tieszen & F.L. Bunnell), pp. 335–410. Dowden, Hutchinson & Ross, Stroudsburg, Penn.

Billings, W.D. (1974). Adaptations and origins of Arctic plants. *Arctic and Alpine Research,* **6,** 129–142.

Billings, W.D. (1987). Constraints to plant growth, reproduction, and establishment in arctic environments. *Arctic and Alpine Research,* **19,** 357–365.

Billings, W.D., Godfrey, P.J., Chabot, B.F. & Borque, D.P. (1971). Metabolic acclimation to temperature in alpine and arctic ecotypes of *Oxyria digyna. Arctic and Alpine Research,* **3,** 277–290.

Bliss, L.C. (1956). A comparison of plant development in microenvironments of Arctic and Alpine plants. *Ecological Monographs,* **26,** 303–337.

Block, W. & Sømme, L. (1982). Cold hardines of terrestrial mites at Signy Island, maritime Antarctic. *Oikos,* **38,** 157–167.

Böcher, T.W. (1983). The allotetraploid *Saxifraga nathorstii* and its probable progenitors *S. azoides* and *S. oppositifolia. Meddelelser om Grønland. Bioscience,* **11,** 1–22.

Brochmann, C. & Elven, R. (1992). Ecological and genetic consequences of polyploidy in arctic *Draba* (Brassiceae). *Evolutionary Trends in Plants,* **6,** 111–124.

Brochmann, C., Soltis, D.E. & Soltis, P.S. (1992a). Electrophoretic relationships and phylogeny of Nordic polyploids in *Draba* (Brassicaceae). *Plant Systematics and Evolution,* **182,** 35–70.

Brochmann, C., Soltis, P.S. & Soltis, D.E. (1992b). Multiple origins of the octoploid Scandinavian endemic *Draba cacuminum:* electrophoretic and morphological evidence. *Nordic Journal of Botany,* **12,** 257–272.

Brochmann, C., Soltis, P.S. & Soltis, D.E. (1992c). Recurrent formation and polyphyly of Nordic polyploids in *Draba* (Brassicaceae). *American Journal of Botany,* **79,** 673–688.

Collier, D.E. & Cummins, W.R. (1989). A field study on the respiration rates in the leaves of temperate plants. *Canadian Journal of Botany,* **67,** 3478–3481.

Crawford, R.M.M. (1989). *Studies in Plant Survival.* Blackwell Scientific Publications, Oxford.

Crawford, R.M.M. (1992). Oxygen availability as an ecological limit to plant distribution. *Advances in Ecological Research,* **23,** 93–185.

Crawford, R.M.M. & Abbott, R.J. (1994). Pre-adaptation of Arctic plants to climate change. *Botanica Acta,* **107,** 271–278.

Crawford, R.M.M. & Balfour, J. (1983). Female predominant sex ratios and physiological differentiation in arctic willows. *Journal of Ecology,* **71,** 149–160.

Crawford, R.M.M. & Balfour, J. (1990). Female-biased sex ratios and differential growth in arctic willows. *Flora,* **183,** 291–302.

Crawford, R.M.M. & Braendle, R. (1996). Oxygen deprivation stress in a changing climate. *Journal of Experimental Botany,* **47,** 145–159.

Crawford, R.M.M., Chapman, H.M., Abbott, R.J. & Balfour, J. (1993). Potential impact of climatic warming on Arctic vegetation. *Flora*, **43**, 367–381.

Crawford, R.M.M., Chapman, H.M. & Hodge, H. (1994). Anoxia tolerance in high Arctic vegetation. *Arctic and Alpine Research*, **26**, 308–312.

Crawford, R.M.M., Chapman, H.M. & Smith, L.C. (1995). Adaptation to variation in growing season length in Arctic populations of *Saxifraga oppositifolia* L. *Botanical Journal of Scotland*, **41**, 177–192.

Danell, K., Elmqvist, T., Ericson, L. & Salomonson, A. (1985). Sexuality in willows and preference by bark-eating voles: defence or not? *Oikos*, **44**, 82–90.

Dowdeswell, J.A. (1995). Glaciers in the High Arctic and recent environmental change. *Philosophical Transactions of the Royal Society of London A*, **352**, 321–334.

Dowdeswell, J.A. & White, J.W.C. (1995). Greenland ice core records and rapid climate change. *Philosophical Transactions of the Royal Society of London A*, **352**, 359–371.

Dunbar, M.J. (1973). Stability and fragility in arctic ecosystems. *Arctic*, **26**, 179–186.

Eck, P. (1990). *The American Cranberry*. Rutgers University Press, New Brunswick.

Elkington, T.T. (1971). Biological flora of the British Isles. *Dryas octopetala* L. *Journal of Ecology*, **59**, 887–905.

Elton, C. (1927). *Animal Ecology*. Sidgwick & Jackson, London.

Forbes, B.C. (1992). Tundra disturbance studies. I. Long-term effects of vehicles on species richness and biomass. *Environmental Conservation*, **19**, 48–58.

Fox, J.F. (1983). Germinable seed banks of interior Alaskan tundra. *Arctic and Alpine Research*, **15**, 405–411.

Grime, J.P. (1979). *Plant Strategies and Vegetation Processes*. Wiley, Chichester.

Grime, J.P. (1993). Stress, competition, resource dynamics and vegetation process. *Plant Adaptation to Environmental Stress* (Ed. by L. Fowden, T. Mansfield & J. Stoddart), pp. 45–63. Chapman & Hall, London.

Grime, J.P., Hodgson, J.G. & Hunt, R. (1988). *Comparative Plant Ecology*. Unwin Hyman, London.

Gudleifsson, B.E. (1994). Metabolite accumulation during ice encasement of timothy grass (*Phleum pratense* L.). *Proceedings of the Royal Society of Edinburgh*, **102B**, 373–380.

Gudleifsson, B.E. & Larsen, A. (1993). Metabolic and cellular impact of ice encasement on herbage plants. *Advances in Plant Cold Hardiness* (Ed. by P.H. Li & L. Christersson), pp. 229–249. CRC Press, Boca Raton, Fla.

Hochachka, P.W. & Somero, G.N. (1973). *Strategies of Biochemical Adaptation*. W.B. Saunders, Philadelphia.

Kinloch, B.B., Westfall, R.D. & Forrest, G.I. (1986). Caledonian Scots pine: origins and genetic structure. *New Phytologist*, **104**, 703–729.

Krupnik, I. (1993). *Arctic Adaptations*. University Press of New England, Hanover.

Larsen, T. (1985). Are arctic ecosystems vulnerable? *Norsk Polarinstitut Rapportserie*, **Nr. 24**, 1–26.

Lid, J. & Lid, D.T. (1994). *Norsk Flora*, 6th edn (Ed. by R. Elven). Det Norske Samlaget, Oslo.

Lincoln, R.J., Boxshall, G.A. & Clark, P.F. (1982). *A Dictionary of Ecology, Evolution and Systematics*. Cambridge University Press, Cambridge.

McGraw, J.B. (1985). Experimental ecology of *Dryas octopetala* ecotypes; relative response to competitors. *New Phytologist*, **100**, 233–241.

McGraw, J.B. (1987). Experimental ecology of *Dryas octopetala* ecotypes. V. Field photosynthesis of reciprocal transplants. *Holarctic Ecology*, **10**, 303–311.

McGraw, J.B. & Antonovics, J. (1983). Experimental ecology of *Dryas octopetala* ecotypes. I. Ecotypic differentiation and life cycle stages of selection. *Journal of Ecology*, **71**, 879–897.

McGraw, J.B. & Vavrek, M.C. (1989). The role of buried viable seeds in arctic and alpine plant communities. *Ecology of Seed Banks* (Ed. by M.A. Leck, V.T. Parker & R.L. Simpson), pp. 91–105. Academic Press, San Diego.

McNulty, A.K. & Cummins, W.R. (1987). The relationship between respiration and temperature in leaves of the arctic plant *Saxifraga cernua*. *Plant, Cell and Environment,* **10,** 319–325.

McNulty, A.K. & Cummins, W.R. (1989). The effect of photoperiod length on respiration in leaves of *Saxifraga cernua* L., an arctic herb. *Plant, Cell and Environment,* **12,** 747–752.

McNulty, A.K., Cummins, W.R. & Pellizari, A. (1988). A field survey of respiration rates in leaves of arctic plants. *Arctic,* **41,** 1–5.

Montgomery, E.G. (1912). Competition in cereals. *Bulletin of the Nebraska Agricultural Station,* **26,** 1–12.

Mooney, H.A. & Billings, W.D. (1960). The annual carbohydrate cycle of alpine plants as related to growth. *American Journal of Botany,* **47,** 594–598.

Mooney, H.A. & Billings, W.D. (1961). Comparative physiological ecology of Arctic and Alpine populations of *Oxyria digyna*. *Ecological Monographs,* **31,** 1–29.

Oksanen, L. (1983). Trophic exploitation and arctic phytomass patterns. *The American Naturalist,* **122,** 45–52.

Oksanen, L. (1990). Predation, herbivory and plant strategies along gradients of primary productivity. *Perspectives on Plant Competition* (Ed. by J.B. Grace & D. Tilman), pp. 445–474. Academic Press, San Diego.

Pielou, E.C. (1994). *A Naturalist's Guide to the Arctic*. University of Chicago Press, Chicago.

Seldal, T., Andersen, K.-J. & Högstedt, G. (1994). Grazing-induced proteinase inhibitors: a possible cause for lemming population cycles. *Oikos,* **70,** 3–11.

Sonesson, M. & Callaghan, T.V. (1991). Strategies of survival in plants of the Fennoscandinavian tundra. *Arctic,* **44,** 95–105.

Stebbins, G.L. (1971). *Chromosomal Evolution in Higher Plants*. Edward Arnold, London.

Stebbins, G.L. (1984). Polyploidy and the distribution of the arctic–alpine flora: new evidence and a new approach. *Botanica Helvetica,* **94,** 1–13.

Teeri, J.A. (1972). *Microenvironmental Adaptations of Local Populations of* Saxifraga oppositifolia *in the High Arctic*. PhD Thesis, Duke University, Durham, NC.

Teeri, J.A. (1973). Polar desert adaptations of a high Arctic plant species. *Science,* **179,** 496–497.

Wadhams, P. (1995). Arctic sea ice extent and thickness. *Philosophical Transactions of the Royal Society of London A,* **352,** 301–309.

Wookey, P.A., Welker, J.M., Parsons, A.N., Press, M.C., Callaghan, T.V. & Lee, J.A. (1994). Differential growth, allocation and photosynthetic responses of *Polygonum viviparum* to simulated environmental change at a high arctic polar semi-desert. *Oikos,* **70,** 131–139.

6. Life strategies of arctic terrestrial arthropods

JEFFREY S.BALE*, IAN D.HODKINSON†, WILLIAM
BLOCK‡, NIGEL R.WEBB§, STEVEN C.COULSON†
AND ANDREW T.STRATHDEE*

**The University of Birmingham, School of Biological Sciences, Edgbaston,
Birmingham B15 2TT, UK*
*†School of Biological and Earth Sciences, Liverpool John Moores University,
Liverpool L3 3AF, UK*
*‡British Antarctic Survey, Natural Environment Research Council, High Cross,
Madingley Road, Cambridge CB3 0ET, UK and*
*§ITE Furzebrook Research Station, Natural Environment Research Council,
Wareham, Dorset BH20 5AS, UK*

INTRODUCTION

The life of arctic arthropods has often been viewed as a struggle for survival, rather than as successful adaptation to, and exploitation of, a harsh and unpredictable physical environment. Survival is, however, merely the preliminary prerequisite for effective adaptation. Despite their comparatively simple structure, arctic habitats present a series of highly variable microenvironments to which the arthropods have successfully developed an equally diverse array of adaptive responses. Here, we review how the arthropods interact with, adapt to, and ultimately exploit this complex environmental templet. We delimit the terrestrial Arctic to include all the treeless habitats, with a growing season of less than 600 day degrees above 0°C, lying to the north of the boreal forest zone (Kauppi & Posch 1988).

Initially, we define the environmental templet for arctic arthropods, identifying the important factors and interactions and emphasizing the significant environmental constraints and opportunities that mould the arthropods' life strategies. We then review the range of adaptive features of arctic arthropods that enable them to survive and succeed in this environment. The manner in which these adaptations are assembled and utilized as 'options for success' is then illustrated for selected species that display strongly contrasting biologies, particularly those with life cycles taking place either above or below ground. Finally, the impacts of climate change on both arthropod species and communities are considered within short- and long-term time perspectives.

When considering adaptation, it is imperative to remember that, in evolutionary terms, arctic faunas are for the most part very young, and comprise a wide spectrum of taxa that are derived from more southerly forms that have adapted (or were

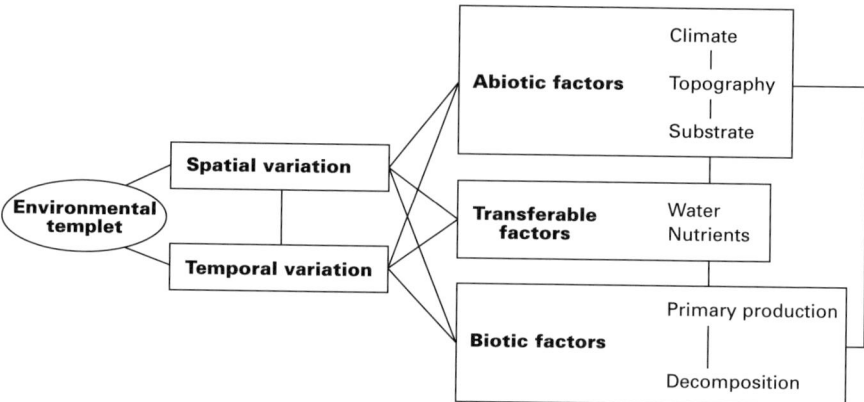

FIG. 6.1. Factor diagram defining the environmental templet of arctic arthropods.

'preadapted') to more extreme conditions (Danks 1990, 1993, 1994). There is little evidence for a uniquely characteristic arctic fauna above the species level: the fauna is primarily derived, by reduction, from those of more temperate regions. Thus, the species are constrained both morphologically and physiologically by prior evolution, yet have successfully adapted in diverse ways to exploit arctic conditions.

The driving process that moulds individual species to the environmental templet is natural selection. Successful adaptation implies that over a complete life cycle the metabolic costs of adaptations are sufficiently offset against energy gains that the organism is able to grow and reproduce successfully. Thus, each adaptational strategy imposes a cost on the organism; for any one species natural selection will act to minimize the total cost *vis-à-vis* the energy gained through feeding. In energetic terms, natural selection acts to maximize the ratio of net secondary production (P) to the cost of maintenance (R).

The environmental templet (*sensu* Southwood 1977) for arctic arthropods, summarized as a factor diagram (Fig. 6.1), comprises both abiotic and biotic elements and the inter-relationships between them. The important driving abiotic variable is climate, which acts either directly on the animal or through topography and substrate to determine the physical environment of the soil. This, in turn, is modified by biotic processes such as the growth and decomposition of plants which transfer water and nutrients between the abiotic and biotic environments. Together these factors act to produce spatial and temporal variation in the environment and determine the stage on which the arctic arthropods must perform. Successful performance demands effective adaptation.

THE ENVIRONMENTAL TEMPLET ILLUSTRATED

The temporal and spatial variation in the microclimates experienced by arctic organisms, and their relationship to the prevailing macroclimate, have been summarized for several low and high arctic sites and the potential for change under global warming reviewed (e.g. French 1970; Corbett 1972; Weller *et al.* 1972; Weller & Holmgren 1974; Courtin & Labine 1977; Rydén & Kostov 1980; Mølgaard 1982; Maxwell 1992; Coulson *et al.* 1993). In these environments, insolation and precipitation are the key templet variables for the arthropods. Extreme climatic events, perhaps occurring infrequently, are probably more significant for survival and adaptation than are the long-term climatic averages.

The Arctic is typically characterized by long, cold winters with continuous darkness, short, cool summers with continuous daylight, low precipitation falling mostly as snow, and continuous snow cover for up to 8 months of the year. These baseline parameters interact with local topography and surface substrate to produce localized variations in microclimate. For example, even small differences in microtopographical relief can produce highly significant differences in microenvironment, creating much spatial heterogeneity of habitat type that different species of arthropod can utilize (see Fig. 6.6). The small size of arctic microarthropods allows them to exploit these favourable microhabitats that are not available to the birds and mammals.

A typical tundra surface is often unstable because of frost heave, resulting in a series of raised hummocks or ridges interspersed with shallow hollows and depressions. The hummocks are often drier, well-drained shedding sites with mineral soils exhibiting restricted organic matter accumulation; the hollows are usually wetter receiving sites with impeded drainage and a tendency to accumulate organic matter, indicating slower rates of decomposition. Snow during winter tends to drift, resulting in shallow accumulation or absence on the ridges and deeper accumulation in the hollows. Lack of snow insulation can result in rapid soil cooling, with soil becoming isothermal with respect to depth. At spring snowmelt the ridges are the first to clear and snow lingers in the hollows, resulting in a shorter effective growing season. The difference is exaggerated by the albedo effect over snow that produces a high reflectance of the warming insolation. By contrast, the dark clear ground absorbs heat and warms rapidly. Thus, the ridges experience a longer growing season and higher temperatures during summer but because of their exposure tend to be colder during the winter. Similarly, sites or microsites with a southern aspect tend to be warmer than those with a northern aspect. These variations in the abiotic templet result in parallel differences in the biotic templet. Different substrate types support different plant communities with different species composition, ground cover, biomass and primary production. Even single plant species can differ significantly in their growth, phenology and flowering success, depending on their precise situation. Their nutritive value to insect herbivores differs seasonally and between tissues (MacLean & Jensen 1985; Danks 1987) and the rates of decomposition of separate species differ widely. Above-ground plant biomass can vary from $5\,\mathrm{g\,m^{-2}}$ in polar desert to $500\,\mathrm{g\,m^{-2}}$ in low shrub tundra (Chapin & Shaver

1985) but the ratio of readily digestible green tissue to woody shoots and stems is often low.

To illustrate the arctic macroclimate and its relationship to the microclimate experienced by the above- and below-ground arthropods we present data for high arctic sites adjacent to Ny-Ålesund (78°30′N, 11°5′E, 8 m asl), on Svalbard (Steffensen 1982; Coulson *et al.* 1995a). Svalbard is warmer than many continental arctic sites because of the warming influence of the Gulf Stream but the climate is particularly variable and displays the full range of climatic stresses to which arthropods are exposed.

At the latitude of Ny-Ålesund, the sun does not set between 18 April and 24 August and does not rise between 25 October and 17 February (Hisdal 1985). Mean monthly temperature is above 0°C for only 3 months of the year (June–August), and does not exceed 5°C in any month (Fig. 6.2). Winter air temperature can fall to −42°C but within the same month rise above 0°C. Frosts can occur in any month but maximum recorded air temperature is 17°C in July. Precipitation is low (371 mm year^{-1}) and falls predominantly as snow, in months when mean air temperature is below 0°C (Fig. 6.3). Precipitation within any one month is highly variable between years but reaches a minimum during the growing season. Often, snowfall is light during early to mid-winter, leaving the soil surface exposed to rapid chilling.

For arctic arthropods the year can be divided into three phases: (i) the winter, where the main climatic constraint is the intensity and duration of subzero temperatures; (ii) the summer, when the main limitations are the high and low

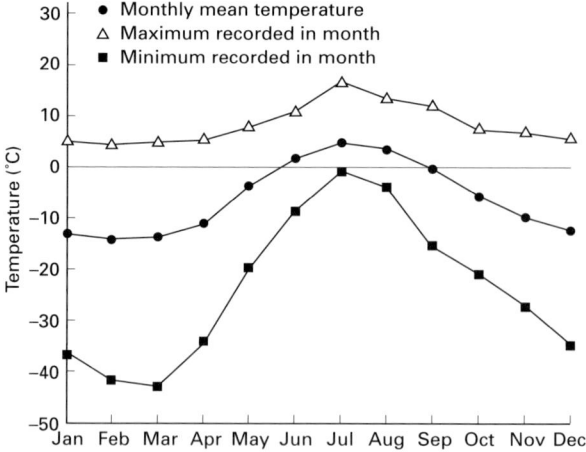

FIG. 6.2. Temperature data for Ny-Ålesund, Svalbard 1969–93 based on screen data provided by Det Norske Meterologiske Institutt.

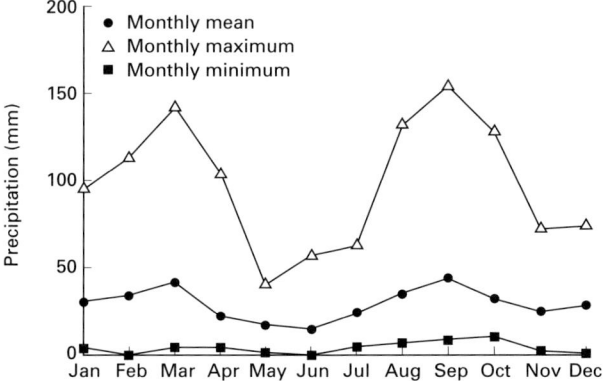

FIG. 6.3. Precipitation data for Ny-Ålesund, Svalbard 1969–93 based on screen data provided by Det Norske Meterologiske Institutt.

temperature extremes coupled with water and total heat availability; and (iii) the shorter transition periods when the arthropods resume or prepare to cease activity. At sites adjacent to Ny-Ålesund in 1993/94 the soil surface vegetation mat was frozen for at least 265 days a year (Table 6.1a). Minimum temperatures at both the soil surface and 3 cm depth varied widely between sites but were higher than minimum air temperature. Temperatures were lowest on exposed polar semi-desert sites and highest in ornithogenic soils at the base of bird cliffs such as at Krykkjefjellet (northerly aspect) and Kjaerstranda (southerly aspect). The frequency of freeze–thaw events was similarly variable with respect to sites and soil depth (7–30 events per year).

The maximum summer temperatures in the surface vegetation/litter mat at the polar semi-desert and tundra heath sites (July and August 1991–93) were 14–17°C above the maximum and 22–25°C above the mean air temperatures respectively (Table 6.1b). Maximum soil temperatures at 3 cm depth were approximately the same as maximum air temperature. The daily frequency with which subzero temperatures occurred at the soil surface and at 3 cm depth of these same sites during July and August (Table 6.1c) showed a rapid decline in summer freezing events with respect to soil depth. Summer soil temperatures of –5°C occur very infrequently and then only at the surface. Looking across these microclimatic datasets, the two aspects which are likely to have the greatest impact on terrestrial arthropods are: (i) the great variation in winter minimum temperatures between sites, related mainly to the timing and depth of snow cover; and (ii) the marked diurnal changes in the temperature of the soil surface and vegetation mat in summer which can fluctuate from –5°C to 30°C.

The overall environmental templet of arctic arthropods is thus complex and variable and the extent to which it is exploited depends on the adaptations of

TABLE 6.1. Microclimatic extremes to which arctic arthropods are potentially exposed, illustrated using data from Ny-Ålesund, Svalbard.

(a) Winter temperature minima and number of yearly freeze–thaw cycles during 1993/94 *

		Min. temp. (°C)	Days < 0°C	Freeze–thaw cycles
Air (screen)		−32.8	245	49
Polar semi-desert	Surface mat	−29.7	272	*
	Soil at 3 cm	−25.0		30
Tundra heath	Surface mat	−20.3	284	14
	Soil at 3 cm	−19.4		13
Krykkjefjellet	Soil at 3 cm	−15.0	265	18
Kjaerstranda	Soil at 3 cm	−8.2	288	7

(b) Summer temperature maxima 1991–93

		Temperature (°C)
Maximum air (screen)		12.6
Mean daily air (July–August)		4.7
Polar semi-desert	Surface mat	27.4
	Soil at 3 cm	12.7
Tundra heath	Surface mat	30.1
	Soil at 3 cm	13.9

(c) Percentage daily frequency of subzero temperatures during July and August

		% Frequency	
		< 0°C	< −5°C
Polar semi-desert	Surface mat	8.1	0.3
	Soil at 3 cm	3.8	0
Tundra heath	Surface mat	9.0	0
	Soil at 3 cm	3.5	0

* Incomplete dataset.

particular species. These have adapted in diverse ways to exploit variations within the templet and communities of complementary species have evolved that fully exploit the range of opportunities available.

CONSTRAINTS AND OPPORTUNITIES

The arctic environment imposes important constraints on the arthropod fauna as well as providing unique opportunities. The primary constraints include abiotic factors such as low temperature, minimal precipitation and highly seasonal light

TABLE 6.2. Constraints, opportunities and their biological impact on arctic terrestrial invertebrates.

	Biological impact
Constraints	
Severe and prolonged winters with subzero temperatures	Potentially lethal or may cause sublethal damage (development and reproduction)
Short, cool summers	Limited thermal budget for development restricting annual powers of increase
Highly variable microclimate	Extreme summer daily temperature fluctuations requiring cold and heat tolerance
Low annual precipitation	Risk of desiccation, varying between microsites and taxa
Highly seasonal photoperiodic regime	Change from continuous light to continuous darkness may not be a reliable environmental trigger
Low-diversity plant communities	Limited range of food plants for herbivores
Soils of poor nutrient content and variable water status	Host plants of low nutritive value
Opportunities	
Species-poor community of arthropods	Limited inter- and intraspecific competition for resources
Low incidence of parasitoids and predators	Reduced pressure from natural enemies
Variable habitat and microclimate	Potential to exploit favourable microsites to maximize annual productivity
Low-diversity plant communities	Requires synchronous phenology and specialist feeding habitats

climate, and biological features such as the low diversity of the plant communities. At the same time, the low incidence of natural enemies and reduced inter- and intraspecific competition for resources create opportunities for various members of the fauna. The main constraints, opportunities and their biological impact are summarized in Table 6.2. In some cases, the same feature can be viewed as both a constraint and an opportunity. For instance, organisms that live on the tops of ridges, which have minimal snow cover, will need to be extremely cold hardy to survive through winter; yet these are the same sites that become clear of snow early in summer thus providing the longest favourable season for growth and development. Similarly, microsites in the vegetative mat where temperatures may reach 30°C, with associated risks of heat stress and desiccation, are also likely to provide the highest annual thermal budget for development and reproduction. On balance, it is evident that constraints outweigh opportunities and to survive and succeed in the arctic climate arthropods need to be highly adapted.

ADAPTATIONS OF ARCTIC TERRESTRIAL INVERTEBRATES

Optimal exploitation of the arctic environment has resulted in the selection of traits that counterbalance each other to maximize the resultant ratio of net secondary productivity (benefit) to respiration (cost) (MacLean 1975). Adaptations of arctic arthropods, often highlighted by comparison with their temperate counterparts, include life cycles and generation times, morphology, behaviour, cold and heat tolerance, desiccation resistance and respiratory metabolism. The adaptations of arctic arthropods, particularly insects, have been reviewed by Downes (1964, 1965), MacLean (1975, 1981), Danks (1981, 1991), Kevan and Danks (1986) and Sømme and Block (1991). Here, we summarize the range of these adaptations, and then consider their mechanisms, functions and integration in more detail in the following section (Options for success), by reference to individual species or taxonomic groups.

Life cycles and generation times

Species living above ground are mainly herbivores where life cycles are closely synchronized with the phenology of their host plants. Many species are univoltine. Whilst it is rare for arctic herbivorous arthropods to have more than one generation a year, there are examples of extended life cycles, in some cases taking 14 years or longer to complete one generation. By contrast, soil-dwelling microarthropods (Collembola and mites) have free-running life cycles usually requiring two or more years per generation (Sømme & Block 1991). Population development is asynchronous and largely independent of season or plant phenology (Addison 1977, 1981).

Morphology and behaviour

Arctic arthropods often show a reduction in body size. This has been viewed as an aid to moisture conservation and location of shelter sites in microhabitats, and may also be a response to limited food resources (Sømme & Block 1991). Reduction or loss of wings is found in a number of arctic insects; selection for brachyptery or aptery is likely to be strong in a climate where temperatures are often below the threshold for flight. Large-bodied arthropods in the Arctic are often dark coloured (melanistic), enabling them to raise their body temperature by heat absorption. Thermoregulation is an important adaptation for flight in adult Lepidoptera and for movement between food sources and digestion in larval stages, for example larvae of *Gynaephora groenlandica* (Kukal, Heinrich & Duman 1988a).

Cold tolerance

The generally low habitat temperatures and frequent freeze–thaw events, increasing the risk of freezing throughout much of the year, have resulted in the development of high levels of cold hardiness in arctic arthropods involving both the freeze-tolerance and freeze-avoidance strategies. In addition, these physiological and biochemical adaptations are accompanied by behavioural responses, in which organisms often migrate to lower soil depths in winter and benefit from the thermal insulation provided by snow cover.

In many freeze-tolerant species, food ingested in summer is evacuated from the digestive system prior to winter, although this does not seem to be essential for winter survival. The crucial feature found in most freeze-tolerant species is the winter synthesis of ice-nucleating agents (INAs), which usually circulate in the haemolymph and function to initiate freezing in safe extracellular areas at high subzero temperatures (−5°C to −10°C). Freezing of water in extracellular areas, such as the haemocoel, cryoconcentrates the unfrozen haemolymph; water then moves from the cells across the cell membrane to the haemocoel to re-establish the osmotic equilibrium. The extracellular ice masses grow in size as temperatures

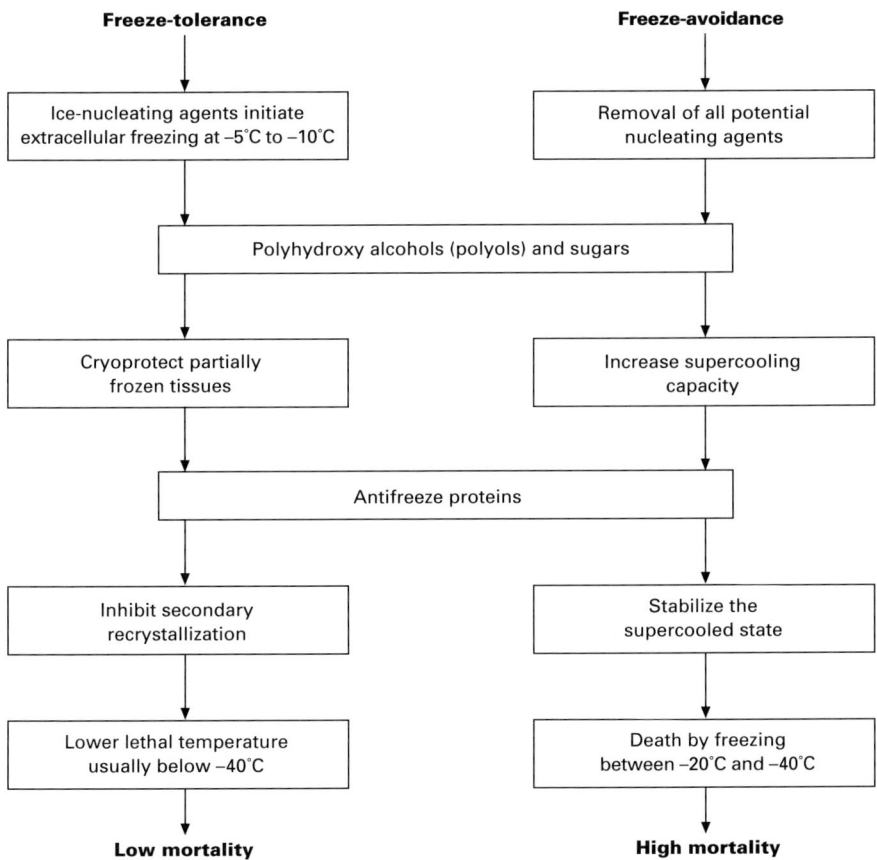

FIG. 6.4. Freeze-tolerance and freeze-avoidance strategies of insect cold hardiness.

become progressively lower in winter, but the critical factor is that the whole process occurs gradually, allowing sufficient time for water to move from the cells where freezing is potentially lethal. Freeze-tolerant insects also synthesize polyols, to cryoprotect the potentially frozen tissues, and antifreeze proteins to inhibit secondary recrystallization. Freeze tolerance is usually regarded as the low-risk/low-mortality overwintering strategy where lower lethal temperatures are often well below −40°C (Fig. 6.4; Leather, Walters & Bale 1993).

For freeze-avoiding arthropods, freezing is fatal. These species have to mask or remove all potential nucleators from their body prior to winter and then synthesize cryoprotectants to lower their body freezing temperature (supercooling point, SCP) (Fig. 6.4). Freeze avoidance is the high-risk/high-mortality strategy, where lower lethal temperatures, as indicated by supercooling points, are usually between −20°C and −40°C. In polar environments, however, soil invertebrates are effectively buffered against severe air temperatures and may have the added protection of accumulated snow cover at the soil surface. While there is increasing evidence that prefreeze mortality is common in insects, particularly after long-term exposure at low subzero temperatures (Bale 1993), it is also evident that polar microarthropods are among the most cold-hardy freeze-avoiding species (Cannon & Block 1988).

Heat tolerance

Tolerance of relatively high temperatures in terrestrial habitats during the arctic summer is advantageous for arthropods in certain situations. Species living in the vegetative mat at the soil surface interface will experience summer temperatures in the region of 25–30°C (Table 6.1b). Recent experiments with soil microarthropods on Svalbard have shown that survival at high temperature is moisture dependent and that Collembola are less able to survive heat stress than oribatid mites (see Options for success).

Desiccation resistance

The response of arctic arthropods to moisture stress has been little studied but clearly heat tolerance and desiccation resistance are likely to be inter-related. Evidence from both field and laboratory experiments has shown that low soil moisture content in summer affects populations of soil microarthropods with a more negative impact on Collembola compared to mites (Hodkinson, Healey & Coulson 1994; Coulson *et al.* 1996).

Development, temperature and respiratory metabolism

The growth and development of arthropods is temperature dependent but in the

Arctic, both the daily mean temperature and the total summer heat budget are low. As an adaptation to this constraint, the threshold temperatures for development and activity are often lower for arctic species than their temperate counterparts, which may be indicative of increased metabolic rates at low temperature (Sømme & Block 1991) as recently demonstrated in the collembolan *Onychiurus arcticus* (Block *et al.* 1994).

OPTIONS FOR SUCCESS

In this section the life-cycle biology and ecophysiology of a range of arctic arthropods are described, showing how the adaptations outlined earlier are successfully combined with a species or species group, while making the important distinction between organisms living above or below ground.

Aphids: Acyrthosiphon svalbardicum

The aphid *Acyrthosiphon svalbardicum* is believed to be endemic to Svalbard (Heikinheimo 1968) where it is monophagous on mountain avens (*Dryas octopetala*) (Strathdee *et al.* 1993a). *A. svalbardicum* feeds on the phloem sap of its host plant which grows as a network of small patches (*Dryas* heaths) with a maximum height of approximately 5 cm. At Ny-Ålesund, following snowmelt and egg hatch, aphids are subject to 24 h daylight throughout the summer season.

The life cycle of *A. svalbardicum* is entirely holocyclic, the aphid overwintering exclusively as eggs produced after mating between oviparae and males. In contrasting temperate holocyclic species (e.g. *Aphis fabae*) (Fig. 6.5a) the fundatrix, which hatches from the overwintered egg, gives rise to a series of parthenogenetic viviparous generations throughout summer. Asexual reproduction is terminated by decreasing autumn photoperiod which induces the production of the sexual morphs, followed by oviposition. The possible premature induction of sexual morphs in the short (but increasing) photoperiod experienced by the fundatrix and immediate successive generations in spring is inhibited by an endogenous mechanism, the interval timer (Marcovitch 1923; Wilson 1938; Lees 1960, 1961).

In the life cycle of *A. svalbardicum* (Fig. 6.5b), the fundatrix gives rise directly to both sexual morphs (oviparae and males) which mate and produce the overwintering eggs after only two generations of active stages. In addition to the sexual morphs, the fundatrices also produce a small number of viviparae (typically four or five in a total reproductive output of about 28 progeny). With continuing favourable conditions, particularly temperature, these viviparae mature and produce a third generation of exclusively sexual morphs, which have the potential to contribute additional overwintering eggs (Strathdee *et al.* 1993a). *A. svalbardicum* thus displays a genetically controlled life cycle in which the dominant production

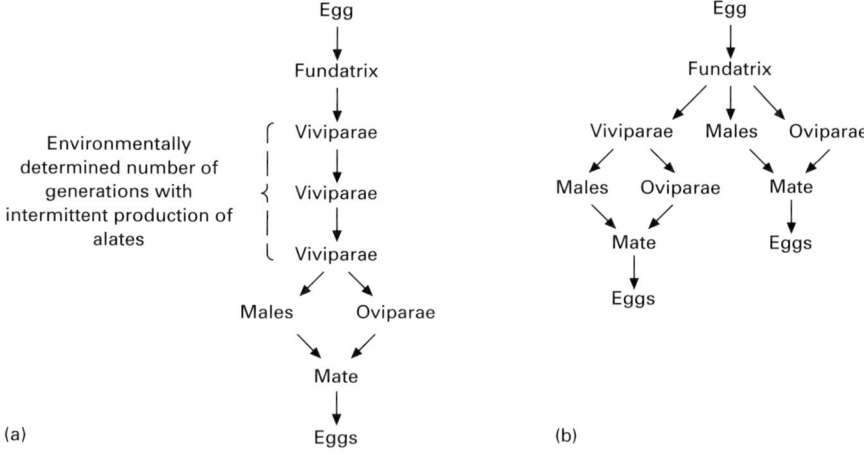

FIG. 6.5. Diagrammatic life cycles of (a) generalized holocyclic temperate species in the Aphidinae and (b) *A. svalbardicum*, a high arctic aphid. (After Strathdee *et al.* 1993a.)

of sexual morphs in the second generation ensures that some eggs are always produced before summer weather conditions deteriorate. Clearly, it requires a longer growing season and more thermal energy to produce three rather than two generations of aphid per year. It seems likely therefore that the egg–fundatrix–sexual morphs–egg route is dominant under the prevailing climate on Svalbard; however, the ability to produce an additional generation of sexual morphs from the second generation viviparae provides a flexibility by which the aphid can exploit favourable microsites, abnormally warm years, or a gradual increase in temperatures through climate warming.

The response of *A. svalbardicum* to temperature elevation is described in the next section, but the importance of natural variation in microclimate is already apparent. The population structure of *A. svalbardicum* was surveyed in patches of *D. octopetala* on a south-facing slope in early July (Fig. 6.6). At the bottom of the slope (patch A) where snow accumulated in winter and was late to clear, late instar and adult fundatrices were present, but there were no second generation nymphs. By contrast, at the top of the slope (patches D & E), where winter snow accumulation was light and cleared early, there were higher numbers of adult fundatrices and increasing numbers of their second generation progeny. The sensitivity of this graded response to the length of the available growing season over a distance of less than 4 m is further illustrated by the 'in between' patches of *D. octopetala* (B, C, F) where population development was more advanced than at the bottom of slope, but less than at the top (Strathdee & Bale 1995).

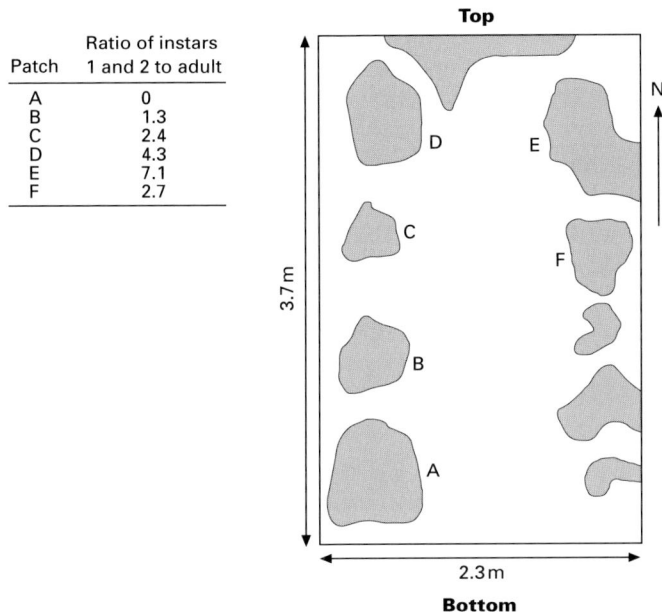

Patch	Ratio of instars 1 and 2 to adult
A	0
B	1.3
C	2.4
D	4.3
E	7.1
F	2.7

Fig. 6.6. Patches of *Dryas octopetala* and ratios of second generation nymphs to adults of *Acyrthosiphon svalbardicum* on a south-facing slope at Ny-Ålesund, Svalbard. (After Strathdee & Bale 1995.)

Psyllids: Cacopsylla palmeni and C. phleobophyllae

Phloem-feeding jumping plant lice or psyllids are the numerically dominant invertebrate herbivores at most low arctic sites. Their annual life cycle emphasizes the significance of precise host–plant synchrony. For example, the successful completion of the life cycle of willow psyllids depends on the species' ability to synchronize with and exploit a short-lived nutrient sink, the willow catkin. This necessitates the psyllid spending over 10 months each year, including winter, in the relatively vulnerable adult stage within the litter layer. Females are sexually mature and active on willow stems before the spring snowmelt is complete and they oviposit on female catkins the moment they appear. Development through five larval instars to adult must take place before the catkin reaches maturity and dehisces. The high degree of the phenological synchrony required for successful psyllid development is emphasized by comparing the relative development time of psyllids vs. catkins at Meade River, Alaska (Fig. 6.7; Hodkinson, Jensen & MacLean 1979). The psyllids *Cacopsylla palmeni* and *C. phlebophyllae* required 40 and 36 days respectively to develop on *Salix* host species that took 40–43 and 39–41 days respectively to produce mature catkins.

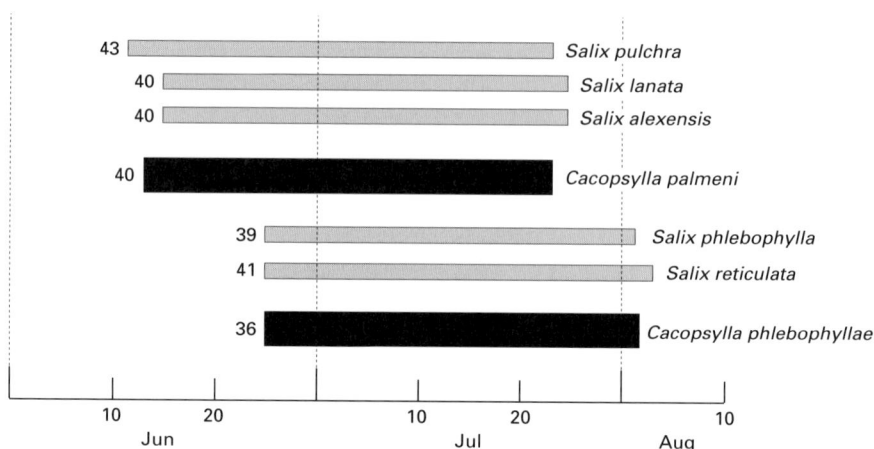

FIG. 6.7. Comparison of development times for psyllid larvae and *Salix* catkins at Meade River, Alaska. Solid bars represent minimum development times for psyllid larvae. Shaded bars above represent minimum development times for the catkins of *Salix* species that act as host plants. (After Hodkinson & Hughes 1982.)

The relative development rates of host plant and insect appear to set the geographical limits of insect distribution. Many arctic psyllid species, for example particularly the *Cacopsylla* species on *Salix*, do not extend over the full geographic range of their host plant and are absent from the northern and/or southern part of the range (MacLean & Hodkinson 1980). The probable mechanism (Fig. 6.8) is that in the northern part of the range the plant grows too slowly to support insect

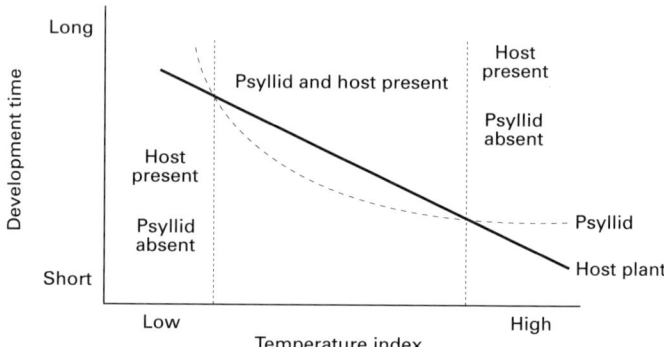

FIG. 6.8. Model to show how the relative development rates of insect and host plant may determine whether the insect life cycle is completed successfully. See text for explanation. (After MacLean 1983.)

development whereas in the south the plant develops too quickly (MacLean 1983). Only over the mid-part of the range is the insect able to match its phenology to that of its host plant.

Gynaephora groenlandica

By contrast with the two or three generations per summer in *A. svalbardicum* and the annual life cycle of arctic psyllids, the high arctic moth *Gynaephora groenlandica* has one of the most protracted life cycles recorded in an insect. *G. gynaephora* is endemic to the Canadian high arctic and Greenland, feeding primarily on buds and leaves of *Salix arctica*. Pupation, adult emergence, mating, oviposition and moulting to the second instar all occur in the first summer. After moulting to the third instar in the second summer, it then takes a further 12 years for the larvae to develop from the third to sixth instar (Fig. 6.9; Kukal & Kevan 1987).

Annual growth and development occurs during a very brief period of activity, mainly in June. At this time, the ambient and ground temperatures are usually below 5°C and 10°C respectively (Kukal & Dawson 1989). Larvae spend approximately

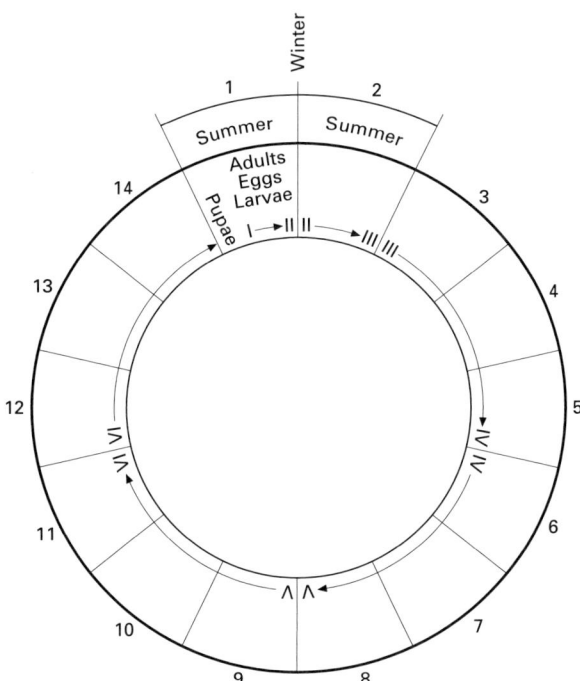

FIG. 6.9. Schematic life cycle of *Gynaephora groenlandica*. (After Kukal & Kevan 1987.)

60% of the time 'basking' in the sun, which raises their body temperature by 25–30°C above ambient (Kukal *et al.* 1988a). Most feeding and movement occurs around midday when temperatures are highest and this is followed by long periods of basking and digestion. The highly restricted period of feeding activity and development in *G. groenlandica* is governed primarily by the limited period of favourable temperatures and the decline in the food quality of its host plant toward the end of June which, in combination with other developmental constraints (low temperature inhibition of moulting, biochemical synthesis of winter cryoprotectants, parasitoid pressure), accounts for the extended 14-year life cycle (Kukal 1991).

In *G. groenlandica*, like *A. svalbardicum*, *C. palmeni* and *C. phlebophyllae*, it is the characteristics of the summer season, rather than the severe winter, which impose the greatest constraint on population abundance. More than 60% of larvae are killed by parasitoids in summer, whereas only 13% of larvae die in winter (Kukal & Kevan 1987). Once feeding ceases at the end of June, the larvae move into sheltered sites, storing glycogen ('voluntary hypothermia'; Regal 1967) as a source for cryoprotectant synthesis, particularly glycerol (Kukal, Serianni & Duman 1988b). In winter, the larvae freeze at about –7°C, although the nucleator is not active in the haemolymph; they remain frozen for up to 9 months during the arctic winter and can survive to –70°C. Interestingly, the larvae are also freeze tolerant during the period of activity and development in summer, although the freezing temperature (supercooling point) is then about –9°C and the larvae die at –15°C (Kukal *et al.* 1988b). Cold acclimation at 5°C leads to mitochondrial degradation, suppressed oxidative metabolism, and a concomitant increase in glycerol content, all of which occur progressively as the larvae enter the frozen overwintering state. Long-term frozen larvae undergo a rapid repair and replication of mitochondria on return to higher temperatures (Kukal, Duman & Serianni 1989).

Soil microarthropods

The predominant arthropod groups in all arctic soil types are oribatid mites (Acari: Cryptostigmata), Collembola (Apterygota) and larval Diptera. Most studies on tundra soil arthropods have been faunistic and few have examined population dynamics (Bohnsack 1971; MacLean 1977; Petersen & Luxton 1982; Coulson *et al.* 1996).

The oribatid mites are a cosmopolitan and very ancient group. Many arctic species appear to have survived Pleistocene glaciations in refugia and have not migrated from lower latitudes (Hammer & Wallwork 1979; Behan-Pelletier, in press). Their ecophysiological characteristics and life-cycle traits preadapt them for arctic conditions (Behan-Pelletier, in press).

Oribatids occur in densities of up to $80\,000\,m^{-2}$, although frequently at less than $20\,000\,m^{-2}$ (Petersen & Luxton 1982). Sampling during the summer months has demonstrated that the mites are concentrated in the top 2.5 cm of the soil with 5% in the 2.5–5.0 cm zone and a few at depths >5.0 cm (Douce & Crossley 1977). In the upper layers the mites are associated with plant material and litter (Bengtson, Fjellberg & Solhoy 1974) and their populations are strongly correlated with the patchiness of the vegetation (Seniczak & Plichta 1978; Hertzberg, Leinaas & Ims 1994). Most arctic oribatids have free-running life cycles in which the adult is preceded by a larva and three nymphal stages. Teletokous parthenogenesis is widespread in genera with circumpolar distributions (Norton & Palmer 1991). Growth and development depend directly on temperature, and life cycles may extend over several years. As a result, most oribatid populations contain 40–50% immatures (Bohnsack 1971; Douce & Crossley 1977; Seniczak & Plichta 1978). There is little apparent synchrony of life cycles with either season or plant phenology, although population densities often increase as summer progresses and decline after mid-summer (Bohnsack 1971, Douce & Crossley 1977).

The Collembola contains many species with circumpolar distributions, and both the richness of the fauna and population densities are generally higher than that of oribatids (Fjellberg 1986). Densities in the active layer of up to 200 000 m^{-2} have been reported, but 10 000–40 000 m^{-2} is more normal. The occurrence of Collembola is often related more closely with soil moisture patterns than is the case with oribatids, with the highest densities occurring in mesic sites (Petersen & Luxton 1982). There is no evidence for seasonal movement of Collembola in the profile of high arctic soils. In subarctic locations with persistent snow cover, Leinaas (1981) has reported upward movement of soil-surface dwelling Collembola into the snow profile during winter.

Collembola have extended life cycles in which a succession of nymphal stages develop slowly to adult without evidence of seasonal synchrony over a period of at least 2 years (Addison 1977, 1981). This results in dense mixed-aged populations which are often highly aggregated (e.g. Hertzberg *et al.* 1994) and may be redistributed over the tundra surface in melt water.

Mites and Collembola show a suite of adaptations that equip them for life in an arctic soil environment. Field samples of the dominant mites (*Diapterobates notatus, Camisia anomia, Ceratoppia hoeli, Hermannia reticulata*) and Collembola (*Onychiurus arcticus, O. groenlandicus* and *Hypogastrura tullbergi*), all of which are freeze avoiding, show similar levels of supercooling from May to August with median SCPs above –10°C, reflecting their summer activity and feeding status (Coulson *et al.* 1995b). Exposure to lower temperatures, cessation of feeding and accumulation of cryoprotectant sugars and polyols depress the winter SCP in these species to below –20°C. As examples, the mid-summer SCP of *D. notatus* (–9°C) decreased to –24°C after starvation and acclimation at 0°C; the concentration of glucose in *O. arcticus* increased from <10 in summer to >30 μg mg^{-1} in the winter cold-hardy condition (Coulson *et al.* 1995b). While acclimation is progressive, probably taking several weeks to achieve maximum winter cold hardiness, by contrast, deacclimation in early summer is rapid. The SCP of *H. tullbergi* increased from −20°C in individuals extracted from frozen soil cores to above −10°C in less than 24h (Fig. 6.10; Coulson *et al.* 1995b).

Ability to survive brief exposure at low subzero temperature (−20°C to −30°C) is essential if temperatures fall this low, but it is equally important for these microarthropods to be able to survive in the supercooled state at higher subzero temperatures for several months, typical of the winter conditions they will experience in their soil environment. In the collembolan *O. arcticus* (Fig. 6.11), which is freeze avoiding, there is a high survival (60%) after 12 weeks at −3°C, decreasing to <40% after the same period at −5°C (Block *et al.* 1994).

If cold tolerance is viewed as an essential prerequisite for survival, then thermal adaptation to low temperature is similarly important for the effective exploitation of the limited opportunities available in the arctic climate. Across a range of soil microarthropod species, 50–70% of individuals are still active at 0°C and locomotory activity is not terminated until −5°C, some 5–10°C lower than in related temperate species (Fig. 6.12; Coulson *et al.* 1995b). In *O. arcticus*, respiratory metabolism, as measured by oxygen consumption, increased markedly between 0°C and 10°C and more gradually up 30°C, before the animals become heat stressed

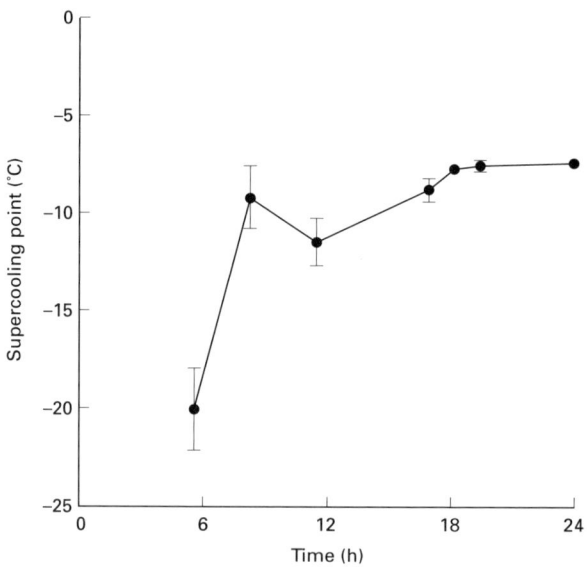

FIG. 6.10. Change in supercooling in *Hypogastrura tullbergi* after extraction from frozen soil.

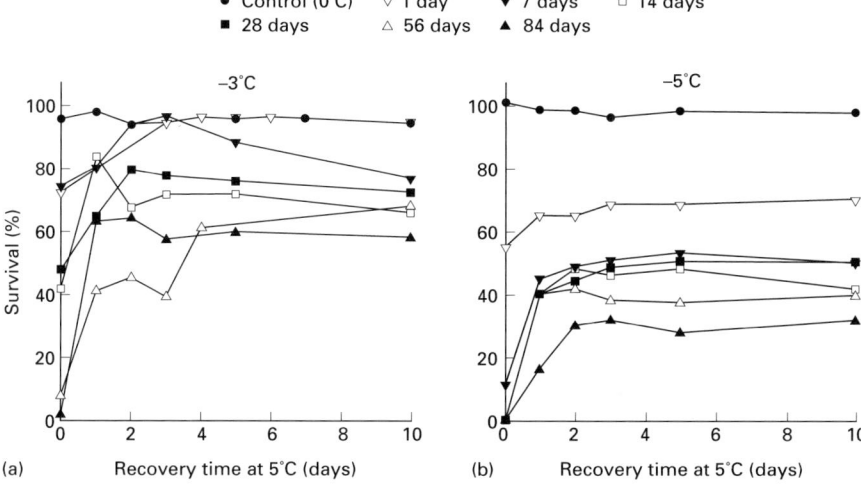

FIG. 6.11. Apparent survival of *Onychiurus arcticus* after exposure to (a) –3°C or (b) –5°C for up to 84 days. After exposure, for a given time period at a particular temperature, animals were allowed to recover for up to 10 days at 5°C and the number apparently surviving on a given day was recorded. This gives rise to an apparent increase in survival up to a plateau as time passes. Thus, for each graph, each line represents recovery after a particular period of exposure to a set temperature. The top control line represents survival of animals that were kept for the full 84 days at 0°C and then allowed to recover. (After Block *et al.* 1994.)

at 35°C (Fig. 6.13; Block *et al.* 1994). Between 0°C and 10°C the Q_{10} was high at 7.0, most probably representing an adaptation to low temperature which enables these organisms to respond rapidly to small thermal increases in their surroundings.

The environmental templet of the soil-inhabiting arthropods also demands the ability to survive high summer temperatures. Studies on Svalbard with seven species of microarthropods have demonstrated remarkable levels of tolerance and survival of supranormal temperatures (Fig. 6.14, Hodkinson *et al.* 1996). None of the species died after 1-h exposure in moist conditions at 30°C. The temperatures required to kill all individuals in 1-h exposures were between 35°C and 40°C (Collembola) and between 40°C and 45°C (oribatid mites) (Fig. 6.14a). In 1-h exposures with dry conditions, there was a higher survival of mites compared to Collembola (Fig. 6.14b). Longer exposures (3 h) in a moist environment shifted the thermal death point downwards by *c.* 2.5°C (Fig. 6.14c) but in dry conditions the Collembola were killed after 3 h at all temperatures, whereas the mites were largely unaffected (Fig. 6.14d). In longer exposures, mature individuals of *O. arcticus* survived for more than 68 days under moist conditions at 25°C, but above 30°C survival was <24 h.

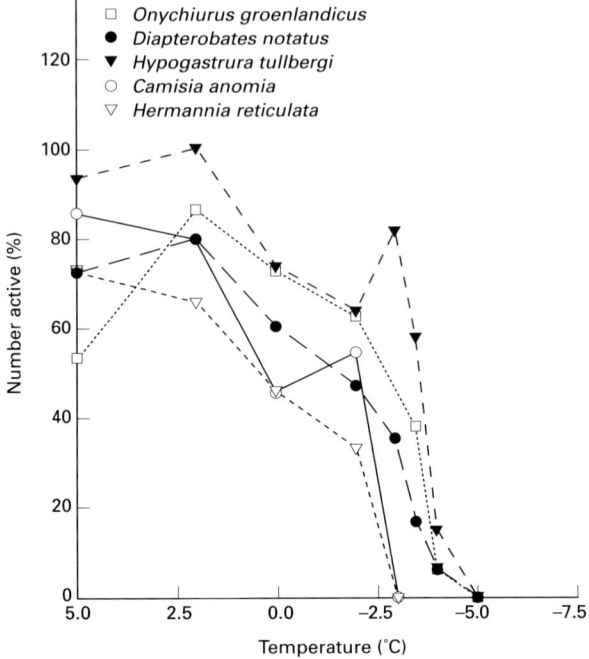

Fig. 6.12. Temperature thresholds for activity of arctic soil microarthropods. (After Coulson *et al.* 1995b.)

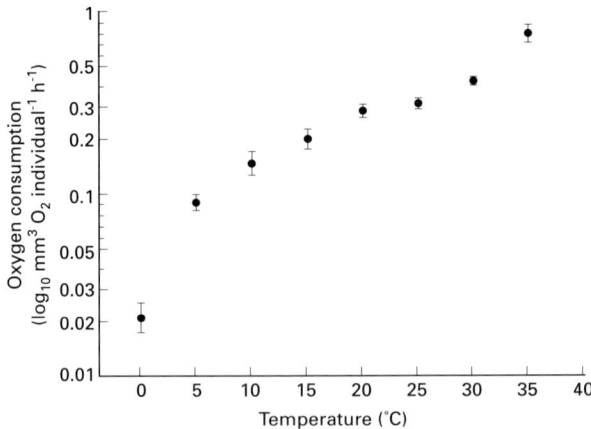

Fig. 6.13. Mean (±SE) oxygen consumption of *Onychiurus arcticus* over temperature range 0–35°C. (After Block *et al.* 1994.)

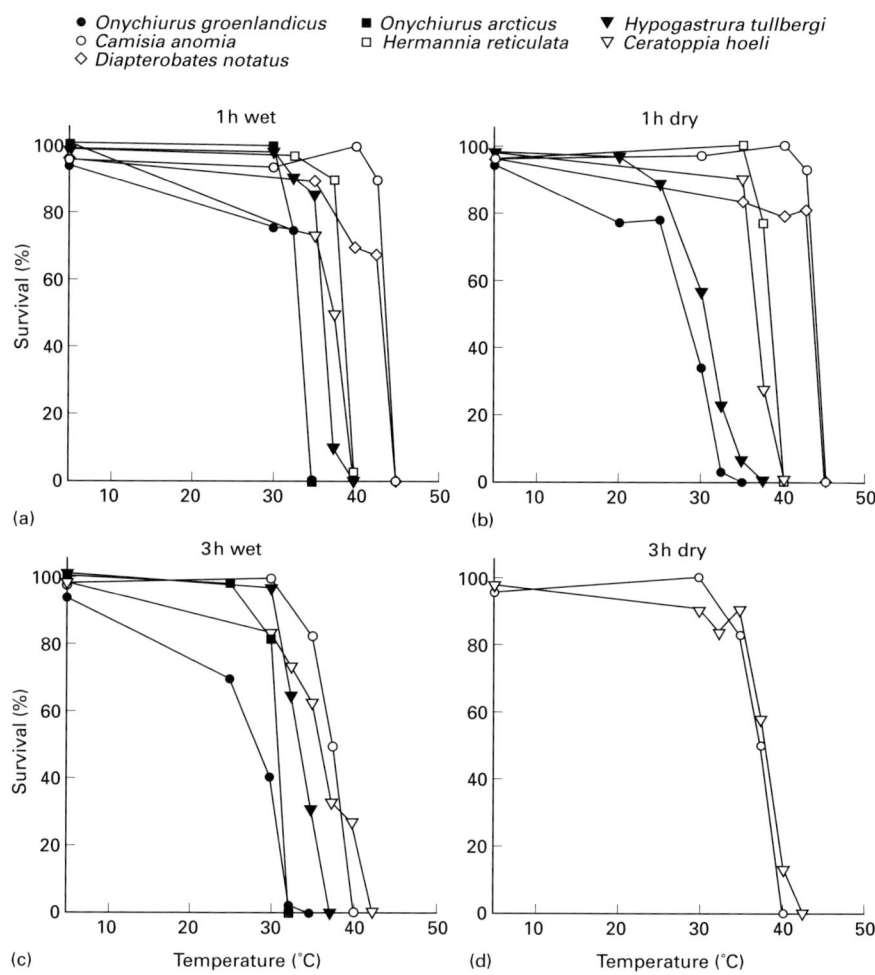

● *Onychiurus groenlandicus*   ■ *Onychiurus arcticus*   ▼ *Hypogastrura tullbergi*
○ *Camisia anomia*   □ *Hermannia reticulata*   ▽ *Ceratoppia hoeli*
◇ *Diapterobates notatus*

FIG. 6.14. Survival of arctic soil microarthropods at a range of temperatures after 1- and 3-h exposures in wet and dry conditions: (a) 1 h wet; (b) 1 h dry; (c) 3 h wet and (d) 3 h dry. Filled symbols, Collembola; open symbols, mites.

IMPACT OF A CHANGING CLIMATE

Above-ground insect herbivore – Acyrthosiphon svalbardicum

The relationship between the aphid *A. svalbardicum* and *D. octopetala* is closely linked through the sap-feeding habit of the herbivore and the absence of alate (winged) morphs which severely constrains the movement of aphids both within and between patches of the host. The seasonal cycle of plant growth and aphid development and reproduction occur simultaneously within a 'window of opportunity' which is initiated at or around the time of snowmelt in early June and terminated by decreasing temperature in late August and September. During this period both the aphid and its host plant are subject to 24 h daylight and prevailing air/vegetation temperature.

The life cycle of *A. svalbardicum* (see Fig. 6.5b) indicates that the second generation of sexual morphs, although likely to fail to mature in most summers because of an inadequate thermal budget, would act as a 'biological amplifier' in a particularly favourable year when they matured and oviposited successfully (Strathdee *et al.* 1993a). When patches of host plant with overwintering eggs of *A. svalbardicum* were enclosed within cloches (Strathdee & Bale 1993) from snowmelt until the end of summer (11 weeks), average leaf surface temperature increased by 2.8°C and total summer thermal budget by 215 day degrees compared with adjacent control areas. This resulted in a marked increase in the population density of *A. svalbardicum* inside the cloches (Fig. 6.15) and an 11-fold increase in egg density ($5769\,\mathrm{m}^{-2}$) at the end of summer compared with controls ($500\,\mathrm{m}^{-2}$) (Strathdee *et al.* 1993a). The main factor accounting for these large differences in aphid and egg density was the advanced phenology of the population in response to warming. Adult fundatrices developed 1 week earlier and second generation viviparae and sexuals matured 2 weeks earlier, in the cloches, an advancement of 18% over the 11-week summer growing season. This advanced phenology provided the third generation sexual morphs (produced by the viviparae in the cloches) with sufficient time to mature and oviposit before the end of the season. The thermal budget

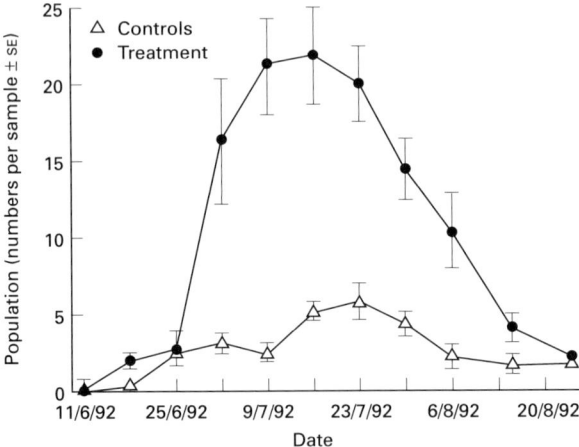

FIG. 6.15. Population densities of *Acyrthosiphon svalbardicum* during summer 1992 in controls and treatment plots. (After Strathdee *et al.* 1993a.)

requirement for the first eggs to be produced each year via the egg–fundatrix–sexual morphs–egg route (see Fig. 6.5b) is 467 day degrees above 0°C, but 710 day degrees are required to complete the egg–fundatrix–vivipara–sexual morphs–egg route, and realize the potential of the extra eggs that are produced if the aphid is able to complete three generations in one summer (Strathdee *et al.* 1993a). Thus, the flexible life cycle of *A. svalbardicum* combines a 'guaranteed' egg production from the first generation of sexual morphs, with the opportunity to increase winter egg density by the exploitation of favourable conditions through the maturation and oviposition of the third generation aphids. Importantly, the response to elevated temperature by aphids is rapid, with major changes observed within a single growing season.

Below-ground soil microarthropods

In field experiments on a polar semi-desert and tundra heath at Ny-Ålesund on Svalbard, Coulson *et al.* (1993, 1996) elevated the thermal budget at the soil surface by approximately 10% (measured as day degrees above 0°C over 3 years). Over the 3-year period of warming there were no significant differences in the densities of either mites or Collembola inside or outside of the polythene tents that were used to elevate temperature until the very last sample date. The data for 'inside' and 'outside' tents were therefore combined as shown in Fig. 6.16. Densities

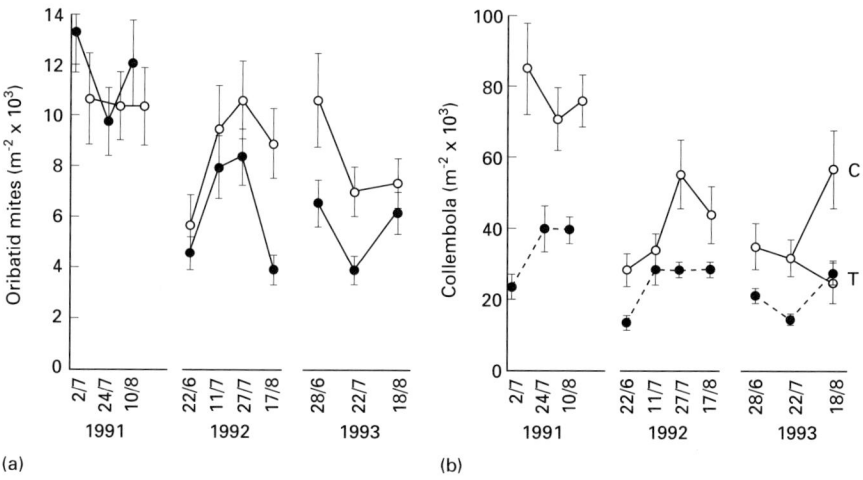

(a)   (b)

FIG. 6.16. Changes in populations of (a) oribatid mites and (b) Collembola at a polar semi-desert (○) and tundra heath (●) 1991–93. Means (± SE) are pooled value of controls (C) and tented plots (T), except August 1993 sample of Collembola where they are plotted separately. (After Coulson *et al.* 1996.)

of oribatid mites were generally higher at the polar semi-desert than the tundra heath (Fig. 6.16a), and showed a declining trend from 1991 to 1993 at both sites. A similar general pattern was observed with the Collembola with significantly higher populations at the polar semi-desert and a marked difference in density between 1991 and 1993 (Fig. 6.16b). The end of summer sample in 1993, however, showed a particularly interesting response where the density of Collembola inside the tents was significantly lower than in the control plots. Precipitation in summer 1993 was low (8 mm) compared with the long-term mean (29 mm). The dry soil conditions at the well-drained polar semi-desert created by an abnormally dry summer and exacerbated by the unusually warm conditions, appeared to create adverse conditions for the Collembola, particularly within the tents, although the mites were much less affected, confirming the differential susceptibility of these two groups of microarthropods to heat stress and desiccation as observed in laboratory experiments (Fig. 6.14).

CONCLUSIONS AND PREDICTIONS

The probable effects of climate change on the ecophysiology and community composition of the arctic primary producers have been discussed in detail (Chapin *et al.* 1992), but despite their important roles as herbivores, pollinators and decomposers, the arthropods have been comparatively little studied. Danks (1992), however, suggested that arctic insects would be excellent organisms for monitoring climate change, proposing a number of key indicator features. These included alterations in faunal composition indices, range limits, presence of marker species, interspecific ratios and the relationships between particular species. Phenological and physiological indicators and the use of key faunal sites could also be significant. Much will depend on the time scale and rate of climate change and the climatic parameters that are modified. It is not temperature alone that will change, patterns of precipitation and cloud cover will alter, leading to significant modification of the whole abiotic and biotic environmental templet. The responses of the different arthropod species will tend to be individualistic (as evidenced by the differences discussed between aphids, mites and Collembola), depending on their life history and physiological characteristics, and will be determined by the type of microhabitat they inhabit.

It is helpful to contrast probable short- and long-term effects on the invertebrates at both the species and community levels. Initially, as we have shown for aphids, mites and Collembola on Svalbard, the thermal environment of the above-ground arthropods will be enhanced proportionately more than that of those living below ground and this, coupled with life cycles that are more responsive to enhanced temperatures (Strathdee *et al.* 1993a,b, 1995), will almost certainly lead to increases in population size and range expansion of above-ground herbivores on a local scale but may not affect soil-dwelling organisms. Most high arctic invertebrate herbivore populations suffer low levels of parasitism and predation but this is likely to increase as temperatures rise, leading to a possible change in the mechanism of population regulation and an effective lengthening of the food chain. For example, *Acyrthosiphon svalbardicum* currently incurs negligible parasitism by a cynipid wasp (*Alloxysta* sp.) and predation by the hoverfly *Syrphus torvus* Osten-Sacken (Hodkinson & Coulson unpublished data). A changing environment will probably lead initially to a redistribution of existing species within the habitat mosaic, as they respond to changing thermal and moisture gradients and modification of the biotic environmental templet. The soil-dwelling Collembola and mites will tend to be favoured in wetter and drier habitats respectively but their population response is likely to be relatively slow. Areas that were metapopulation sinks (*sensu* Watkinson & Sutherland 1995) may be sufficiently ameliorated to support population sources. On a local scale there will probably be a gradual colonization of newly exposed sites by existing species, most of which are likely to show improved performance at slightly enhanced temperatures. Rates of colonization will depend on dispersal abilities which are likely to be low for flightless soil-dwelling species but more rapid for winged above-ground forms. There is little to suggest that increased temperature alone will lead to species loss.

In the longer term there is likely to be a redistribution of species along the latitudinal (thermal) gradient. As conditions in the High Arctic improve, the arthropod community will be subject to competition from colonizing species from further south. The establishment of new host plants will be followed by new host-specific herbivores. Survival of existing arthropod species will depend on their ability to compete with incoming species, by avoiding competition by rapid adaptation and specialization or escaping competition by shifting their general distribution northwards. There is, however, an ultimate northern land limit beyond which these northern species cannot be pushed. Some changes are likely to be subtle with host-specific insects, such as the psyllids, shifting their distribution patterns within the overall geographical range of their host plant. Where there is continuous land connection, such as in the Low Arctic regions of the northern continental land masses, northward spread of both animals and plants will probably be a gradual process. However, most of the High Arctic comprises islands that differ in their degree of physical isolation. Colonization of an island group like Spitsbergen, which is well isolated from both Greenland and Europe, will depend on the dispersal abilities of the organisms. For most flying insects this does not present a problem. Elton (1925) observed live aphids on Spitsbergen that had their probable origin in mainland Europe. The present fauna, under prevailing conditions, is clearly able to withstand repeated invasion by these less well-adapted species from the south but this balance will gradually change as temperatures increase and the frequency of successful establishment increases. The soil faunas experience different problems. Being flightless, they are dependent on alternative agencies for dispersal and thus rates of dissemination are likely to be low. The most likely mechanism for introducing southern soil-dwelling species into the High Arctic is to be carried in soil attached to the many species of migratory tundra-breeding bird species that move into the Arctic during summer. Here, basic physiology is likely to be important, with desiccation-resistant species, such as mites, having a greater chance of survival during dispersal compared with the Collembola.

To conclude, the response of arctic arthropod communities to climate change will be complex and dependent on the adaptive responses of the individual species to a shifting environmental templet. The processes of change are dynamic and without a fixed end point. Different elements of the arthropod communities will respond at varying rates and with different degrees of success, depending on their existing life histories and physiologies. Community change will reflect the incremental changes occurring at the species level.

ACKNOWLEDGEMENTS

The research on Svalbard was funded by Grant GST/02/534 from the Arctic Terrestrial Ecology Special Topic Programme of the UK Natural Environment Research Council. We are grateful for field and technical assistance from Roger Worland, Chris Wooley, Chris Wright and Nick Cox.

REFERENCES

Addison, J.A. (1977). Population dynamics and biology of Collembola on Truelove Lowland. *Truelove Lowland Devon Island, Canada: A High Arctic Ecosystem* (Ed. by L.C. Bliss), pp. 363–382. University of Alberta Press, Edmonton.

Addison, J.A. (1981). Biology of *Hypogastrura tullbergi* (Collembola) at a high Arctic site. *Holarctic Ecology*, **4**, 49–58.

Bale, J.S. (1993). Classes of insect cold hardiness. *Functional Ecology*, **7**, 751–753.

Behan-Pelletier, V.M. Oribatid mite fauna of northern ecosystems: a product of evolutionary adaptations or physiological constraints? *Proceedings of IX International Congress of Acarology, Columbus, Ohio*, in press.

Bengtson, S.-A., Fjellberg, A. & Solhoy, T. (1974). Abundance of tundra arthropods in Spitsbergen. *Entomologica Scandinavica*, **5**, 137–142.

Block, W., Webb, N.R., Coulson, S., Hodkinson, I.D. & Worland, M.R. (1994). Thermal adaptation in the Arctic collembolan *Onychiurus arcticus*. *Journal of Insect Physiology*, **40**, 715–722.

Bohnsack, K.K. (1971). Distribution of oribatids near Barrow, Alaska. *Proceedings 3rd International Congress of Acarology, Prague* (Ed. by M. Daniel & B. Rosicky), pp. 71–74. W. Junk, The Hague.

Cannon, R.J.C. & Block, W. (1988). Cold tolerance of microarthropods. *Biological Reviews*, **63**, 23–77.

Chapin, S.F., Jeffries, R.L., Reynolds, J.F., Shaver, G.R. & Svoboda, J. (1992). *Arctic Ecosystems in a Changing Climate: An Ecophysiological Perspective*. Academic Press, San Diego.

Chapin, F.S. & Shaver, G.R. (1985). Arctic. *Physiological Ecology of North American Plant Communities* (Ed. by B.F. Chabot & H.A. Mooney), pp. 16–40. Chapman & Hall, New York.

Corbet, P.S. (1972). The microclimate of arctic plants and animals, on land and in fresh water. *Acta Arctica*, **18**, 1–43.

Coulson, S.J., Hodkinson, I.D., Block, W., Webb, N.R. & Worland, M.R. (1995b). Low summer temperatures: a potential mortality factor for high Arctic soil arthropods? *Journal of Insect Physiology*, **41**, 783–792.

Coulson, S., Hodkinson, I.D., Strathdee, A.T., Bale, J.S., Block, W., Worland, M.R. & Webb, N.R. (1993). Simulated climate change: the interaction between vegetation type and microhabitat temperatures at Ny Alesund, Svalbard, *Polar Biology*, **13**, 67–70.

Coulson, S.J., Hodkinson, I.D., Strathdee, A.T., Block, W., Webb, N.R., Bale, J.S. & Worland, M.R. (1995a). Thermal environments of Arctic soil organisms during water. *Arctic and Alpine Research*, **27**, 365–371.

Coulson, S.J., Hodkinson, I.D., Webb, N.R., Block, W., Bale, J.S., Strathdee, A.T., Worland, M.R. & Wooley, C. (1996). Effects of experimental temperature elevation on high arctic soil micro-arthropod populations. *Polar Biology*, **16**, 147–153.

Courtin, G.M. & Labine, C.L. (1977). Microclimatological studies on Truelove Lowland. *Truelove Lowland, Devon Island, Canada: A High Arctic Ecosystem* (Ed. by L.C. Bliss), pp. 73–106. University of Alberta Press, Edmonton.

Danks, H.V. (1981). *Arctic Arthropods*. Entomological Society of Canada, Ottawa.

Danks, H.V. (1987). Insect–plant interactions in arctic regions. *Revue d'Entomologie du Quebec*, **31**, 52–75.

Danks, H.V. (1990). Arctic insects: instructive diversity. *Canada's Missing Dimension: Science and History in the Canadian Arctic Islands*, Vol. II (Ed. by C.R. Harrington), pp. 444–470. Canadian Museum of Nature, Ottawa.

Danks, H.V. (1991). Winter habitats and ecological adaptations for winter survival. *Insects at Low Temperature* (Ed. by R.E. Lee & D.L. Denlinger), pp. 231–259. Chapman & Hall, London.

Danks, H.V. (1992). Arctic insects as indicators of environmental change. *Arctic*, **45**, 159–166.

Danks, H.V. (1993). Patterns of diversity in the Canadian insect fauna. Systematics and entomology: diversity distribution, adaptation and application. *Memoirs of the Entomological Society of Canada* (Ed. by G.E. Ball & H.V. Danks), **165**, 272 pp.

Danks, H.V. (1994). Regional diversity of insects in North America. *American Entomologist*, **40**, 50–55.

Douce, G.K. & Crossley, D.A. (1977). *Acarina* abundance and community structure in an arctic coastal tundra. *Pedobiologia*, **17**, 32–42.

Downes, J.A. (1964). Arctic insects and their environment. *Canadian Entomologist*, **96**, 279–307.

Downes, J.A. (1965). Adaptations of insects in the Arctic. *Annual Review of Entomology*, **10**, 257–274.

Elton, C.S. (1925). The dispersal of insects to Spitsbergen. *Transactions of the Royal Entomological Society of London* (1925), 289–299.

Fjellberg, A. (1986). Collembola from the Canadian high arctic. Review and additional records. *Canadian Journal of Zoology*, **64**, 2386–2390.

French, H.M. (1970). Soil temperatures in the active layer, Beaufort Plain. *Arctic*, **23**, 229–239.

Hammer, M. & Wallwork, J.A. (1979). A review of the world distribution of oribatid mites (Acari: Cryptostigmata) in relation to continental drift. *Biologiske Skrifter det kongelige Danske videnskabernes Selskab, København*, **22**, 3–31.

Heikinheimo, O. (1968). The aphid fauna of Spitzbergen. *Annales Entomologici Fennici*, **34**, 82–93.

Hertzberg, K., Leinaas, H.P. & Ims, R.A. (1994). Patterns of abundance and demography: Collembola in a habitat patch gradient. *Ecography*, **17**, 349–359.

Hisdal, V. (1985). *Geography of Svalbard*. Norsk Polarinstitutt, Oslo.

Hodkinson, I.D., Jensen, T.S. & MacLean, S.F. (1979). The distribution, abundance and host plant relationships of *Salix*-feeding psyllids (Homoptera: Psylloidea) in arctic Alaska. *Ecological Entomology*, **4**, 119–132.

Hodkinson, I.D., Coulson, S., Webb, N.R. & Block, W. (1996). Can high Arctic soil microarthropods survive elevated summer temperatures? *Functional Ecology,* **10,** 314–321.

Hodkinson, I.D., Healey, V. & Coulson, S. (1994). Moisture relationships of the high Arctic collembolan *Onychiurus arcticus. Physiological Entomology,* **19,** 109–114.

Hodkinson, I.D. & Hughes, M.K. (1982). *Insect Herbivory.* Chapman & Hall, London.

Kauppi, P. & Posch, M. (1988). A case study of the effects of CO_2-induced climatic warming on forest growth and the forest sector. A. Productivity reactions of northern boreal forests. *Assessments in Cool Temperate and Cold Regions. The Impact of Climatic Variations on Agriculture,* Vol. 1 (Ed. by M.L. Parry, T.R. Carter & N.T. Konijn), pp. 183–195. Kluwer Academic, Dordrecht.

Kevan, P.G. & Danks, H.V. (1986). Adaptations of Arctic insects. *The Arctic and its Wildlife* (Ed. by B. Sage), pp. 55–57. Croom Helm, London.

Kukal, O. (1991). Behavioural and physiological adaptations to cold in a freeze tolerant Arctic insect. *Insects at Low Temperature* (Ed. by R.E. Lee & D.L. Denlinger), pp. 276–300. Chapman & Hall, New York.

Kukal, O. & Dawson, T.E. (1989). Temperature and food quality influences on feeding behaviour, assimilation efficiency and growth rate of arctic woolly-bear caterpillars. *Oecologia,* **79,** 526–532.

Kukal, O., Duman, J.G. & Serianni, A.S. (1989). Cold-induced mitochondrial degradation and cryoprotectant synthesis in freeze-tolerant arctic caterpillars. *Journal of Comparative Physiology B,* **158,** 661–671.

Kukal, O., Heinrich, B. & Duman, J.G. (1988a). Behavioural thermoregulation in the freeze tolerant arctic caterpillar, *Gynaephora groenlandica. Journal of Experimental Biology,* **138,** 181–193.

Kukal, O. & Kevan, P.G. (1987). The influence of parasitism on the life history of a high arctic insect, *Gynaephora groenlandica* (Wocke) (Lepidoptera: Lymantriidae). *Canadian Journal of Zoology,* **65,** 156–163.

Kukal, O., Serianni, A.S. & Duman, J.G. (1988b). Glycerol metabolism in a freeze-tolerant arctic insect: An *in vivo* 13-C NMR study. *Journal of Comparative Physiology B,* **158,** 175–183.

Leather, S.R., Walters, K.F.A. & Bale, J.S. (1993). *The Ecology of Insect Overwintering.* Cambridge University Press, Cambridge.

Less, A.D. (1960). The role of photoperiod and temperature in the determination of the parthenogenetic and sexual forms in the aphid *Megoura vicia* Buckton. 2. The operation of the interval timer in young clones. *Journal of Insect Physiology,* **9,** 153–164.

Lees, A.D. (1961). Clonal polymorphism in aphids. *Insect Polymorphism* (Ed. by J.S. Kennedy), pp. 68–78. Symposium No. 1, Royal Entomological Society.

Leinaas, H.P. (1981). Activity of arthropoda in snow within a coniferous forest soil with special reference to Collembola. *Holarctic Ecology,* **4,** 127–138.

MacLean, S.F. (1975). Ecological adaptations of tundra invertebrates. *Physiological Adaptations to the Environment* (Ed. by F.J. Vernberg), pp. 269–300. Intext Educational Publishers, New York.

MacLean, S.F. (1977). Community organisation in the soil invertebrates of Alaskan arctic tundra. *Soil Organisms as Components of Ecosystems* (Ed. by U. Lolim & T. Persson), *Proceedings VI International Colloquium Soil Zoology; Ecological Bulletin, Stockholm,* **25,** 90–101.

MacLean, S.F. (1981). Introduction: invertebrates. *Tundra Ecosystems: A Comparative Analysis* (Ed. by L.C. Bliss, O.W. Heal & J.J. Moore), pp. 509–516. Cambridge University Press, Cambridge.

MacLean, S.F. (1983). Life cycles and the distribution of psyllids (Homoptera) in arctic and subarctic Alaska. *Oikos,* **40,** 445–451.

MacLean, S.F. & Hodkinson, I.D. (1980). The distribution of psyllids (Homoptera: Psylloidea) in arctic and subarctic Alaska. *Arctic & Alpine Research,* **12,** 369–376.

MacLean, S.F. & Jensen, T.S. (1985). Food plant selection by insect herbivores in Alaskan arctic tundra: the role of plant life form. *Oikos,* **44,** 211–221.

Marcovitch, S. (1923). The migration of the Aphididae and the appearance of the sexual forms as affected by the relative length of daily exposure. *Journal of Agricultural Research,* **28,** 513–522.

Maxwell, B. (1992). Arctic climate: potential for change under global warming. *Arctic Ecosystems in a Changing Climate: An Ecophysiological Perspective* (Ed. by S.F. Chapin, R.L. Jeffries, J.F. Reynolds, G.R. Shaver & J. Svoboda) pp. 11–34. Academic Press, San Diego.

Mølgaard, P. (1982). Temperature observations in high arctic plants in relation to microclimate in the vegetation of Peary Land, North Greenland. *Arctic and Alpine Research*, **14**, 105–115.

Norton, R.A. & Palmer, S.C. (1991). The distribution, mechanisms and evolutionary significance of parthenogenesis in oribatid mites. *The Acari: Reproduction, Development and Life-history Strategies* (Ed. by R. Schuster & P.W. Murphy), pp. 107–136. Chapman & Hall, London.

Petersen, H. & Luxton, M. (1982). A comparative analysis of soil fauna populations and their role in decomposition processes. *Oikos*, **39**, 287–388.

Regal, P.J. (1967). Voluntary hypothermia in reptiles. *Science*, **155**, 1551–1553.

Rydén, B.E. & Kostov, L. (1980). Thawing and freezing in tundra soils. *Ecology of a Subarctic Mire* (Ed. by M. Sonesson). *Ecological Bulletin, Stockholm*, **30**, 251–281.

Seniczak, S. & Plichta, W. (1978). Structural dependence of moss mite populations (Acari, Oribatei) and patchiness of vegetation in moss–lichen-tundra at the north coast of Hornsund, West Spitsbergen. *Pedobiologia*, **18**, 145–152.

Sømme, L. & Block, W. (1991). Adaptations to alpine and polar environments in insects and other terrestrial arthropods. *Insects at Low Temperature* (Ed. by R.E. Lee & D.L. Denlinger), pp. 318–359. Chapman & Hall, New York.

Southwood, T.R.E. (1977). Habitat, the templet for ecological strategies? *Journal of Animal Ecology*, **46**, 337–365.

Steffensen, E.L. (1982). The climate of Norwegian Arctic stations. *Klima*, **5**, 3–43.

Strathdee, A.T. & Bale, J.S. (1993). A new cloche design for elevating temperature in polar terrestrial ecosystems. *Polar Biology*, **13**, 577–580.

Strathdee, A.T. & Bale, J.S. (1995). Factors affecting the distribution of *Acyrthosiphon svalbardicum* (Hemiptera: Aphididae) on Spitsbergen. *Polar Biology*, **15**, 375–380.

Strathdee, A.T., Bale, J.S., Block, W.C., Coulson, S.J., Hodkinson, I.D. & Webb, N.R. (1993a). Effects of temperature elevation on a field population of *Acyrthosiphon svalbardicum* (Hemiptera: Aphididae) on Spitsbergen. *Oecologia*, **96**, 457–465.

Strathdee, A.T., Bale, J.S., Block, W.C., Webb, N.R., Hodkinson, I.D. & Coulson, S.J. (1993b). Extreme adaptive life-cycle in a high arctic aphid, *Acyrthosiphon svalbardicum*. *Ecological Entomology*, **18**, 254–258.

Strathdee, A.T., Bale, J.S., Strathdee, F.C., Block, W.C., Coulson, S.J., Webb, N.R. & Hodkinson, I.D. (1995). Climatic severity and the response to temperature elevation of Arctic aphids. *Global Change Biology*, **1**, 23–28.

Watkinson, A.R. & Sutherland, W.J. (1995). Sources, sinks and pseudosinks. *Journal of Animal Ecology*, **64**, 126–130.

Weller, G., Cubley, S., Parker, S., Trabant, D. & Benson, C. (1972). The tundra microclimate during snow-melt at Barrow, Alaska. *Arctic*, **25**, 291–300.

Weller, G. & Holmgren, B. (1974). The microclimates of the arctic tundra. *Journal of Applied Meteorology*, **13**, 854–862.

Wilson, F. (1938). Some experiments on the influence of environment upon forms of *Aphis chloris* Koch. *Transactions of the Royal Entomological Society of London*, **87**, 168–188.

7. Environmental fluctuations in arctic marine ecosystems as reflected by variability in reproduction of polar bears and ringed seals

IAN STIRLING*† AND NICHOLAS J. LUNN†*

*Canadian Wildlife Service, 5320 122 Street, Edmonton, Alberta,
Canada T6H 3S5 and
†Department of Biological Sciences, University of Alberta, Edmonton, Alberta,
Canada T6G 2E7

INTRODUCTION

There is considerable interest in the extent and duration of natural environmental fluctuations and in unidirectional environmental change caused by human activities. Consequently, monitoring natural ecological variation in marine ecosystems is an important area of research (e.g. Croxall *et al.* 1988). For example, the reproductive performance of albatrosses, penguins and fur seals tends to track the distribution, abundance and availability of antarctic krill (*Euphausia superba* Dana) which is one of the key species in the antarctic food web (Croxall 1992). Although little is known about interannual variation in productivity in arctic marine ecosystems, opportunities exist to examine this phenomenon through analysis of long-term data on reproduction of apical predators, including seals and polar bears (*Ursus maritimus* Phipps).

Polar bears are distributed throughout the circumpolar Arctic in relatively discrete subpopulations. The ecological factors that influence their seasonal movements, distribution and behaviour vary widely between different geographic areas. Throughout their range, polar bears feed predominantly on ringed seals (*Phoca hispida* Schreber) and, to a lesser degree, on bearded seals (*Erignathus barbatus* Erxleben) (Stirling & Archibald 1977; Smith 1980). Most important, however, is that although polar bears are capable of catching seals of all age classes, young-of-the-year form the bulk of their diet (Stirling & Archibald 1977; Smith 1980; Hammill & Smith 1991). Recent analyses using predation and energy matrices confirm that the high levels of polar bear predation sustained by ringed seal populations are only possible because a large proportion of the total number of animals taken are young-of-the-year (Stirling & Øritsland 1995). Consequently, fluctuations in the productivity of ringed seal pups will be reflected immediately on polar bear reproduction and cub survival.

Ringed seal pups are born in early April and are weaned at 6 weeks of age (McLaren 1958) by which time they are approximately 50% fat by wet weight (Stirling & McEwan 1975; Lydersen, Hammill & Ryg 1992). Polar bears prefer fat to other parts of a seal. From shortly after they are born until break-up of the annual ice in early summer, when they become less accessible to polar bears, ringed seal pups are abundant, probably easier to catch because they are less experienced, and represent a high caloric return per unit of energy expended by a hunting polar bear. Polar bears become inactive during periods when seals are unavailable, such as during the open water season in western Hudson Bay from late July to early November, or simply less accessible, such as during intensely cold and inclement weather in mid-winter (Latour 1981; Messier, Taylor & Ramsay 1994). During these periods they fast, relying on their fat reserves in a hibernation-like physio-logical state for up to several months at a time (Nelson *et al.* 1983; Derocher *et al.* 1990). Polar bears reach their lightest weights of the year in late March, just prior to the birth of the next cohort of ringed seal pups, which also suggests it is the success of their hunting in spring and early summer that maximizes the body reserves

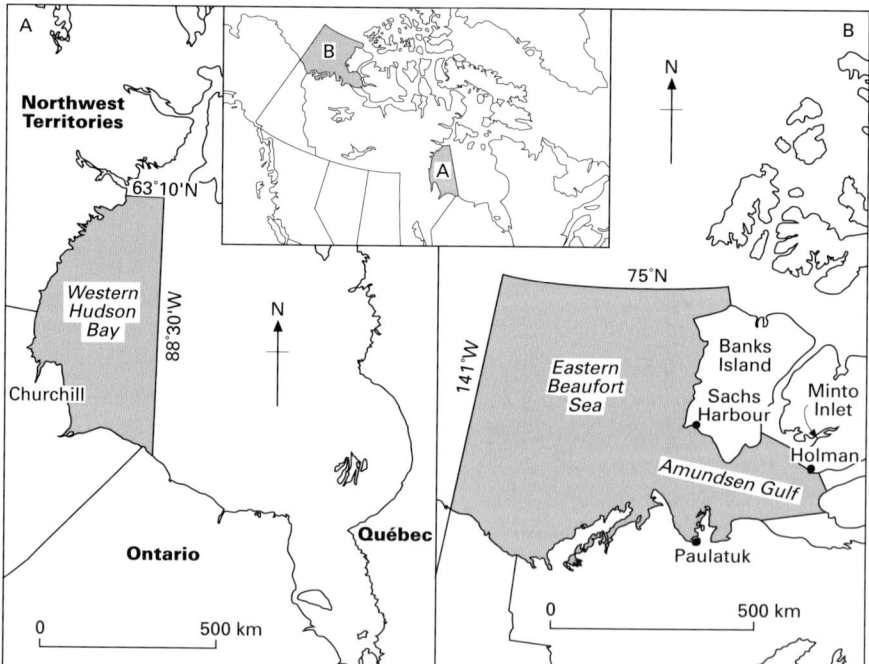

FIG. 7.1. Location of study areas in western Hudson Bay and the eastern Beaufort Sea in the Canadian Arctic.

necessary for survival, reproduction and nursing of cubs through the rest of the year. Thus, if major fluctuations in the biological productivity of arctic marine ecosystems occur, they should be reflected in the reproductive performance of polar bears and their principal prey, the ringed seal.

In this chapter, we examine the reproductive success of polar bears in western Hudson Bay and of polar bears and ringed seals in the eastern Beaufort Sea in the Canadian Arctic (Fig. 7.1) to provide insight into: (i) the amount of variability that occurs in the reproductive parameters of these species; and (ii) what such variations tell us about the ecological fluctuations experienced in the marine ecosystems of these two areas.

MATERIALS AND METHODS

Study areas and data collection

Ecologically, our two study areas differed in several important ways. Hudson Bay is shallow (average depth 125 m). The annual ice melts completely every summer by about late July and refreezes in November. Most marine water enters the bay from the north after passing through the interisland channels and polynyas of the High Arctic Archipelago, although some Atlantic water comes from the Labrador Sea. Fresh water enters through several large rivers. Although polar bears in several populations fast on land during the open-water period, no bears do so for as long as those in Hudson Bay. Polar bears there hunt on the sea ice through the winter and spring but the whole population must fast for about 4 months during the open-water period. Pregnant females remain ashore for approximately 8 months. Most female polar bears in western Hudson Bay breed for the first time at 4 years of age and, in some years, 40% or more of the litters are weaned at 1.5 years of age which is 1 year earlier than in other areas (Stirling, Pearson & Bunnell 1976; Stirling *et al.* 1977a; Ramsay & Stirling 1988). The reasons for these differences are not known. The study area in western Hudson Bay included the coastal areas of Ontario, Manitoba and the Northwest Territories, bounded by 63°10′N to the north and 88°30′W to the east (Fig. 7.1a). Field methods, and data collected during population studies of polar bears from 1965 to 1994, are summarized by Derocher and Stirling (1992). There are few data on seals in Hudson Bay.

The Beaufort Sea is partially ice covered throughout the year. To the north, multi-year pack ice from the polar basin predominates over water several hundred metres deep. Along the mainland coast and west coast of Banks Island, the annual ice over the continental shelf melts by late July and refreezes in most years by early October. Most water enters the Beaufort Sea from the cold, relatively unproductive, abyss of the polar basin via the west coast of Banks Island in the southerly flow of the Beaufort Gyre, although some also enters through Amundsen

Gulf. Polar bears there move north to the multi-year ice when the annual ice melts, south again when it refreezes (Stirling 1990), and thus are not forced to fast for an extended period of time. Most female polar bears in the Beaufort Sea do not breed for the first time until they are 5 years old and cubs are weaned at 2.5 years of age, although a few litters in the most northerly areas are not weaned until they are 3.5 years old. In the eastern Beaufort Sea, the study area lay east of 141°W and south of 75°S, including Amundsen Gulf (Fig. 7.1b). Field methods, and data collected during population studies of polar bears from 1971 to 1979, 1985 to 1987 and 1992 and 1994, are summarized by Stirling, Calvert and Andriashek (1980) and Stirling, Andriashek and Calvert (1993). Field methods and data collected on seals from 1971 through 1994 are summarized by Stirling, Kingsley and Calvert (1982), Kingsley (1986), Smith (1987), Kingsley and Byers (1990) and the Fisheries Joint Management Committee, Inuvik, NWT (unpublished data).

Definitions and calculation of reproductive parameters

All bears were assumed to be born on 1 January. COY (cub-of-the-year) are bears less than 1 year of age. Yearling cubs are between 1 and 2 years of age and 2-year-old cubs are between 2 and 3 years of age but still with their mother. Subadults were independent bears 2 and 3 years old in Hudson Bay and from 2 to 4 years of age in the Beaufort Sea. Females 4 years and older in Hudson Bay and 5 years and older in the Beaufort Sea were defined as adults because those are the ages at which most breed for the first time in the two areas respectively.

We used data from captured female polar bears, alone or accompanied by cubs of different ages, to estimate their age-specific natality. Because some cubs are weaned at 1.5 years of age in western Hudson Bay, while no cubs are weaned at less than 2.5 years of age in the Beaufort Sea, and because data were collected in different seasons, natalities in those two areas were calculated as per Ramsay and Stirling (1988) and Stirling et al. (1980) respectively. Three-year running means were calculated to estimate natality in western Hudson Bay from 1980 to 1994 and in the Beaufort Sea from 1971 to 1979 and from 1992 to 1994. From 1985 to 1987 in the Beaufort Sea, sample sizes were large enough to facilitate annual estimations of natality so that annual variability through that short period could still be evaluated.

Reproduction of ringed seals can be monitored cost effectively from the annual harvest taken during the open-water season by Inuit hunters. In a normal population, a minimum of 30–40% of the seals taken are young-of-the-year and ovulation rates of adult females normally exceed 80% (McLaren 1958; Smith 1987). Data on ringed seal abundance and reproduction in the Beaufort Sea were taken from Stirling and Archibald (1977), Stirling, Archibald and DeMaster (1977b), Smith and Stirling (1978), Stirling et al. (1982), Kingsley (1986), Smith (1987), Kingsley and Byers (1990), Harwood and Stirling (1992) and unpublished data from the Fisheries Joint Management Committee, Inuvik, NWT.

Comparison of condition of adult polar bears in western Hudson Bay between years

To evaluate variability in condition of adult males and females fasting on land in western Hudson Bay in fall, we used the formula $C = (W/L^2) \times 100$ where C = index of condition (Quetelet Index), W = weight in kilograms, and L = body length in metres (Ganong 1991) and calculated 3-year running means. To control for variation in the timing of sampling periods between years, weights were scaled to a constant capture date of 21 September by adding or subtracting 0.85 kg to or from the weights of all bears for each day they were caught before or after that day (Derocher & Stirling 1992).

RESULTS AND DISCUSSION

Western Hudson Bay

Population studies of polar bears in western Hudson Bay have been ongoing from 1965 through 1994 although not all the data collected prior to 1980 are directly comparable to those recorded afterward (Stirling *et al.* 1977a; Derocher & Stirling 1992). In the early to mid-1980s, the natality of female polar bears in western Hudson Bay was the highest recorded anywhere in polar bear range, and nowhere else did females successfully wean cubs at 1.5 years of age instead of at the normal age of 2.5 years. Subsequently, a long-term decline in condition of adult female polar bears and survival of their cubs was documented from the 1970s through the late 1980s (Derocher & Stirling 1992), as reflected by a significant decline in condition indices ($r = -0.245$, $n = 310$, $P < 0.001$) (Fig. 7.2). This decline did not constitute a threat to the population because even when natality was at its lowest in the late 1980s, the rates were still higher than the upper range of values for bears elsewhere in the Arctic (e.g. Stirling *et al.* 1976, 1980). For the past 12 years, estimates of population size have remained relatively constant (Fig. 7.2) (Derocher & Stirling 1995; Canadian Wildlife Service unpublished data), indicating that the declines in condition and natality have not influenced the population's ability to maintain itself. Thus, the more important (but unanswered) question is probably not why natality declined from the early 1980s but how could natality have been sustained at a level so much higher than other polar bear populations in the first place, what facilitated the successful weaning of yearlings there but nowhere else in their range, and how could females manage these physiological feats in a habitat where pregnant females must also fast for 8 months or more?

It is unknown if overall biological productivity has declined, the abundance or accessibility of seals has changed, or other factors are involved which could account for the long-term changes in natality and condition of adult polar bears in western Hudson Bay. However, the cause of the increase in condition in 1992 (Fig. 7.2) may have resulted from a climatic perturbation. Mt Pinatubo erupted in the Philippines and released a large amount of particulate matter into the atmosphere, which resulted in significant climatic cooling over the northern hemisphere (McCormick, Thomason & Trepte 1995). In Hudson Bay, break-up was 3 or more weeks later in 1992 than 1991. The bears were heavier when they came ashore in 1992, probably because they had been able to feed on seals for longer (Stirling & Derocher 1993), and natality rose in 1993 (Fig. 7.2). In 1993, the temperature was still cooler than normal but not as cold as in 1992 (McCormick et al. 1995). Although the timing of break-up appeared to return to the normal range in 1993, the cooler temperatures probably still delayed break-up and allowed bears to feed longer before coming ashore than they were able to in 1991. Consequently, the condition of adult females did not decline to pre-1992 levels although natality began to decline again in 1994.

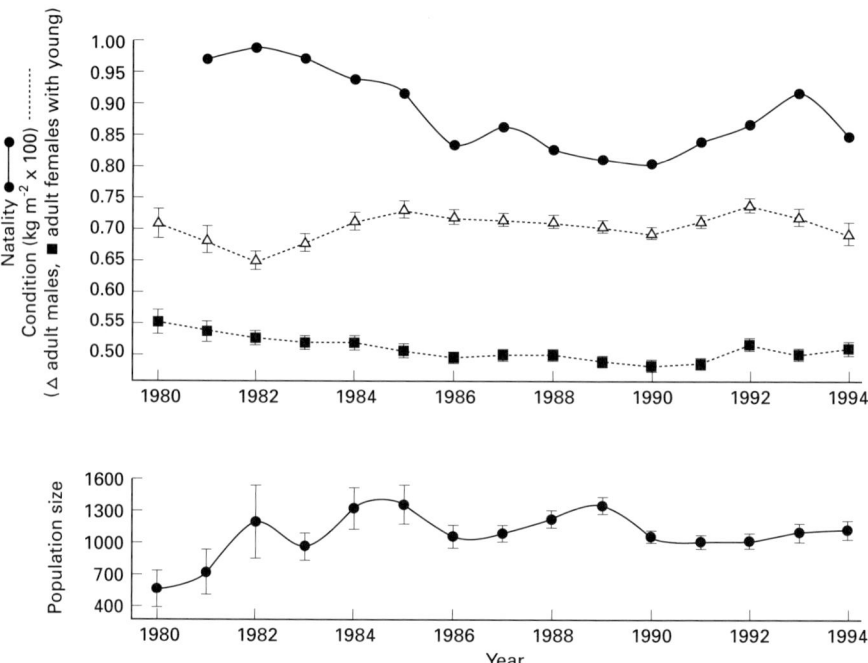

FIG. 7.2. Changes in natality, condition of adult male and female polar bears, and population size in western Hudson Bay from 1980 through 1994, expressed as 3-year running means. (Estimates of population size are from Derocher & Stirling 1995 and Canadian Wildlife Service unpublished data.)

Although the condition indices of adult females declined through the late 1980s, there was no corresponding decline in adult males ($r = 0.023$, $n = 446$, $P > 0.2$). Similarly, Atkinson, Stirling and Ramsay (1995) demonstrated a small but statistically significant decline in total body length of adult females from the late 1960s through the mid-1980s, which was absent in adult males, but proposed no explanation for the difference. Behavioural and physiological differences that may make adult males less vulnerable to losing physical condition than females include that they do not use their own fat reserves to nurse young and they steal and consume seals killed by bears of all other age and sex classes (Stirling 1974), as well as killing seals themselves. In addition, we note this is the only population of polar bears known to have an adult sex ratio significantly skewed in favour of adult females as a result of sex-selective harvesting (Derocher & Stirling 1995). In the eastern Beaufort Sea, Stirling *et al.* (1993) showed that females accompanied by young, especially those with COY, showed different habitat preferences from adult males during the most important feeding period in the spring. If adult females with COY in western Hudson Bay show a similar pattern and degree of habitat segregation from adult males in the spring, to that demonstrated in the eastern Beaufort Sea (Stirling *et al.* 1993), then it is possible that the level of intrasexual competition between females has increased while that between males has not. At present, this hypothesis remains untested.

Beaufort Sea

Polar bears

Between 1971 and 1987, the natality of female polar bears fluctuated widely with the maximum recorded rates being double the minimum (Fig. 7.3). Of particular interest are the large-scale changes that took place between 1971 and 1979. Natality was high from 1971 through 1973, declined sharply in 1974, remained low through 1976, and then recovered rapidly again from 1977 through 1979. When compared in 3-year blocks, the mean natality of adult females, 6 years of age or greater, was high from 1971 to 1973 (0.552, $n = 69$), low from 1974 to 1976 (0.360, $n = 90$), and high again from 1977 to 1979 (0.607, $n = 61$).

Other parameters related to polar bear reproduction and weights also varied markedly through the same periods in the 1970s. The age-specific weights of male and female bears were significantly lighter in 1974 and 1975 than they were from 1971 to 1973 and the differences were greatest in the subadult age classes (Kingsley 1979). Similarly, the greatest changes in natality occurred in the youngest breeding females. Although most females in the Beaufort Sea breed for the first time at 5 years of age, when conditions are particularly good, some 4-year-olds may also breed (Stirling *et al.* 1976; Lentfer *et al.* 1980). Between 1971 and 1973 and 1974

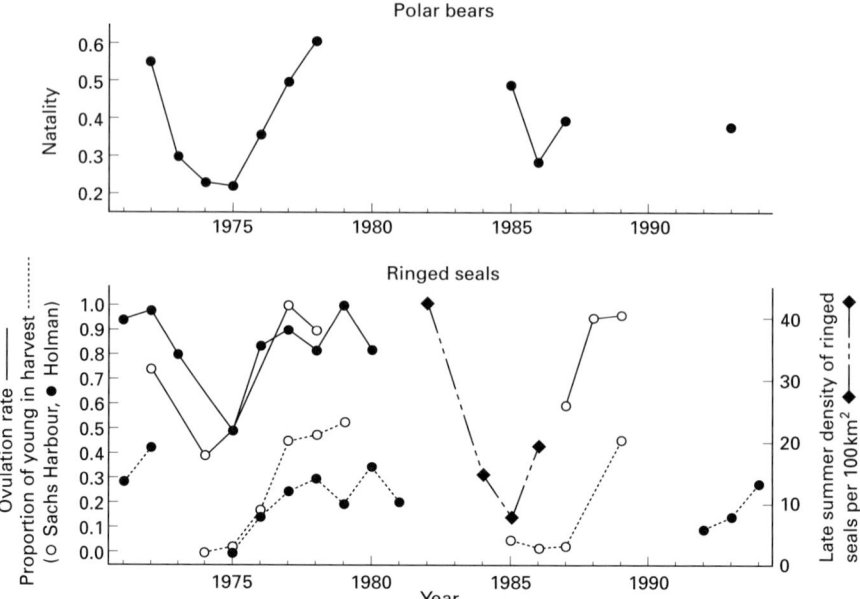

Fig. 7.3. Changes in natality of polar bears and indices of productivity of ringed seals in the eastern Beaufort Sea from 1971 through 1994. Natalities in 1971–79 and 1992–94 are expressed as 3-year running means. (Data on seals taken from Stirling & Archibald 1977; Smith & Stirling 1978; Stirling *et al.* 1982; Smith 1987; Kingsley & Byers 1990; Harwood & Stirling 1992; L.A. Harwood & T.G. Smith unpublished data.)

and 1976, the natality of 5-year-olds dropped 23.6% from 0.462 to 0.353 and that of 4-year-old females dropped 95% from 0.666 to 0.033. In addition, there was a significant decline in the proportion of adult females (≥ 5 years) accompanied by cubs of any age in 1971–73 (38/44 = 86.4%) compared to 1974–76 (37/62 = 59.7%) (*G*-test of independence with Williams' correction, $G_{adj} = 9.28$, $P < 0.005$) which suggests that cub mortality increased in the latter period as well.

The mean litter size declined from 1.73 ± 0.07 ($n = 41$) in 1971–73, to 1.59 ± 0.08 ($n = 39$) in 1974–76, and increased again to 1.80 ± 0.10 ($n = 25$) in 1977–79. However, these differences were not statistically significant ($F_{2,102} = 1.657$, $P > 0.15$), suggesting that females that were able to support cubs as well as themselves kept their complete litters, but females that lost cubs prior to weaning probably lost whole litters.

Although natality of polar bears also showed a marked decline in the mid-1980s (Fig. 7.3), it is more difficult to determine when changes began and how long they may have persisted because we have only 3 years of data. Nevertheless, between 1985 and 1986, natality of females aged 6 years and older declined from

0.490 to 0.287 (41.4%), with a partial recovery to 0.380 in 1987. Between 1985 and 1987, the mean litter size of cubs declined (1985: 1.71 ± 0.11, $n = 24$; 1986: 1.55 ± 0.09, $n = 33$; 1987: 1.53 ± 0.09, $n = 30$) although the difference was not statistically significant ($F_{2,84} = 0.923$, $P > 0.4$). Natality of 5-year-old females showed the greatest change, declining from 0.857 in 1985 to 0.209 in 1986 (75.6%) and remained essentially unchanged in 1987 at 0.222. There was a significant decline in the weights of the COY of females of all ages when captured on the sea ice in April and May ($F_{2,48} = 8.82$, $P < 0.001$). The mean weights of cubs in 1986 (12.6 ± 0.7, $n = 15$) and 1987 (11.7 ± 0.5, $n = 24$) were not significantly different (multiple pairwise comparisons, Tukey's method, $P > 0.6$) but both cohorts were significantly lighter than cubs in 1985 (15.6 ± 0.8, $n = 12$; 1985 vs. 1986, $P < 0.05$; 1985 vs. 1987, $P < 0.001$).

The natality of polar bears from 1992 to 1994 (Fig. 7.3) is not completely comparable with those from earlier years because the animals were all captured in the northernmost part of the study area where multi-year ice is more abundant and some females keep their cubs for 3.5 years before weaning (I. Stirling & N.J. Lunn unpublished data).

Ringed seals

In 1971 and 1972, the proportions of pups in the open-water harvest were 0.29 and 0.42 respectively, while the ovulation rate of adult females ranged between 0.74 and 0.97 at Sachs Harbour and Holman in 1971–72 (Fig. 7.3), values which indicate strong reproduction. Surveys of ringed seal birth lairs in 1973 (Smith & Stirling 1978), and the proportion of young-of-the-year in the seals killed by polar bears in the springs of 1971–73, also indicated high pup production (Stirling & Archibald 1977). From 1974 through 1976, the proportions of pups in the open-water harvest at Sachs Harbour were 0.00, 0.02, and 0.17 respectively and the ovulation rates of adult females in those samples were 0.39 and 0.49 respectively (Fig. 7.3). Similarly, at Holman, there was only one pup in an open-water sample of 391 in 1975 and only 0.14 in 1976 (Fig. 7.3). Surveys of ringed seal birth lairs indicated similar declines in 1974 and 1975, with only a small increase in 1976 (Smith & Stirling 1978). Between 1974 and 1975, the estimated total numbers of ringed seals hauled out on the ice over the eastern Beaufort Sea (excluding Amundsen Gulf), during the moult in late June, dropped by about half and remained relatively low until 1978 (Stirling *et al.* 1982). The proportion of pups in the open-water harvest, and the ovulation rates of adult females in 1976, began to increase at both Holman and Sachs Harbour and appeared back to the normal range for a healthy ringed seal population from 1977 through 1980 (Fig. 7.3). Finally, the unsmoothed age frequencies from the open-water harvest at Sachs Harbour and Holman clearly show that the cohorts from 1974 to 1976 are under-represented (Stirling *et al.* 1982; Smith 1987).

In the mid-1980s, studies of polar bears, like those of ringed seals during the same period, were not as extensive as both were in the 1970s. Consequently, the timing of the changes in productivity that occurred during that period is less distinct. In 1986, Inuit hunters at Sachs Harbour reported that no young-of-the-year seals were taken during the open-water hunt and that only reduced numbers had been seen in 1984 and 1985. From the age composition and ovulation rates of ringed seals collected from the open-water sample at Sachs Harbour in 1987–89, Kingsley and Byers (1990) confirmed that pup production was low in 1987 and, on the basis of missing cohorts, concluded that recruitment had also been low from 1984 through 1986. In 1988 and 1989, reproduction recovered to normal levels (Fig. 7.3).

Densities of ringed seals in the open water of the southern Beaufort Sea in the late summer and fall showed a similar pattern of change in relative abundance (Fig. 7.3). Densities (ringed seals per $100 \, km^2$), which declined from 1982 (42.2) through 1984 (14.7) and 1985 (7.9), recovered partially in 1986 (19.4) (Harwood & Stirling 1992). Regardless of difficulties associated with interpreting the reproductive data from 1984 through 1986, the results described above suggest that following the high densities of ringed seals in the open water in 1982, and unknown reproductive success and population densities in 1983, both parameters declined markedly through 1984 to very low levels in 1985–87, followed by recovery in 1988 and 1989.

From 1992 through 1994, the proportion of pups in the combined open-water harvest of ringed seals at Minto Inlet, Holman and Paulatuk were 0.10 (13/133), 0.15 (23/153) and 0.28 (55/195) respectively (L.A. Harwood & T.G. Smith personal communication). These data are not as extensive as those from the 1970s and 1980s (Fig. 7.3), or yet supported by data on ovulation rates, but they suggest a decline in ringed seal productivity occurred again in the early 1990s.

Effects of fluctuations in ringed seal reproduction on polar bears

From the results presented above, it seems clear that the most critical factor affecting reproductive success, subsequent condition and probably survival of polar bears is the availability of ringed seal pups from about mid-April through to break-up sometime in July. For example, during 1971–73, when ringed seal productivity was high, 54.8% (17/31) of the seals found killed by polar bears in the study area and that could be aged were pups (Stirling & Archibald 1977). In contrast, in 1974–75 when productivity was low, none of 35 seals found killed by polar bears was a pup, and the per cent utilization of carcasses was significantly higher. Finally, if older seals were equally vulnerable to capture by polar bears as weaned pups appear to be, then condition and natality of the bears would probably not decline in years when reproduction of ringed seals declines.

The availability of pups to prey upon in spring is especially important for adult female polar bears with COY for two reasons. First, by the time they return to the sea ice from their maternity dens, they have depleted most of their stored fat and depend on an immediate and abundant supply of ringed seal pups with which to replenish themselves and feed their cubs. Second, when females with COY first leave their maternity dens, they appear to select ringed seal pupping habitat in the fast ice. Although ringed seals are abundant in this habitat, they are less accessible than seals in moving-ice and floe-edge habitats until after the snow covering birth lairs and breathing holes melts in late spring (Stirling *et al.* 1993). Although subadult seals are more abundant and accessible in moving-ice and floe-edge habitats, densities of adult male polar bears are also higher there in spring and they sometimes prey on young cubs (Taylor, Larsen & Schweinsburg 1985). Thus, it appears that while high productivity of ringed seal pups is essential to bears of all age classes to maintain condition and high overwinter survival, the availability of pups to adult females with COY in the weeks immediately following departure from the maternity den is critical. For example, in the spring of 1974, when ringed seal pups first became scarce, we captured two very thin lone adult female polar bears that had nursed very recently, from which we deduced they had already lost their litters. A third emaciated female was accompanied by two cubs which were so thin that one could barely walk. We have not seen females with cubs in this condition in the Beaufort Sea, or elsewhere in the Arctic, before or since. Only in 1975 were females with COY found in disproportionately high numbers in habitats normally preferred by adult males, apparently in response to the paucity of ringed seal pups in their normal pupping habitat (Stirling *et al.* 1993).

In comparison to the rapid increase in the natality of adult female polar bears after 1974–76, the recovery of natality of 4- and 5-year-old females in 1977–79 was modest. We suggest the consequences of lower food availability were probably greater for younger females because they are less experienced hunters, less dominant, and therefore less able to retain seals they have killed or scavenged (Stirling 1974; I. Stirling unpublished data). Similarly, younger and lighter female polar bears probably had limited reserves to support body growth and reproduction until ringed-seal reproduction returned to normal in the late 1970s.

Although the data are not as extensive in the 1980s as in the 1970s, it seems clear that declines and subsequent increases in ringed seal productivity were again reflected by polar bears (Fig. 7.3). The relatively low natality of polar bears in 1992–94 (Fig. 7.3) may be because the sample was taken further to the north where productivity is lower. Alternatively, if the lower reproduction of ringed seals recorded in the main study area extended far enough north, it might also have influenced natality of polar bears.

The environmental factors that cause periods of either high or low productivity of ringed seals in the eastern Beaufort Sea, that last for 3 or more years (Fig. 7.3), are unknown. However, particularly heavy ice in the winters of 1973–74 and 1984–85 coincided with the onset of the decline in productivity of ringed seals, and consequently polar bears, in those decades. Stirling *et al.* (1977b, 1982) and Kingsley and Byers (1990) speculated that the heavy ice might have stimulated a decline in primary productivity, resulting in further changes in the ecosystem that culminated in the decline in reproduction of ringed seals. Tummers (1980) demonstrated that the maximum surface sea temperature was 0.62°C cooler in the heavy ice year of 1974 than in the much lighter ice year of 1975. Whether colder water in 1974 resulted from the presence of heavier ice or could have been part of the overall change that stimulated its occurrence is unknown. Although primary productivity could have recovered in the 2–3 years of normal ice conditions, populations of crustaceans and fish would have taken longer to return to their former abundance. In the above context, it is also interesting to note that in 1966, following the abnormally heavy ice conditions of 1964, only 8% (15) of a sample of 190 seals taken at Sachs Harbour during the open-water sealing period were young-of-the-year (Stirling *et al.* 1977b), suggesting that some of the changes we documented during the 1970s, 1980s, and probably 1990s, have occurred before.

CONCLUSIONS

Taken together, the data on polar bears and ringed seals from the eastern Beaufort Sea and polar bears from western Hudson Bay indicate that significant variation occurs in both natality and body condition, but over different time frames. In similar studies, fluctuations in the reproductive rates of sea birds and seals have been demonstrated over periods of several years duration and hypotheses of causal mechanisms have been suggested. For example, Polovina *et al.* (1994) presented data indicating declines in productivity of three marine species, including Hawaiian monk seals (*Monachus schauinslandi* Matschie), beginning in the early 1980s and continuing to the present, possibly as a consequence of a large-scale and long-term climate event that altered deep-level oceanic mixing with consequent effects on ecosystem productivity. In comparison, in the Scotia Sea region of the Southern Ocean, decreases in the reproductive performance of albatrosses, penguins and fur seals that last 3–4 years have been documented (Croxall *et al.* 1988) and are thought to be due to the redistribution of krill outside of the normal foraging range of these predators by oceanographic and atmospheric forces (Priddle *et al.* 1988). Similarly, Testa *et al.* (1991) demonstrated 4–5 year fluctuations in the reproduction of three antarctic phocids over periods of 20–40 years and suggested the possibility that they were influenced by the El Niño Southern Oscillation.

To date, we have no data with which to speculate on the causes of the fluctuations in natality and condition of polar bears and ringed seals documented in western Hudson Bay and the eastern Beaufort Sea. The contrasting patterns of change in the two areas suggests that different environmental processes may be operating but neither appears to simply represent the periodic extremes of annual variation most often reported from population studies (e.g. Weatherhead 1986). The similarities of these two, relatively longer- and shorter-term, patterns to those reported from seals in the northwestern Hawaiian Islands and in the Antarctic respectively suggest the possibility that the fluctuations in Hudson Bay are caused by a longer-term climatic event and those in the eastern Beaufort Sea by shorter-term oceanographic factors. Nevertheless, it is clear that long-term changes are occurring and can be monitored through species at the top of the ecological pyramid. Over the long term, developing an understanding of the processes involved and a predictive capability of their effects on birds and mammals will be of significant benefit to their conservation and management as well as to the ecosystems of which they are a part.

ACKNOWLEDGEMENTS

We are particularly grateful to the Canadian Wildlife Service, the Polar Continental Shelf Project, the Natural Sciences and Engineering Research Council, the Manitoba Department of Natural Resources, and the NWT Department of Renewable Resources, for their long-term support of our research on polar bears and seals throughout the Canadian Arctic. Additional financial assistance for individual projects was received from the Arctic Petroleum Operators Association, Beaufort Sea Project, Department of Indian and Northern Affairs, Dome Petroleum, Malden Mills, Inuvialuit Game Council, Northern Oil and Gas Assessment Project, World Wildlife Fund (Canada) and the World Society for the Protection of Animals. For access to unpublished data on ringed seals from the eastern Beaufort Sea and polar bears from western Hudson Bay, we thank Lois Harwood, T.G. Smith, the Fisheries Joint Management Committee, Inuvik, NWT, and M.A. Ramsay. Finally, we thank D.S. Andriashek, Wendy Calvert, A.E. Derocher, S. Miller, N.A. Øritsland, M.A. Ramsay, D.B. Siniff, T.G. Smith and Cheryl Spencer for assistance in the field, the laboratory, and discussions of ideas presented in this chapter.

REFERENCES

Atkinson, S.N., Stirling, I. & Ramsay, M.A. (1995). The effect of growth in early life on adult body size in polar bears (*Ursus maritimus*). *Journal of Zoology, London*, **239**, 225–234.

Croxall, J.P. (1992). Southern Ocean environmental changes: effects on seabird, seal and whale populations. *Philosophical Transactions of the Royal Society of London Series B*, **338**, 319–328.

Croxall, J.P., McCann, T.S., Prince, P.A. & Rothery, P. (1988). Reproductive performance of seabirds and seals at South Georgia and Signy Island, South Orkney Islands, 1976–1987: implications for Southern Ocean monitoring studies. *Antarctic Ocean and Resources Variability* (Ed. by D. Sahrhage), pp. 261–285. Springer, Berlin.

Derocher, A.E., Nelson, R.A., Stirling, I. & Ramsay, M.A. (1990). Effects of fasting and feeding on serum urea and serum creatinine levels in polar bears. *Marine Mammal Science*, 6, 196–203.

Derocher, A.E. & Stirling, I. (1992). The population dynamics of polar bears in western Hudson Bay. *Wildlife 2001: Populations* (Ed. by D.R. McCullough & R.H. Barrett), pp. 1150–1159. Elsevier Applied Science, London.

Derocher, A.E. & Stirling, I. (1995). Estimation of polar bear population size and survival in western Hudson Bay. *The Journal of Wildlife Management*, 59, 215–221.

Ganong, W.F. (1991). *Review of Medical Physiology*, 15th edn. Appleton & Lange, Norwalk, Conn.

Hammill, M.O. & Smith, T.G. (1991). The role of predation in the ecology of the ringed seal in Barrow Strait, Northwest Territories. *Marine Mammal Science*, 7, 123–135.

Harwood, L.A. & Stirling, I. (1992). Distribution of ringed seals in the southeastern Beaufort Sea in late summer. *Canadian Journal of Zoology*, 70, 891–900.

Kingsley, M.C.S. (1979). Fitting the von Bertalanffy growth equation to polar bear age–weight data. *Canadian Journal of Zoology*, 57, 1020–1025.

Kingsley, M.C.S. (1986). *Distribution and Abundance of Seals in the Beaufort Sea, Amundsen Gulf, and Prince Albert Sound, 1984*. Environmental Studies Revolving Funds Report, No. 25, Ottawa.

Kingsley, M.C.S. & Byers, T. (1990). *Status of the Ringed Seal in Thesiger Bay, N.W.T., 1987–89*. Report prepared for the Fisheries Joint Management Committee, Inuvik.

Latour, P.B. (1981). Spatial relationships and behaviour of polar bears (*Ursus maritimus* Phipps) concentrated on land during the ice-free season of Hudson Bay. *Canadian Journal of Zoology*, 59, 1763–1774.

Lentfer, J.W., Hensel, R.J., Gilbert, J.R. & Sorensen, F.E. (1980). Population characteristics of Alaskan polar bears. *International Conference on Bear Research and Management*, 4, 109–115.

Lydersen, C., Hammill, M.O. & Ryg, M.S. (1992). Water flux and mass gain during lactation in free-living ringed seal (*Phoca hispida*) pups. *Journal of Zoology, London*, 228, 361–369.

McCormick, P.M., Thomason, L.W. & Trepte, C.R. (1995). Atmospheric effects of the Mt. Pinatubo eruption. *Nature*, 373, 399–404.

McLaren, I.A. (1958). The biology of the ringed seal (*Phoca hispida* Schreber) in the Eastern Canadian Arctic. *Journal of the Fisheries Research Board of Canada Bulletin*, 118, 97 pp.

Messier, F., Taylor, M.K. & Ramsay, M.A. (1994). Denning ecology of polar bears in the Canadian Arctic Archipelago. *Journal of Mammalogy*, 75, 420–430.

Nelson, R.A., Folk, G.E. Jr, Pfeiffer, E.W., Craighead, J.J., Jonkel, C.J. & Steiger, D.L. (1983). Behavior, biochemistry, and hibernation in black, grizzly, and polar bears. *International Conference on Bear Research and Management*, 5, 284–290.

Polovina, J.J., Mitchum, G.T., Graham, N.E., Craig, M.P., Demartini, E.E. & Flint, E.N. (1994). Physical and biological consequences of a climate event in the central North Pacific. *Fisheries Oceanography*, 3, 15–21.

Priddle, J., Croxall, J.P., Everson, I., Heywood, R.B., Murphy, E.J. & Sear, C.B. (1988). Large-scale fluctuations in distribution and abundance of krill – a discussion of possible causes. *Antarctic Ocean and Resources Variability* (Ed. by D. Sahrhage), pp. 169–182. Springer, Berlin.

Ramsay, M.A. & Stirling, I. (1988). Reproductive biology and ecology of female polar bears (*Ursus maritimus*). *Journal of Zoology, London*, 214, 601–634.

Smith, T.G. (1980). Polar bear predation of ringed and bearded seals in the land-fast sea ice habitat. *Canadian Journal of Zoology*, 58, 2201–2209.

Smith, T.G. (1987). The ringed seal, *Phoca hispida*, of the Canadian western Arctic. *Bulletin of the Canadian Journal of Fisheries and Aquatic Sciences*, 216, 81 pp.

Smith, T.G. & Stirling, I. (1978). Variation in the density of ringed seal (*Phoca hispida*) birth lairs in the Amundsen Gulf, Northwest Territories. *Canadian Journal of Zoology*, 56, 1066–1070.

Stirling, I. (1974). Midsummer observations on the behavior of wild polar bears (*Ursus maritimus*). *Canadian Journal of Zoology*, 52, 1191–1198.

Stirling, I. (1990). Polar bears and oil: ecologic perspectives. *Sea Mammals and Oil: Confronting the Risks* (Ed. by J.R. Geraci & D.J. St Aubin), pp. 223–234. Academic Press, New York.

Stirling, I., Andriashek, D. & Calvert, W. (1993). Habitat preferences of polar bears in the Western Canadian Arctic in late winter and spring. *Polar Record*, **29**, 13–24.

Stirling, I. & Archibald, W.R. (1977). Aspects of predation of seals by polar bears in the eastern Beaufort Sea. *Journal of the Fisheries Research Board of Canada*, **34**, 1126–1129.

Stirling, I., Archibald, W.R. & DeMaster, D.P. (1977b). Distribution and abundance of seals in the eastern Beaufort Sea. *Journal of the Fisheries Research Board of Canada*, **34**, 976–988.

Stirling, I., Calvert, W. & Andriashek, D. (1980). Population ecology studies of the polar bear in the area of southeastern Baffin Island. *Canadian Wildlife Service Occasional Paper*, **44**, 33 pp.

Stirling, I. & Derocher, A.E. (1993). Possible impacts of climatic warming on polar bears. *Arctic*, **46**, 240–245.

Stirling, I., Jonkel, C., Smith, P., Robertson, R. & Cross, D. (1977a). The ecology of the polar bear (*Ursus maritimus*) along the western coast of Hudson Bay. *Canadian Wildlife Service Occasional Paper*, **33**, 64 pp.

Stirling, I., Kingsley, M.C.S. & Calvert, W. (1982). The distribution and abundance of seals in the eastern Beaufort Sea, 1974–79. *Canadian Wildlife Service Occasional Paper*, **47**, 23 pp.

Stirling, I. & McEwan, E.H. (1975). The caloric value of whole ringed seals (*Phoca hispida*) in relation to polar bear (*Ursus maritimus*) ecology and hunting behaviour. *Canadian Journal of Zoology*, **53**, 1021–1027.

Stirling, I. & Øritsland, N.A. (1995). Relationships between estimates of ringed seal and polar bear populations in the Canadian Arctic. *Canadian Journal of Fisheries and Aquatic Sciences*, **52**, 2594–2612.

Stirling, I., Pearson, A.M. & Bunnell, F.L. (1976). Population ecology studies of polar and grizzly bears in northern Canada. *Transactions of the North American Wildlife and Natural Resources Conference*, **41**, 421–429.

Taylor, M.K., Larsen, T. & Schweinsburg, R.E. (1985). Observations of intraspecific aggression and cannibalism in polar bears (*Ursus maritimus*). *Arctic*, **38**, 303–309.

Testa, J.W., Oehlert, G., Ainley, D.G., Bengtson, J.L., Siniff, D.B., Laws, R.M. & Rounsvell, D. (1991). Temporal variability in Antarctic marine ecosystems: periodic fluctuations in the phocid seals. *Canadian Journal of Fisheries and Aquatic Sciences*, **48**, 631–639.

Tummers, E.L. (1980). *Heat Budgets of the Southeast Beaufort Sea for the Years 1974 and 1975.* MSc Thesis, Naval Postgraduate School, Monterey, Calif.

Weatherhead, P.J. (1986). How unusual are unusual events? *The American Naturalist*, **128**, 150–154.

8. Trophic interactions in arctic ecosystems and the occurrence of a terrestrial trophic cascade

DAWN R.BAZELY* AND ROBERT L.JEFFERIES†

*Department of Biology, York University, 4700 Keele Street, North York, Ontario, Canada M3J 1P3 and

†Department of Botany, University of Toronto, 25 Willcocks Street, Toronto, Ontario, Canada M5S 3B2

INTRODUCTION

A short growing season with few days above 10°C, restricted availability of nitrogen, phosphorus (and water) and extensive permanent ice cover over much of the central basin of the Arctic Ocean, all combine to limit biological production in arctic terrestrial, fresh-water and marine ecosystems (McCoy 1983; Bliss 1986; Nadelhoffer et al. 1992; Welch et al. 1992). Yet there is considerable abiotic and biotic diversity within each of these systems which is associated with striking differences in primary and secondary productivity.

Where resource turnover and availability are sufficiently high, the systems may be relatively productive and sustain multitrophic food webs, in accordance with food-chain theory, which predicts longer food chains in more productive habitats (Fretwell 1977; Oksanen et al. 1981). Polynyas (leads of open water) in sea ice and marine ice-edge systems are regions where enhanced biological productivity occurs across trophic levels, from phytoplankton to sea birds and mammals (Dunbar 1981; Alexander 1995), some consisting of at least five trophic levels of surprising complexity (Hobson & Welch 1992). In contrast, mean annual net primary production for high arctic polar desert communities is about $1 \, g \, m^{-2} \, year^{-1}$ with low species diversity (Billings 1992). Food chains in two-dimensional habitats, such as polar deserts, tend to be shorter than those of relatively productive three-dimensional habitats, such as low arctic tall-shrub tundra (Lawton 1989; Bliss & Matveyeva 1992) which has 10 times the vascular plant species of polar deserts (Rannie 1986).

Within northern ecosystems stochastic events, which include variability in weather patterns (e.g. the position of the arctic front in summer; Scott 1992), fire, freeze–thaw cycles, snow-banks and storm surges, bring about fluctuations in numbers of primary and secondary producers at different spatial and temporal scales (Walker & Walker 1991). Frequent physical disturbance tends to shorten food chains (Menge & Sutherland 1976), precluding the development of species-rich, highly differentiated, reticulate food webs in which the overall effects of disturbances are buffered (Lawton 1989; Strong 1992) (but see Power, Marks & Parker 1992 for a discussion of successional time scales). In some cases where resource limitation occurs, species diversity is low, food chains are short and trophic relationships can be represented as a ladder rather than a web comprising autotrophs, herbivores, and primary and secondary carnivores (cf. Strong 1992). Terrestrial arctic food chains lend themselves to schematic representation as short trophic ladders, where one trophic level is occupied by one species or a small guild of species with shared susceptibility. These chains may be vulnerable to the effects of 'trophic cascades' (*sensu* Carpenter, Kitchell & Hodgson 1985).

In this chapter we briefly review the concept of a trophic cascade, establish a set of criteria that we consider necessary to demonstrate its presence, and we discuss not only trophic interactions in arctic ecosystems, in general, but also the impact of herbivory by lesser snow geese on arctic coastal ecosystems and the relevance of the trophic cascade model to this particular interaction.

TROPHIC CASCADES: GENERAL CONSIDERATIONS

Trophic cascades have been described as 'runaway consumption and downward dominance through the food chain' (Strong 1992). The biomass of primary producers is severely reduced or destroyed when a herbivore increases dramatically in numbers in the absence of control by species at the next trophic level. A chain-like response follows perturbation of this upper level, leading to strong effects on populations lower down the trophic ladder (Power 1992). The concept of trophic cascade is controversial, since it impinges on the wider question of the extent to which populations are regulated by the availability of nutrients and other resources ('bottom-up' effects) and predation ('top-down' effects). This question has been addressed mostly in aquatic, rather than terrestrial systems. Kitchell and Carpenter (1992) describe the trophic cascade hypothesis as applied to lakes, recognizing potential lake productivity as being set by nutrient inputs, and deviations are hypothesized to stem from food-chain effects, in which the activities of organisms at the top trophic level radiate or 'cascade' down through all trophic levels (Fig. 8.1). Even nutrient availability may be influenced by top-down effects (Kitchell & Carpenter 1992). To date, limnological research into top-down and bottom-up effects has identified many methodological problems and the realization that species interactions extend beyond a particular trophic level resulting in a multiplicity of feedback control processes between different levels.

Trophic cascades are usually considered to result from top-down effects (i.e. predator controlled), but cascades also may operate from the bottom-up (Hunter & Price 1992). For example, interactions between primary producers and their

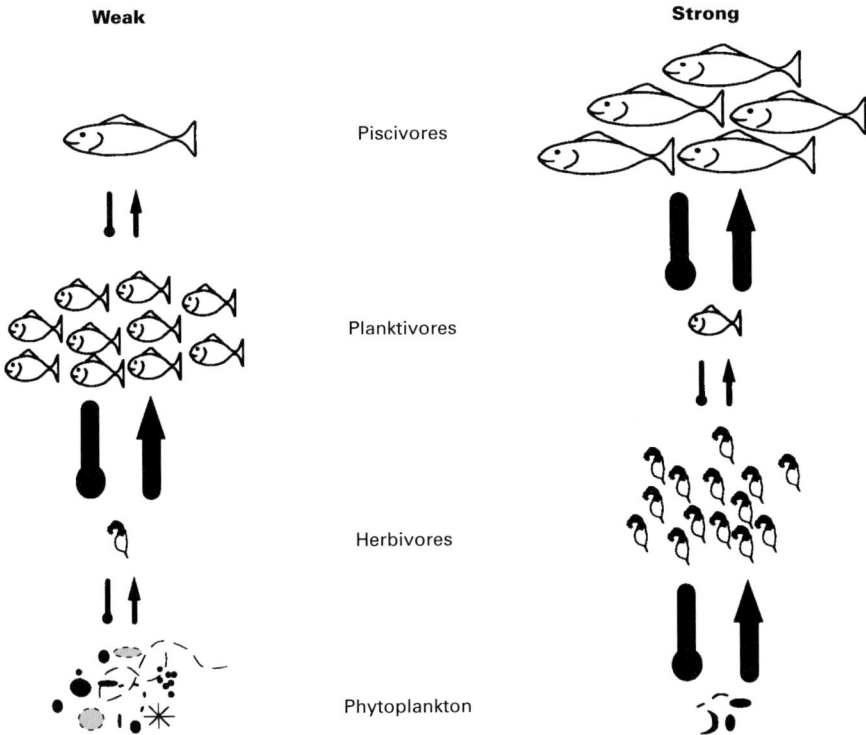

Weak **Strong**

Piscivores

Planktivores

Herbivores

Phytoplankton

FIG. 8.1. Diagrammatic representation of the effects of strong and weak trophic cascades on populations of organisms at different trophic levels in an aquatic ecosystem. Where a strong trophic cascade occurs, 'top-down' effects from predation by piscivores dominate the trophic ladder. In contrast, in a weak cascade, the top-down effects are limited to only part of the trophic ladder and the effects are dissipated over time. In biomanipulation experiments, populations of piscivores are frequently manipulated. (Reproduced by kind permission of C.W. Ramcharan and D.J. McQueen.)

symbionts (e.g. nitrogen fixation), may directly affect species diversity and population turnover at higher trophic levels (Strong *et al.* 1995). Trophic cascades have frequently been investigated by manipulation of piscivorous lake fish populations (Fig. 8.1) (DeMelo, France & McQueen 1992). The effects of nutrient availability and predation are invariably confounded, with both contributing to interactions between organisms, although their relative contributions under a given circumstance are rarely measured (McQueen 1990; DeMelo *et al.* 1992). Increased zooplankton predation of algae (a top-down effect) and decreased nutrient concentrations in the water column (a bottom-up effect) both reduce algal biomass. In addition, nutrients such as nitrogen and phosphorus may be locked up for a number of algal generations in longer-lived zooplankton (Elser 1992). Removal of bream (*Abramis brama*), which cause bioturbation of sediments, reduces nutrient availability in the water through the absence of faeces and suspended particulate matter (Lammens 1988; McQueen 1990). Other confounding effects include differential selectivity of primary producers by herbivores (Haney 1987), shifts in zooplankton size classes leading to changes in feeding behaviour, and omnivory of many secondary producers (Polis 1991). These factors dampen the effects of initially strong top-down processes on lower trophic levels (McQueen 1990; DeMelo *et al.* 1992). Short-term, top-down manipulation of aquatic trophic dynamics can rarely be sustained in the longer term (Mills & Forney 1988). In addition, externally-driven environmental stochasticity, which may be manifested as temporal or spatial heterogeneity, generates variance in primary productivity which may mask the effects of a trophic cascade.

Many studies have tested predictions from both the trophic cascade and bio-manipulation models in aquatic ecosystems (Kitchell 1992; reviewed in DeMelo *et al.* 1992), but few experimental studies have addressed trophic cascades in terrestrial ecosystems (e.g. Kajak, Andrzejewska & Wojcik 1968; Spiller & Schoener 1990; Rosenheim, Wilhoit & Armer 1993; Marquis & Whelan 1994), partly because experimental studies in terrestrial systems tend to explore specific animal–plant interactions (Hunter & Price 1992). In addition, the clear identification of different trophic levels is often problematic, as easy assignment of species to a level may be obscured by omnivory (Spiller & Schoener 1990; Power 1992), although omnivory does not necessarily preclude the occurrence of trophic cascades (Power 1990; Power *et al.* 1992).

In a trophic cascade, occurrence of sustained, chain-like responses by organisms at different trophic levels depends on positive-feedback processes which bring about sudden and pronounced self-amplifying changes of matter; the system acts to reinforce change in the direction of the deviation (Maruyama 1963; DeAngelis, Post & Travis 1986). When such a sustained deviation occurs, negative feedback processes, which decelerate rates of change, either fail to act or respond weakly so that they are unable to restore the system to the original equilibrium. By definition, feedback processes represent effects that are delayed in time (i.e. there is a lag-phase). If the time delays are of sufficient length, the system moves to an alternative state more rapidly than negative feedbacks can respond and a new equilibrium is established (cf. Oksanen 1990).

In the absence of negative feedbacks, chain-like responses associated with a major perturbation generate instability within the system; that is, the system lacks resilience and moves to an alternative state and populations or communities characteristic of the former state disappear or are much reduced in number. These characteristics appear applicable to trophic cascades. A major perturbation, such as that created in a biomanipulation experiment in which piscivores are added to lakes, is associated with a strong positive feedback that sets in motion a series of other positive and negative feedbacks that operate in parallel and are dependent on this primary positive feedback. The feedbacks produce a continuum of responses of organisms which are directly or indirectly affected by the primary perturbation. The responses become more and more attenuated the further processes are distanced spatially and temporally from the primary feedback.

Three of the principles governing positive feedback responses (DeAngelis *et al.* 1986) are particularly relevant to a discussion of trophic cascades. Positive feedbacks accelerate, as a result of the self-amplification of changes in matter, unless a negative feedback operates (e.g. exponential growth). An example is the increasing rate of destruction of the upper layer of soil in the Sahel by wind and water erosion, a consequence of overgrazing and removal of plant cover by domestic livestock (Graetz 1991). As plant cover decreases, increased rates of erosion occur, and plants are unable to re-establish in the impoverished soil, hence the feedback cycle accelerates and the rates of wind and water erosion increase exponentially (Graetz 1991). This example also demonstrates the second principle associated with positive feedbacks, the non-linearity of responses. The outcome of accelerating rates of soil erosion is that the system collapses, which is the third principle (i.e. the system moves to an alternative state). Predictions of the effects of a trophic cascade on different trophic levels based on principles governing positive feedback responses are given in Table 8.1. These predictions are examined further on page 195, where the effects of a terrestrial trophic cascade on a herbivore population (lesser snow geese, *Anser caerulescens caerulescens*) and the coastal vegetation of the Hudson Bay are discussed.

TABLE 8.1. Predictions of the effects of a trophic cascade on different trophic levels based on principles governing positive feedback responses.

- Interactions between organisms can be represented as a trophic ladder as a first approximation where there is only one species or where there is shared vulnerability of species at each trophic level

- A major perturbation driven by a strong positive feedback which affects the organisms in the upper trophic levels is sustained over ecological time

- Effects of the perturbation cascade down a trophic ladder altering population and community structure at lower trophic levels

- The primary positive feedback that initiates the perturbation is likely to be coupled to other secondary positive feedbacks that amplify the overall effects of the cascade. Counteracting negative feedbacks are weak or absent. There are accelerating rates of change, involving both abiotic and biotic threshold responses

- The cascade leads to the destruction or severe modification of the system, and the establishment of an alternative state with the associated changes in community structure

ARCTIC TROPHIC RELATIONSHIPS

Terrestrial ecosystems

Factors limiting secondary production include the amount, availability and suitability of food (McNaughton *et al.* 1991). In arctic tundra ecosystems, which may be relatively diverse (Bliss & Matveyeva 1992), spatial and temporal heterogeneity in the distribution of energy and materials place major constraints upon the transfer of these entities between trophic levels. Plant–herbivore interactions in northern latitudes are driven by nutritional pulses in space and time (Jefferies *et al.* 1992; Jefferies, Klein & Shaver 1994). The mobility of herbivores allows them to exploit such varied food pulses as the inflorescences of *Pedicularis* and *Oxytropis* and the fruiting bodies of fungi. Thus, the degree to which suitable forage species can be exploited influences secondary production. Fluctuations in herbivore populations arise, in part, from such temporal and spatial variation in forage availability, which in turn is frequently influenced by the slow rates of nitrogen turnover in northern soils (Chapin, Johnson & McKendrick 1980; Giblin *et al.* 1991; Nadelhoffer *et al.* 1992). Following defoliation, these slow rates limit regrowth of vegetation. Nevertheless, shoots of some long-lived perennials such as *Eriophorum vaginatum* are able to regrow rapidly after three or four episodes of grazing within a season, because of the presence of large storage reserves (Shaver, Chapin & Gartner 1986) and tussocky growth forms which protect these reserves.

In contrast to habitats where high nutrient turnover rates and regrowth of vegetation cannot be sustained, herbivores are able to influence turnover rates at some sites (Jefferies *et al.* 1994). The likelihood of observing such intervention by herbivores is greatest in early successional communities. The presence of enhanced nitrogen turnover via positive feedbacks characterizes strong plant–herbivore interactions. Such processes are recipient (herbivore)-driven and depend on intensive foraging which accelerates nutrient turnover by maintaining high faecal inputs. In some cases increased nitrogen-fixation in grazed sites also contributes to increased turnover and availability (Schultz 1968; Jefferies, Bazely & Cargill 1986; Pastor, Naiman & Dewey 1987; Jefferies 1988a,b; Pastor *et al.* 1988). Such interactions involving positive feedbacks are inherently unstable and do not continue indefinitely (DeAngelis *et al.* 1986).

Bliss (1986) has estimated the herbivore-carrying capacity of arctic tundra at the landscape scale ($1000\,\mathrm{km}^{-2}$ units). The major herds of caribou (*Rangifer tarandus*) in the western and central North American Arctic occur at average densities varying from 0.50 to 0.35 animals km^{-2}. Comparable values for Peary caribou (*R. t. pearyi*) in the Canadian Arctic Archipelago are between 0.04 and 0.30 animals km^{-2}. Estimates of muskoxen (*Ovibos moschatus*) densities range from 0.03 to 0.30 animals km^{-2}. In contrast, the calculated carrying capacity of herbivores for different plant communities in the High and Low Arctic is between five and ten times the above values, based on net above-ground primary production, rates of forage consumption and body weight of animals (Bliss 1986). Differences between the observed and estimated values may be attributed to density-independent winter mortality caused by factors such as deep snow and ice crusts. The net effect is to reduce these herbivore populations far below the carrying capacity of tundra habitats.

Three trophic levels may be recognized in some arctic terrestrial ecosystems where carnivores exploit increased herbivore populations arising from transient bursts of primary production. However, since interactions among organisms are so numerous, the trophic levels are rarely discrete, and most animals are generalists in their feeding habits and are omnivorous. For example, top-level carnivores such as polar bears (*Ursus maritimus*) eat a variety of different types of vegetation including fruits and berries; likewise, herbivores such as caribou eat bird eggs and geese eat insects. Additionally, in these pulse-regulated systems, where grazing is episodic and may be coupled with a flowering event, flows of energy and nutrients alternate between a detritivore food web (in which plant tissue dies and decomposes) and a herbivore-based food web. For much of the time there may be only one well-represented trophic level (excluding the soil microbial trophic level). In polar desert communities, for example, where net primary production is as low as $1\,g\,m^{-2}$ per year, grazers are limited or absent. Hence, arctic ecosystems are basically plant-decomposer systems (Bliss 1986). A similar conclusion was reached by Oksanen *et al.* (1996) in their study of grazing webs in the Scandinavian subarctic.

There have been few formal studies of energy transfer among trophic levels in arctic tundra ecosystems (Whitfield 1977; MacLean 1980; Chapin *et al.* 1980; Bliss 1986). Results from different plant communities on Devon Island, Canada (Whitfield 1977; Bliss 1986) indicated that 98–99% of the standing crop comprised plant material and that 90–91% of net production was attributable to plants (Fig. 8.2). Saprovores contributed up to 0.6% of the standing crop and up to 2.7% of net production, while comparable figures for the microflora were 0.9% and 8.3% respectively. The cushion-plant subsystem gave some of the highest production figures (Bliss 1986). Herbivores and carnivores comprised only 0.1% of the standing crop, but accounted for up to 1.6% of the net production in sedge–moss meadow and cushion-plant communities (Whitfield 1977; Bliss 1986). Above-ground vertebrate herbivores and carnivores were responsible for most of the net secondary production (Bliss 1977). Hence, the saprovores (Protozoa, Rotifera, Tardigrada, Enthytraeidae, Crustacea, Collembola and Diptera) were a much larger trophic unit than either vertebrates or carnivores. Invertebrates as a whole assimilated about 5% of the total production of these communities, which was about four times greater than assimilation by all vertebrates (Ryan 1977). Although they were inactive for up to 10 months a year, the below-ground invertebrates were the major animal

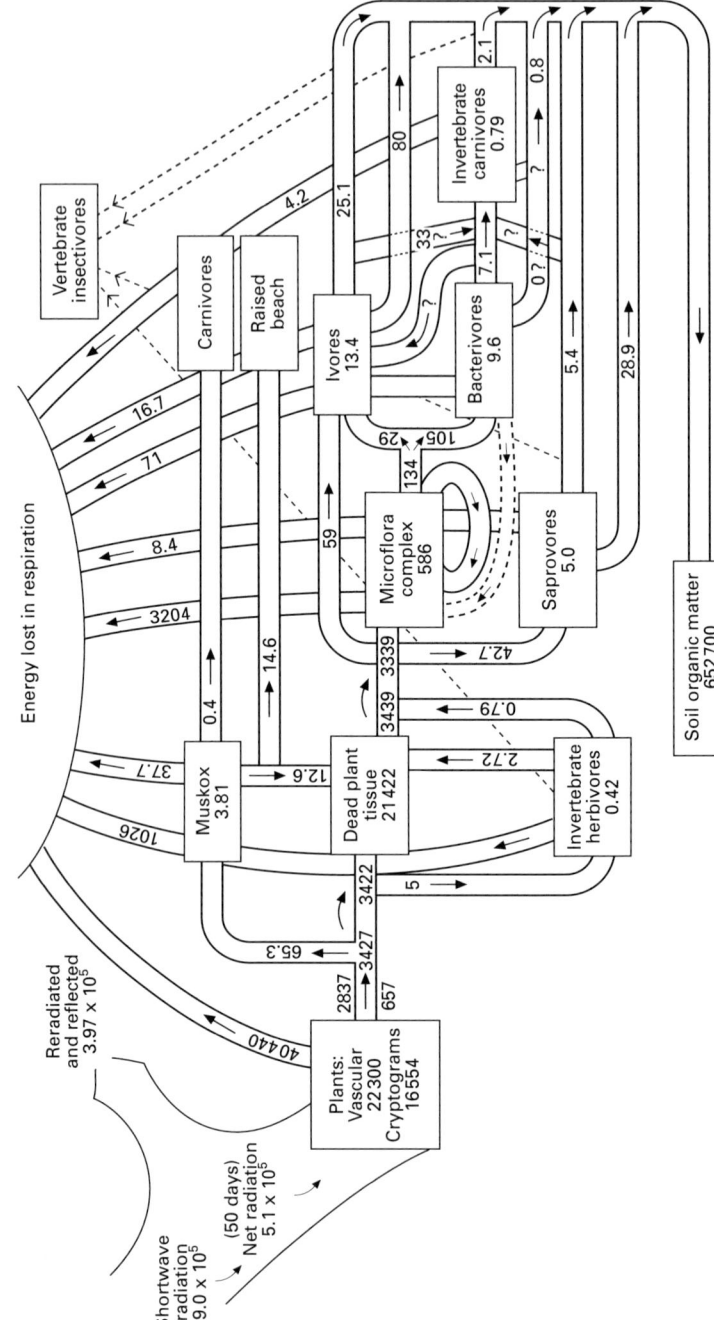

FIG. 8.2. Energy flow diagram for the wet sedge–moss ecosystem, Truelove Lowland, Devon Island, Canada. Standing crop (boxes) and energy flow (arrows) are expressed in kJ m^{-2}. Incoming energy is for the length of the growing season. Ivores are fungal, bacteria and protozoan feeders. Vertebrate insectivores are seed and insect feeding birds. (After Whitfield 1977 and Bliss 1986. Published with permission from L.C. Bliss and Plenum Press.)

group involved in energy transfer (Ryan 1977). At Barrow, Alaska, data for plant standing crop and net production were generally similar to that described above, although the microflora was the second most important group, followed by the saprovores and soil invertebrate herbivores (Chapin *et al.* 1980; MacLean 1980). The combined invertebrates assimilated only 1.4% of the total net production, a value three times greater than that of vertebrates.

Aquatic ecosystems

Establishing trophic relationships within aquatic communities is 'a daunting task' (Hobson & Welch 1992). This is especially so when it is not possible to sample food webs throughout the year because of ice cover, and when infrequent sampling renders trophic models susceptible to the limitations of both temporal and spatial scaling. The use of stable isotope analysis, in which the abundances of $\delta^{13}C$ and $\delta^{15}N$ are measured, allows determination of trophic relationships and the resolution of some of these problems; there have been a number of such studies in both fresh-water and marine arctic ecosystems (Fry & Sherr 1988; Owens 1988). Under certain circumstances the ^{15}N isotope undergoes a predictable stepwise enrichment between prey and consumer tissues with increasing trophic level (Miyake & Wada 1967; Owens 1988). It is thus possible to determine relative trophic positions within aquatic (and terrestrial) ecosystems by measuring isotopic abundances in prey and consumers. In addition, stable-isotope measurements of carbon, nitrogen and sulphur have been used to determine the strength of food-web connections (DeNiro & Epstein 1978; Peterson & Fry 1987). The trophic positions established from these data represent long-term averages of food assimilated (Tieszen *et al.* 1983; Hobson & Welch 1992).

In arctic fresh-water ecosystems few investigators have used multiple species of isotopes to analyse trophic relationships. Kling, Fry and O'Brien (1992) examined planktonic food-web structure in eight oligotrophic lakes in Alaska using carbon and nitrogen isotope analyses of algae and zooplankton. The dominant zooplankton were the herbivorous copepod, *Diaptomus pribilofensis*, and the large predaceous copepod, *Heterocope septentronatis*. *Diaptomus* is exclusively herbivorous but *Heterocope* may eat both phytoplankton and *Diaptomus* (Luecke & O'Brien 1983). The general enrichment of ^{15}N in a predator relative to its prey was found to average 3.4‰ across a variety of systems, and was caused by the preferential ex-cretion of the lighter ^{14}N as a by-product of protein synthesis (Minagawa & Wada 1984).

Kling *et al.* (1992) found that realized trophic interactions varied among lakes. In some lakes, the planktonic food web was based on the classic three-tiered trophic structure (phytoplankton, *Diaptomus*, *Heterocope*); $\delta^{15}N$ values were *c.* 3‰ enriched at each trophic level. The analyses excluded fish and microheterotrophs which, if included, would have extended the trophic structure to at least four levels in some lakes. In other lakes, the step-wise enrichment of ^{15}N was less clear-cut. There was a continuum in the relative enrichment of ^{15}N between *Diaptomus* and *Heterocope* that was caused by *Heterocope* foraging on organisms, such as phytoplankton, lower down the food chain. Thus, in these situations, the realized, as opposed to potential, food web had only slightly more than two trophic levels and the δ^{15}N values indicated the degree of omnivory.

The presence of only two or three realized planktonic trophic levels in arctic lakes (Kling *et al.* 1992) is supported by the findings of Gu, Schell and Alexander (1994) for an Alaskan subarctic lake, but contrasts greatly with results from recent studies of temperate lakes in which five to six potential trophic levels have been observed (excluding fish and microheterotrophs) (Sprules & Bowerman 1988). However, since detailed data on realized trophic interactions are sparse, generalizations are difficult to make (Kling *et al.* 1992). There were no obvious patterns in δ^{15}N values either within season or among animal species in two of the lakes studied. Elsewhere, in high latitude humic lakes, inputs of terrestrial detritus are an important source of carbon for zooplankton (Hessen, Andersen & Lyche 1990) but this was not significant in the lakes described above. As Kling *et al.* (1992) emphasize, their studies raise the question of why food webs in similar lakes containing the same species are structured differently.

Results from experimental and observational studies of arctic lake ecosystems have provided considerable evidence for bottom-up control of phytoplankton density and production (Whalen & Alexander 1984; O'Brien *et al.* 1992) and the heterotrophic microplankton community (Rublee 1992). Increased production from addition of nutrients stimulated zooplankton after 2 years but had little effect on the benthos or sediments (O'Brien *et al.* 1992). Fish manipulations influenced the densities of large zooplankton but had little effect on small-bodied zooplankton (Kling *et al.* 1992; O'Brien *et al.* 1992).

Stable isotope analyses have been used to examine food-web patterns in a number of arctic marine food webs (McConnaughey & McRoy 1979; Dunton *et al.* 1989). However, only one study, conducted in Lancaster Sound in the Canadian High Arctic, included higher-level carnivores such as whales, seals, sea birds and polar bears and indicated that there was a considerable range in both carbon and nitrogen stable-isotope ratios throughout the food web (Fig. 8.3). (Hobson & Welch 1992; Welch *et al.* 1992). Their isotopic model (Fig. 8.3) confirms previous suggestions of the occurrence of five trophic levels at such sites (Bradstreet & Cross 1982), although an analysis of the benthic food web may reveal a longer food chain (McConnaughey & McRoy 1979).

The mean δ^{15}N value of 5.4‰ for particulate organic matter was similar to average values found at other northern ocean sites (Minagawa & Wada 1984; Dickson 1986). Overall, stable-isotope ratios of nitrogen varied from 4.9‰ in

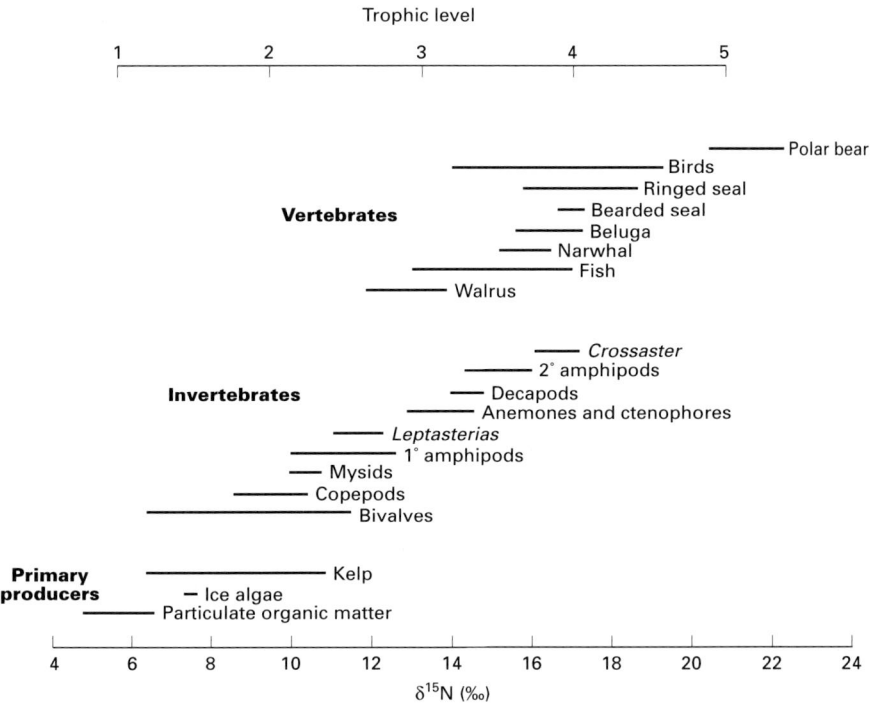

FIG. 8.3. Ranges of δ^{15}N values for marine organisms from the Barrow Strait–Lancaster Sound region (Northwest Territories, Canada) and their associated trophic positions according to an isotopic model using a trophic enrichment value of +3.8‰ (not applicable to marine birds). (Published with permission from H.E. Welch and *Marine Ecology Progress Series.*)

particulate organic matter to 21.9‰ for polar bear muscle. Patterns varied between stable isotopes: δ^{15}N values showed a steady enrichment with increasing trophic position, unlike δ^{13}C values. In the Sound, polar bears which feed almost exclusively on ringed seals (*Phoca hispida*) (Stirling & McEwan 1975) showed at +3.8‰ enrichment in δ^{15}N compared with the corresponding value for seals (Hobson & Welch 1992). A similar difference was found between the copepod, *Calanus hyperboreus*, and particulate organic matter, its primary food. In the case of piscivorous birds, the δ^{15}N isotopic fractionation factor between bird diet and muscle tissue was less than the stepwise increment of 3.8‰, possibly because the birds excrete uric acid rather than urea (Mizutani, Kabaya & Wada 1991; Hobson & Clark 1992). In all other components of the marine food web, an enrichment factor of 3.8‰ was found between prey and predator, although invertebrates such as copepods, barnacles and bivalves, and echinoderms and amphipods, which are predators, occupied a broad range of isotopic values, as they also feed on suspended particles in the water column (Hobson & Welch 1992).

Some organisms in this food web have very varied diets, as confirmed by the isotopic analysis. For example, arctic cod (*Boreogadus saida*) feed on a mixed diet of copepods and amphipods (Bradstreet *et al.* 1986) and although age–class differences are evident, the isotopic data indicate that the primary diet is amphipods. Beluga whales feed on cod (Bradstreet *et al.* 1986), but the isotopic data indicate that there is a substantial component of the diet that consists of prey of lower trophic levels. A mixed-diet strategy appears to be characteristic of narwhal (*Monodon monoceros*), ringed seals and sea birds (Gaston & Nettleship 1981; Bradstreet & Cross 1982; Bradstreet *et al.* 1986).

These $\delta^{15}N$ studies (Hobson & Welch 1992) highlighted the importance of both invertebrates and arctic cod in nutrient (and energy) transfer. These aggregations of organisms from different trophic levels, which transfer energy to higher trophic levels, reflect the intra-annual cycle of energy flow caused by pulsed primary production and subsequent respiration (Welch *et al.* 1992). Welch *et al.* (1992) concluded that there was more than adequate primary production to supply the food web and support sea mammals and birds. Where higher vertebrates were present, food chains were generally long (five trophic levels), which may explain the relatively high concentrations of organochlorine contaminants in arctic marine mammals compared with concentrations in water and plankton (Muir *et al.* 1992; Alexander 1995).

Various gaps remain in our understanding of marine trophic relationships. Some aspects of the planktonic and benthos food webs have not been investigated, and the role of water currents in regulating autochthonous and allochthonous fluxes of materials is poorly understood (Welch *et al.* 1991, 1992). Finally, the transfer of energy between organisms in the upper trophic levels (i.e. cod, sea birds and sea mammals) requires additional studies (but see Stirling, Archibald & DeMaster 1977; Stirling, Kingsley & Calvert 1982).

In conclusion, arctic aquatic food chains are long, especially in marine environments. This is unusual, given the overall low productivity per unit area (Welch *et al.* 1992). This apparent paradox may reflect the pulse-regulated nature of both these aquatic ecosystems, and their terrestrial counterparts, whereby seasonal resource acquisition and population growth are highly restricted in time. During unfavourable periods for growth and reproduction, low maintenance costs (or migration) enable populations to survive. For example, the weights of adult male polar bears peak in late spring or early summer when they feed on ringed seal pups, thereafter declining for the remainder of the annual weight cycle (M.A. Ramsay personal communication). This 'idling' survival strategy, which is shown by a large number of northern organisms (Jefferies *et al.* 1992), may allow extended food chains to occur because high-energy demands by organisms do not occur year round. In addition, Oksanen *et al.* (1996) have pointed out that many top-level predators utilize resources from different biomes (e.g. polar bears). The pulse-regulated feeding strategy is analogous to the pulse-regulated primary production of arid/semi-arid environments (Westoby 1979/80). There is a phasing of feeding bouts with events in order to meet dietary requirements of animals at specific stages of development (Jefferies, Klein & Shaver 1994).

ARCTIC BREEDING GEESE AND THE OCCURRENCE OF A TERRESTRIAL TROPHIC CASCADE

Population size and habitat use

McNaughton (1984) suggested that flocking behaviour in terrestrial vertebrate herbivores maintains intense levels of herbivory which influences the growth of forage plants and leads to the production of grazing lawns. Many species of arctic breeding geese feed in flocks, either on their breeding, wintering or staging grounds (Cramp & Simmons 1977). While intense herbivory does not result necessarily in the initiation of trophic cascades, the foraging activities of large colonial populations of geese may be expected to exert strong top-down effects (Cargill & Jefferies 1984b).

Population growth rates of many species of arctic breeding geese (but not all) have increased significantly over the past 30 years (Ogilvie & St Joseph 1976; CWS, USFWS & Atlantic Flyway Council 1981; Boyd & Pirot 1989; Owen & Black 1991; Madsen 1991; Fox, Boyd & Warren 1992). A number of factors have contributed to this increase: decreased hunting pressure, the provision of reserves and refuges, changes in agricultural land-use patterns resulting in increased forage availability on wintering and staging grounds (an energy and nutrient subsidy), and the recent exploitation of additional natural habitats by geese (Ogilvie & St Joseph 1976; CWS *et al.* 1981; Owen & Black 1991; Fox *et al.* 1992; Fox 1993). While the relative importance of these different factors varies among goose populations and species, many of them are ultimately the result of human activities. Their combined effect represents a biomanipulation of arctic breeding goose populations on a massive scale. In the past, European and North American goose populations were most likely limited both by food availability on their wintering grounds and by hunting pressure (Ogilvie & St Joseph 1976; Owen & Black 1991). In addition, populations of geese have exploited alternative forage sources. During the 1970s in North America, greater snow geese (*Anser caerulescens atlanticus*) started to forage in *Spartina* marshes and farm fields on their staging grounds (CWS *et al.* 1981; Bédard, Nadeau & Gauthier 1986).

The burgeoning numbers of geese have attracted great interest with respect to their impact on agricultural land, where they have come into conflict with farmers (Bédard *et al.* 1986; Owen 1990) and in wetland habitats in both their wintering grounds and arctic breeding grounds (Lieff 1973; Giroux & Bédard 1987; Kerbes,

Kotanen & Jefferies 1990). Studies have examined the impact of goose grazing (on above-ground plant parts) and grubbing (of roots and rhizomes) on both standing crop and on net primary production (e.g. Smith & Odum 1981; Cargill & Jefferies 1984b; Bazely & Jefferies 1986; Giroux & Bédard 1987; Belanger & Bédard 1994). From the perspective of trophic cascades, interactions of interest between geese and their forage species are those in which geese may be acting as keystone herbivores (Kerbes, Kotanen & Jefferies 1990; Strong 1992). In addition, the growth responses of forage plants to different grazing intensities are invariably non-linear. In these cases the potential arises for the establishment of positive and negative feedbacks (DeAngelis *et al.* 1986). These types of interactions can only be identified from studies examining the effects of different foraging intensities on above- or below-ground primary production of plant communities.

Impact of goose foraging in natural habitats

Natural habitats grazed by geese include arctic sedge meadows (Kotanen & Jefferies 1987; Jefferies 1988a; Gauthier 1991), arctic and temperate salt marshes (Prins, Ydenberg & Drent 1980; Smith & Odum 1981; Cargill & Jefferies 1984a,b; Madsen & Mortensen 1987; Madsen 1989; Bazely, Ewins & McCleery 1991) and estuarine marshes (Giroux & Bédard 1987; Belanger & Bédard 1994). The species richness of these arctic salt-marsh communities tends to be low, and with the exception of the geese there are no other major herbivores. Hence, one of the criteria for the initiation of trophic cascade, namely the presence of simplified food webs or trophic ladders, appears to be met. In addition, amounts of plant biomass in some of these habitats are low, and the geese frequently remove large portions of the annual primary production. For example, greater snow geese grazing on wet-meadow vegetation on Bylot Island, Canada removed from 30 to 64% of the net above-ground primary production (NAPP) of *Dupontia fisheri* and from 65 to 85% of *Eriophorum scheuchzeri* vegetation (Gauthier 1991). Lesser snow geese feeding on salt marshes dominated by a grass, *Puccinellia phryganodes*, and a sedge, *Carex subspathacea*, on the Hudson Bay Coast, near Churchill, Canada, removed up to 90% of NAPP (Cargill & Jefferies 1984b). Staging brent geese (*Branta bernicla*) on the Danish island of Langli were estimated to have consumed from 47 to 97% of NAPP of salt-marsh swards dominated by *Puccinellia maritima* (Madsen 1989). Grubbing by geese also results in the removal of large amounts of below-ground biomass. Greater snow geese grubbing for roots and rhizomes in North Carolina marshes dominated by *Spartina alterniflora, S. patens, Distichlis spicata* and *Scirpus robustus*, removed 50–64% of the below-ground biomass (Smith & Odum 1981). In the St Lawrence estuary geese removed 19–60% of net below-ground primary production (NBPP) of rhizomes of *S. americanus* (Giroux & Bédard 1987).

The impact of high levels of off-take on primary production by foraging geese is variable. Grazing by lesser snow geese on salt marshes on Hudson Bay removed up to 90% of NAPP, resulting in very low biomass levels, but overall, the NAPP of grazed swards dominated by *Puccinellia phryganodes* increased (Cargill & Jefferies 1984b; Hik & Jefferies 1990; Hik, Sadul & Jefferies 1991). In experiments with *Puccinellia phryganodes*, compensatory growth following a grazing episode by captive goslings resulted in higher levels of above-ground biomass compared with that of ungrazed swards (Hik & Jefferies 1990). In contrast, compensatory growth in *Carex subspathacea*, across a range of grazing intensities, resulted in similar biomass levels to those of ungrazed swards (Zellmer *et al.* 1993). In the same area, when geese grazed on swards of *Festuca rubra* and *Calamagrostis deschampsioides* there was little, if any, compensatory growth within the season, so that 44 days after the grazing episode in plots of *Festuca rubra*, above-ground biomass was negatively correlated with grazing pressure (Zellmer *et al.* 1993). Grazing by greater snow geese on Bylot Island also had no effect on NAPP, except it was depressed at very high grazing pressures (Gauthier 1991). In temperate salt marshes on the Dutch island of Schiermonnikoog, grazing by barnacle geese (*Branta leucopsis*) on *Festuca rubra* swards did not affect NAPP in some sites but at others NAPP increased (D.R. Bazely unpublished data). In some of the same sites, growth rates of *Plantago maritima* plants grazed by brent geese were significantly greater than those of ungrazed plants (Prins, Ydenberg & Drent 1980).

Positive feedbacks initiated by lesser snow geese

The intense foraging activities of both greater and lesser snow geese exert strong top-down effects which alter plant community structure (Smith & Odum 1981; Bazely & Jefferies 1986; Giroux & Bédard 1987; Belanger & Bédard 1994). Grubbing by greater snow geese significantly increased the heterogeneity of *Scirpus* marshes in South Carolina, and destroyed large portions of the habitat (Smith & Odum 1981). Lesser snow geese have been described as a keystone herbivore in their arctic breeding grounds (Kerbes *et al.* 1990). In both fresh-water and salt marshes, herbivory by lesser snow geese has not only altered plant community composition (Bazely & Jefferies 1986; Hik, Jefferies & Sinclair 1992) but also has modified rates of nutrient cycling and other soil processes (Jefferies *et al.* 1986, 1995; Wilson 1993). Top-down effects, initiated directly and indirectly by human management, have allowed the lesser snow goose population to grow, resulting in major changes in populations of organisms at lower trophic levels.

The number of lesser snow geese breeding at La Pérouse Bay has increased from 1800 pairs in 1968 to over 8000 pairs in 1988 (Cooch & Cooke 1991) and by 1991 it reached an estimated 23000 pairs (R.H. Kerbes unpublished data). Lesser snow geese are colonial breeders and following hatch, in early- to mid-June, they

Positive feedback leading to increased primary production of forage plants

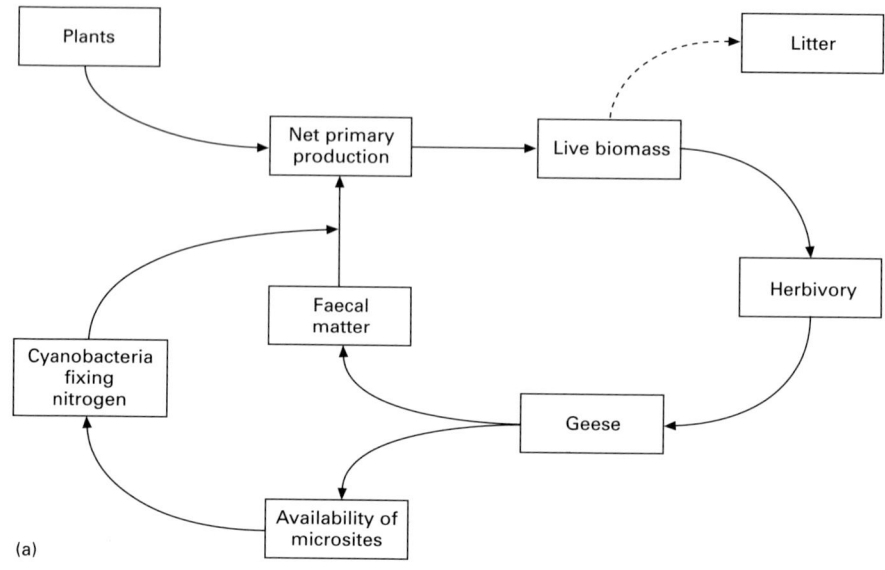

(a)

Positive feedback leading to decreased primary production of forage plants

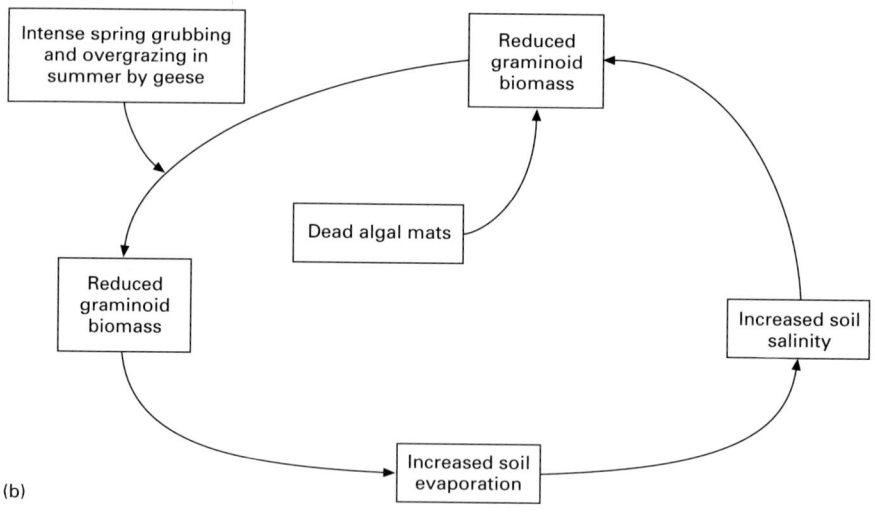

(b)

forage intensively. Goslings increase in weight from 80 g to 1500 g (fledging mass) in 7 weeks (Cooke, Rockwell & Lank 1995). During the posthatch period, goose grazing leads to increased biomass production and tissue turnover of preferred forage species via a positive feedback (Cargill & Jefferies 1984b; Kotanen & Jefferies 1987; Bazely & Jefferies 1989b). In the absence of grazing, growth of these salt-marsh swards dominated by *Puccinellia phryganodes* and *Carex subspathacea* is nitrogen limited (Cargill & Jefferies 1984a). Goose grazing increases nitrogen availability and NAPP in grazed compared with ungrazed swards via this positive feedback that involves two mechanisms (Fig. 8.4). First, within a season, availability is increased by rapid nitrogen recycling in goose droppings (Bazely & Jefferies 1985; Ruess, Hik & Jefferies 1989). Passage of forage through the gut is fast; geese defecate every 4–5 min and fresh goose droppings contain high levels of soluble nitrogen. Experimental addition of goose droppings to grazed and ungrazed swards results in increased biomass production (Bazely & Jefferies 1985). Intense grazing also maintains open swards with bare microsites which are colonized by cyanobacteria that fix nitrogen, particularly early in the season (Bazely & Jefferies 1989a). Thus, goose grazing initiates a positive feedback in which the increased growth of swards of *Puccinellia* under grazed conditions is driven by increased nitrogen availability. The onset of fall and the departure of the geese from the area act as negative feedbacks that restrict further plant growth within the season.

A series of late spring thaws during the 1980s in the northern Hudson Bay region, coincident with the increasing goose population, has resulted in the presence of large numbers of staging snow geese at La Pérouse Bay during the pre- and early nesting period (Jefferies *et al.* 1995). At this time, just prior to above-ground plant growth, both staging (birds *en route* to northern breeding sites) and nesting geese grub for roots and rhizomes of preferred salt-marsh graminoids. This initiates another positive feedback process (Fig. 8.4) that has led to the destruction of the salt-marsh swards (Jefferies 1988a,b; Kerbes *et al.* 1990; Iacobelli & Jefferies 1991; Srivastava & Jefferies 1996). This particular series of trophic interactions meets all the criteria for a trophic cascade in which top-down effects, initiated by human management, have allowed the lesser snow goose population to grow, resulting in major changes in populations at lower trophic levels. The destruction of salt-marsh

---

FIG. 8.4. (*facing page*) The interactions between lesser snow geese and the coastal salt-marsh vegetation on which they forage at La Pérouse Bay, Manitoba can be represented as two positive feedbacks. In the first positive feedback (a), grazing by geese results in increased net above-ground primary production of forage species as a result of the availability of faecal nitrogen for plant growth and enhanced rates of nitrogen fixation by cyanobacteria growing on sediments where swards are heavily grazed. In the second positive feedback (b), removal of the vegetation cover by grubbing exposes the sediments leading to increased evaporation and hypersalinity in the soil which decreases plant growth. Summer grazing and algal mats which develop in spring both reduce available forage and accelerate the process.

swards dominated by *Puccinellia* and *Carex* has escalated, due to the establishment of a positive feedback associated with the development of hypersaline soil conditions following removal of surface vegetation by grubbing (Srivastava & Jefferies 1995a, 1996). The development of hypersaline conditions has led also to the death of willows (Iacobelli & Jefferies 1991). Once the surface vegetation is stripped away, increased evaporation from the bare soil occurs, which draws inorganic salts to the surface from underlying marine clays deposited when this region was the Tyrell Sea. The hypersaline soil conditions depress the growth of *Puccinellia* and *Carex*, which fail to recolonize patches of bare, hypersaline soil, at least in the short term (< 10 years) (Srivastava & Jefferies 1995b; R.L. Jefferies unpublished data). Hence, over this time patches of remaining vegetation continue to diminish in size as a result of grubbing each spring, heavy grazing in summer and the inability of plants to re-establish in hypersaline conditions. Ultimately, an alternative state of bare mud is reached, which acts as an increasingly strong negative feedback on plant and animal growth at the local level.

Thus, at La Pérouse Bay the early spring grubbing activities of both the nesting and staging populations of large numbers of lesser snow geese have resulted in the destruction of the preferred posthatch foraging habitats. Soil salinity levels are higher in areas denuded of vegetation and rates of soil nitrogen mineralization are significantly lower (Wilson 1993). This 'degenerative' positive feedback with negative consequences for vegetation growth is in operation at La Pérouse Bay and elsewhere along the coasts of Hudson and James Bays.

The immediate impact of the feedback on the lesser snow goose population is that during the posthatch period some goose families walk up to 35 km from La Pérouse Bay to marshes where food is still available and goslings can grow to 1500 g at fledging (Cooch *et al.* 1993). In contrast, the body mass of goslings reared within the bay has decreased significantly over the past 15 years. Cooch *et al.* (1991) and Francis *et al.* (1992) have shown that the reduced gosling size leads to reduced gosling survivorship and increased mortality between fledging and the time the birds are hunted. The poor growth of goslings may also be a consequence of the birds feeding on alternative, less nutritious forage in the vicinity of the bay (Gadallah & Jefferies 1995a,b). Negative feedbacks are operating to restrict population growth, at least at the local level.

The vegetation at La Pérouse Bay is unlikely to re-establish in the foreseeable future. Isostatic uplift, soil erosion, hypersalinity and the renewed grubbing each year limit the development of swards. Some exclosures have been devoid of vegetation for 12 years (R.L. Jefferies unpublished data). There is anecdotal evidence that coincident with these changes there have been declines in populations of soil invertebrates, and some species of shore birds and ducks such as wigeon, which are grazers. As a result of similar processes to those occurring at La Pérouse Bay, sites at McConnell River (NWT) on the west coast of Hudson Bay have been stripped of vegetation and the soil has been eroded to reveal underlying glacial debris (Kerbes *et al.* 1990).

CONCLUSIONS

There are relatively few documented cases of trophic cascades operating in terrestrial habitats which involve herbivores feeding on vascular plants. However, the positive feedbacks initiated by lesser snow geese on their subarctic breeding grounds are examples of self-amplifying changes which are resulting in the destruction of the existing plant communities. This plant–herbivore interaction is unstable in the long term and is the result of the biomanipulation of goose populations arising from human management. As pointed out by McQueen *et al.* (1992), those biomanipulations which resulted in strong trophic cascades in lake communities were massive, and completely restructured the aquatic communities. The observations for lesser snow geese constitute a terrestrial parallel to the limnological studies of biomanipulation and trophic cascades.

ACKNOWLEDGEMENTS

We thank D. McQueen, M. Power, D. Strong, P. Lundberg, J. van der Koppel and B. Welch for help, good advice and discussions during the preparation of this manuscript. In addition we thank two anonymous referees for their comments. Both the authors also thank the Natural Sciences and Engineering Research Council of Canada for financial support for these studies. Finally, many thanks to M. Stasiuk and C. Siu who typed various versions of the manuscript in record time.

REFERENCES

Alexander, V. (1995). The influence of the structure and function of the marine food web on the dynamics of contaminants in Arctic Ocean ecosystems. *The Science of the Total Environment*, **160/161**, 593–603.

Bazely, D.R., Ewins, P.J. & McCleery, R.H. (1991). Possible effects of local enrichment by gulls on feeding-site selection by wintering barnacle geese *Branta leucopsis*. *Ibis*, **133**, 111–114.

Bazely, D.R. & Jefferies, R.L. (1985). Goose faeces: a source of nitrogen for plant growth in a grazed salt marsh. *Journal of Applied Ecology*, **22**, 693–703.

Bazely, D.R. & Jefferies, R.L. (1986). Changes in the composition and standing crop of salt marsh communities in response to the removal of a grazer. *Journal of Ecology*, **74**, 693–706.

Bazely, D.R. & Jefferies, R.L. (1989a). Lesser snow geese and the nitrogen economy of a grazed salt marsh. *Journal of Ecology*, **77**, 24–34.

Bazely, D.R. & Jefferies, R.L. (1989b). Leaf and shoot demography of an arctic stoloniferous grass, *Puccinellia phryganodes*, in response to grazing. *Journal of Ecology*, **77**, 811–822.

Bédard, J., Nadeau, A. & Gauthier, G. (1986). Effects of spring grazing by greater snow geese on hay production. *Journal of Applied Ecology*, **23**, 65–75.

Belanger, L. & Bédard, J. (1994). Role of ice scouring and goose grubbing in marsh plant dynamics. *Journal of Ecology*, 82, 437–445.

Billings, W.D. (1992). Phytogeographic and evolutionary potential of the arctic flora and vegetation in a changing climate. *Arctic Ecosystems in a Changing Climate: An Ecophysiological Perspective* (Ed. by F.S. Chapin III, R.L. Jefferies, J.F. Reynolds, G.R. Shaver & J. Svoboda), pp. 91–109. Academic Press, San Diego.

Bliss, L.C. (1977). General summary, Truelove Lowland ecosystem. *Truelove Lowland, Devon Island, Canada: A High Arctic Ecosystem* (Ed. by L.C. Bliss), pp. 657–675. University of Alberta Press, Edmonton.

Bliss, L.C. (1986). Arctic ecosystems: their structure, function, and herbivore carrying capacity. *Grazing Research at Northern Latitudes* (Ed. by O. Gudmundsson), pp. 5–26. Plenum Press, New York.

Bliss, L.C. & Matveyeva, N.V. (1992). Circumpolar Arctic vegetation. *Arctic Ecosystems in a Changing Climate: An Ecophysiological Perspective* (Ed. by F.S. Chapin III, R.L. Jefferies, J.F. Reynolds, G.R. Shaver & J. Svoboda), pp. 59–89. Academic Press, San Diego.

Boyd, H. & Pirot, J.Y. (Eds) (1989). *Flyways and Reserve Networks for Water Birds.* Canadian Wildlife Service: IWRB Special Publication No. 9. IWRB, Slimbridge.

Bradstreet, M.S.W. & Cross, W. (1982). Trophic relationships at high Arctic ice edges. *Arctic*, 35, 1–12.

Bradstreet, M.S.W., Finely, K.J., Sekerak, A.D., Griffiths, W.B., Evans, C.R., Fabijan, M. & Stallard, H.E. (1986). *Aspects of the Biology of Arctic Cod* (Boreogadus saida) *and its Importance in Arctic Marine Food Chains.* Canadian Technical Report, Fisheries & Aquatic Science, No. 1491. Department of Fisheries & Oceans, Ottawa.

Canadian Wildlife Service, US Fish and Wildlife Service & Atlantic Flyway Council (1981). *A Greater Snow Goose Management Plan.* Environment Canada, Ottawa.

Cargill, S.M. & Jefferies, R.L. (1984a). Nutrient limitation of primary production in a sub-arctic salt-marsh. *Journal of Applied Ecology*, 21, 657–668.

Cargill, S.M. & Jefferies, R.L. (1984b). The effects of grazing by lesser snow geese on the vegetation of a sub-arctic salt marsh. *Journal of Applied Ecology*, 21, 669–686.

Carpenter, S.R., Kitchell, J.R. & Hodgson, J.R. (1985). Cascading trophic interactions and lake productivity. *BioScience*, 35, 634–639.

Chapin, F.S., Johnson, D.A. & McKendrick, J.D. (1980). Seasonal movements of nutrients in plants of differing growth form in an Alaskan tundra ecosystem: implications for herbivory. *Journal of Ecology*, 68, 189–209.

Chapin, F.S., Miller, P.C., Billings, W.D. & Coyne, P.I. (1980). Carbon and nutrient budgets and their control in coastal tundra. *An Arctic Ecosystem, the Coastal Tundra at Barrow, Alaska* (Ed. by J. Brown, P.C. Miller, L.L. Tieszen & F.L. Bunnell), pp. 458–482. Dowden, Hutchinson & Ross, Stroudsburg, Penn.

Cooch, E.G. & Cooke, F. (1991). *Demographic Changes in a Snow Goose Population: Biological and Management Implications* (Ed. by C.M. Perrins, J.-D. Lebreton & G.J.M. Hirons), pp. 168–189. Bird Population Studies. Oxford University Press, Oxford.

Cooch, E.G., Jefferies, R.L, Rockwell, R.F. & Cooke, F. (1993). Environmental change and the cost of philopatry: an example in the lesser snow goose. *Oecologia*, 93, 128–138.

Cooch, E.G., Lank, D.B., Rockwell, R.F. & Cooke, F. (1991). Long-term decline in body size in a snow goose population: evidence of environmental degradation? *Journal of Animal Ecology*, 60, 483–496.

Cooke, F., Rockwell, R.F. & Lank, D.B. (1995). *The Snow Geese of La Pérouse Bay. Natural Selection in the Wild.* Oxford University Press, Oxford.

Cramp, S. & Simmons, K.E.L. (1977). *Handbook of the Birds of Europe, the Middle East, and North Africa: The Birds of the Western Palearctic.* Vol. 1, Ostrich–Ducks. Oxford University Press, Oxford.

DeAngelis, D.L., Post, W.M. & Travis, C.C. (1986). *Positive Feedback in Natural Systems.* Springer, Berlin.

DeMelo, R., France, R. & McQueen, D.J. (1992). Biomanipulation: hit or myth? *Limnology and Oceanography*, 37, 192–207.

DeNiro, M.J. & Epstein, S. (1978). Influence of diet on the distribution of carbon isotopes in animals. *Geochimica et Cosmochimica Acta*, 42, 495–506.

Dickson, M.L. (1986). *A Comparative Study of the Pelagic Food Chains in Two Newfoundland Fjords using Stable Carbon and Nitrogen Isotope Tracers*. MSc Thesis, Memorial University, St John's, Newfoundland.

Dunbar, M.J. (1981). *Physical Causes and Biological Significance of Polynyas and Other Open Water in Sea Ice. Polynyas in the Canadian Arctic*, pp. 29–43. Occasional Papers, No. 45. Canadian Wildlife Service, Ottawa.

Dunton, K.H., Saupe, S.M., Golikov, A.N., Schell, D.M. & Schonber, S.V. (1989). Trophic relationships and isotopic gradients among arctic and subarctic marine fauna. *Marine Ecology Progress Series*, 56, 89–97.

Elser, J. (1992). Phytoplankton dynamics and the role of grazers in Castle Lake, California. *Ecology*, 73, 887–902.

Fox, A.D. (1993). Pre-nesting feeding selectivity of pink-footed geese *Anser brachyrhynchus* in artificial grasslands. *Ibis*, 135, 417–423.

Fox, A.D., Boyd, H. & Warren, S.M. (1992). The phenology of spring staging of pre-nesting geese in Iceland. *Ecography*, 15, 289–295.

Francis, C.M., Richards, M.H., Cooke, F. & Rockwell, R.F. (1992). Long-term changes in survival rates of lesser snow geese. *Ecology*, 73, 1346–1362.

Fretwell, S.D. (1977). The regulation of plant communities by food chains exploiting them. *Perspectives in Biology and Medicine*, 20, 169–185.

Fry, B. & Sherr, E.B. (1988). $\delta^{13}C$ measurements as indicators of carbon flow in marine and freshwater ecosystems. *Stable Isotopes in Ecological Research* (Ed. by P.W. Rundel, J.R. Ehleringer & K.A. Nagy), pp. 196–229. Springer, New York.

Gadallah, F.L. & Jefferies, R.L. (1995a). Comparison of the nutrient contents of the principal forage plants utilized by lesser snow geese on summer breeding grounds. *Journal of Applied Ecology*, 32, 263–275.

Gadallah, F.L. & Jefferies, R.L. (1995b). Forage quality in brood rearing areas of the lesser snow goose and the growth of captive goslings. *Journal of Applied Ecology*, 32, 276–287.

Gaston, A.J. & Nettleship, D.N. (1981). *The Thick-billed Murres of Prince Leopold Island*. Canadian Wildlife Service Monograph No. 6. Ottawa, Canada.

Gauthier, G. (1991). *Impact of Grazing by Greater Snow Goose* (Chen caerulescens atlantica) *on Vegetation on Bylot Island, NWT*. Unpublished Report for Environment Canada. Dep't de biologie & Centre d'Etudes Nordiques, Université Laval, Quebec.

Giblin, A.E., Nadelhoffer, K.J., Shaver, G.R., Laundre, J.A. & McKerrow, A.J. (1991). Biogeochemical diversity along a riverside toposequence in Arctic Alaska. *Ecological Monographs*, 61, 415–435.

Giroux, J.F. & Bédard, J. (1987). Effects of grazing by greater snow geese on the vegetation of tidal marshes in the St Lawrence estuary. *Journal of Applied Ecology*, 24, 773–788.

Graetz, R.D. (1991). Desertification: a tale of two feedbacks. *Ecosystem Experiments, Scope, 45* (Ed. by H.A. Mooney, E. Medina, D.W. Schindler, E.-D. Schulze & B.H. Walker), pp. 59–87. Wiley, New York.

Gu, B., Schell, D.M. & Alexander, V. (1994). Stable carbon and nitrogen isotopic analysis of the plankton food web in a subarctic lake. *Canadian Journal of Fisheries and Aquatic Science*, 51, 1338–1344.

Haney, J. (1987). Field studies on zooplankton-cyanobacteria interactions. *New Zealand Journal of Marine and Freshwater Research*, 21, 467–475.

Hessen, D.O., Andersen, T. & Lyche, A. (1990). Carbon metabolism in a humic lake: pool sizes and cycling through zooplankton. *Limnology and Oceanography*, 35, 84–99.

Hik, D. & Jefferies, R.L. (1990). Increases in the net above-ground primary production of a salt-marsh forage grass: a test of the predictions of the herbivore-optimization model. *Journal of Ecology*, 78, 180–195.

Hik, D.S., Jefferies, R.L. & Sinclair, A.R.E. (1992). Foraging by geese, nostatic uplift and asymmetry in the development of salt-marsh plant communities. *Journal of Ecology*, **80**, 395–406.

Hik, D.S., Sadul, H.A. & Jefferies, R.L. (1991). Effects of the timing of multiple grazings by geese on net above-ground primary production of swards of *Puccinellia phryganodes*. *Journal of Ecology*, **79**, 715–730.

Hobson, K.A. & Clark, R.G. (1992). Assessing avian diets using stable isotope analysis. II. Factors influencing diet-tissue fractionation. *Condor*, **94**, 189–197.

Hobson, K.A. & Welch, H.E. (1992). Determination of trophic relationships within a high arctic marine food web using $\delta^{13}C$ and $\delta^{15}N$ analysis. *Marine Ecology Progress Series*, **84**, 9–18.

Hunter, M.D. & Price, P.W. (1992). Playing chutes and ladders: heterogeneity and the relative roles of bottom-up and top-down forces in natural communities. *Ecology*, **73**, 724–732.

Iacobelli, A. & Jefferies, R.L. (1991). Inverse salinity gradients in coastal marshes and the death of stands of *Salix*: the effects of grubbing by geese. *Journal of Ecology*, **79**, 61–73.

Jefferies, R.L. (1988a). Pattern and process in arctic coastal vegetation in response to foraging by lesser snow geese. *Plant Form and Vegetation Structure* (Ed. by M.J.A. Werger, P.J.M. van der Art, H.J. During & J.T.A. Verhoeven), pp. 281–300. SPB Academic Publishing, The Hague.

Jefferies, R.L. (1988b). Vegetational mosaics, plant–animal interactions and resources for plant growth. *Plant Evolutionary Biology* (Ed. by L.D. Gottlieb & S.K. Jain), pp. 341–369. Chapman & Hall, London.

Jefferies, R.L., Bazely, D.R. & Cargill, S.M. (1986). Effects of grazing on tundra vegetation – a positive feedback model. *Rangelands: A Resource Under Siege* (Ed. by P.J. Joss, P.W. Lynch & O.B. Williams), p. 50. Australian Academy of Science, Canberra.

Jefferies, R.L., Gadallah, F.L., Srivastava, D.R. & Wilson, D.J. (1995). Desertification and trophic cascades in arctic coastal ecosystems: a potential climatic change scenario? *Global Change and Arctic Terrestrial Ecosystems* (Ed. by T.V. Callaghan), pp. 201–206. European Commission, Ecosystems Research Report 10. Luxembourg.

Jefferies, R.L., Klein, D.R. & Shaver, G.R. (1994). Vertebrate herbivores and northern plant communities: reciprocal influences and responses. *Oikos*, **71**, 193–206.

Jefferies, R.L. & Svoboda, J., Henry, G., Raillard, M. & Ruess, R. (1992). Tundra grazing systems and climatic change. *Arctic Ecosystems in a Changing Climate: An Ecophysiological Perspective* (Ed. by F.S. Chapin III, R.L. Jefferies, J.F. Reynolds, G.R. Shaver & J. Svoboda), pp. 391–412. Academic Press, San Diego.

Kajak, A., Andrzejewska, L. & Wojcik, Z. (1968). The role of spiders in the decrease of damages caused by *Acridoidea* on meadows—experimental investigations. *Ekologia Polska*, Series A, **16**, 755–764.

Kerbes, R.H., Kotanen, P.M. & Jefferies, R.L. (1990). Destruction of wetland habitats by lesser snow geese: a keystone species on the west coast of Hudson Bay. *Journal of Applied Ecology*, **27**, 242–258.

Kitchell, J.F. (1992). *Food Web Management: A Case Study of Lake Mendota*. Springer, New York.

Kitchell, J.F. & Carpenter, S.R. (1992). Cascading trophic interactions. *The Trophic Cascade in Lakes* (Ed. by S.R. Carpenter & J.F. Kitchell), pp. 1–14. Cambridge University Press, Cambridge.

Kling, G.W., Fry, B. & O'Brien, W.J. (1992). Stable isotopes and planktonic trophic structure in arctic lakes. *Ecology*, **73**, 561–566.

Kling, G.W., O'Brien, W.J., Miller, M.C. & Hershey, A.E. (1992). The biogeochemistry and zoogeography of lakes and rivers in arctic Alaska. *Hydrobiologia*, **240**, 1–14.

Kotanen, P.M. & Jefferies, R.L. (1987). Responses of arctic sedges to release from grazing: leaf elongation in two species of *Carex*. *Canadian Journal of Botany*, **67**, 1414–1419.

Lammens, E.H.R.R. (1988). Trophic interactions in the hypertrophic Lake Tjeikemeer: top-down and bottom-up effects in relation to hydrology, predation and bioturbation during the period 1974–1985. *Limnologica*, **19**, 81–85.

Lawton, J.H. (1989). Food webs, *Ecological Concepts* (Ed. by J.M. Cherrett), pp. 43–78. Blackwell Scientific Publications, Oxford.

Lieff, B.C. (1973). *The Summer Feeding Ecology of Blue and Canada Geese at the McConnell River, NWT*. PhD Thesis, University of Western Ontario.

Luecke, C. & O'Brien, W.J. (1983). The effect of *Heterocope* predation on zooplankton communities in arctic ponds. *Limnology and Oceanography*, **28**, 367–377.

MacLean, S.F. (1980). The detritus-based trophic system. *An Arctic Ecosystem. The Coastal Tundra at Barrow, Alaska* (Ed. by J. Brown, P.C. Miller, L.L. Tieszen & F.L. Bunnell), pp. 411–457. Dowden, Hutchinson & Ross, Stroudsburg, Penn.

Madsen, J. (1989). Spring feeding ecology of brent geese, *Branta bernicla*: Annual variation of salt marsh food supplies and effects of grazing on growth of vegetation. *Danish Review of Game Biology*, **13**(7). Communication No. 224. Vildtbiologisk St. Rønde, Denmark.

Madsen, J. (1991). Status and trends of goose population in the Western Palearctic in the 1980s. *Ardea*, **79**, 113–122.

Madsen, J. & Mortensen, C.E. (1987). Habitat exploitation and interspecific competition of moulting geese in East Greenland. *Ibis*, **129**, 25–44.

Marquis, R.J. & Whelan, C.J. (1994). Insectivorous birds increase growth of white oak through consumption of leaf-chewing insects. *Ecology*, **75**, 2007–2014.

Maruyama, M. (1963). The second cybernetics: deviation-amplifying mutual causal processes. *American Scientist*, **51**, 164–179.

McConnaughey, T. & McRoy, C.P. (1979). Food-web structure and the fractionation of carbon isotopes in the Bering Sea. *Marine Biology*, **53**, 257–262.

McCoy, G.A. (1983). Nutrient limitation in two arctic lakes, Alaska. *Canadian Journal of Fisheries and Aquatic Science*, **40**, 1195–1202.

McNaughton, S.J. (1984). Grazing lawns: animals in herds, plant form, and coevolution. *American Naturalist*, **124**, 863–886.

McNaughton, S.J., Oesterheld, M., Frank, D.A. & Williams, K.J. (1991). Primary and secondary production in terrestrial ecosystems. *Comparative Analyses of Ecosystems* (Ed. by J. Cole, G. Lorett & S. Findlay), pp. 120–139. Springer, Berlin.

McQueen, D.J. (1990). Manipulating lake community structure: where do we go from here? *Freshwater Biology*, **23**, 613–620.

McQueen, D.J., Mills, E.L., Forney, J.L., Johannes, M.R.S. & Post, J.R. (1992). Trophic level relationships in pelagic food webs: comparisons derived from long-term data sets for Oneida Lake, New York (USA) and Lake St George, Ontario (Canada). *Canadian Journal of Fisheries and Aquatic Sciences*, **49**, 1588–1596.

Menge, B.A. & Sutherland, J.P. (1976). Species diversity gradients: synthesis of the roles of predation, competition, and temporal heterogeneity. *American Naturalist*, **130**, 730–757.

Mills, E.L. & Forney, J.L. (1988). Trophic dynamics and development of pelagic food webs. *Complex Interactions in Lake Communities* (Ed. by S.R. Carpenter), pp. 14–27. Springer, Berlin.

Minagawa, M. & Wada, E. (1984). Stepwise enrichment of ^{15}N along food chains: further evidence and the relation between ^{15}N and animal age. *Geochimica et Cosmochimica Acta*, **48**, 1135–1140.

Miyake, Y. & Wada, E. (1967). The abundance ratio of ^{15}N/^{14}N in marine environments. *Records of Oceanographic Works in Japan*, **9**, 32–53.

Mizutani, H., Kabaya, Y. & Wada, E. (1991). Nitrogen and carbon isotope compositions relate linearly in cormorant tissues and its diet. *Isotopenpraxis*, **4**, 166–168.

Muir, D.C.G., Wagemann, R., Hargrave, B.T., Thomas, D.J., Peakall, D.B. & Norstrom, R.J. (1992). Arctic marine ecosystem contamination. *Science of the Total Environment*, **122**, 75–134.

Nadelhoffer, K.J., Giblin, A.E., Shaver, G.R. & Linkins, A.E. (1992). Microbial processes and plant nutrient availability in arctic soils. *Arctic Ecosystems in a Changing Climate: An Ecophysiological Perspective* (Ed. by F.S. Chapin III, R.L. Jefferies, J.F. Reynolds, G.R. Shaver & J. Svoboda), pp. 281–300. Academic Press, San Diego.

O'Brien, W.J., Hershey, A.E., Hobbie, J.E., Hullar, M.A., Kipphut, G.W., Miller, M.C., Mollar, B. & Vestal, J.R. (1992). Control mechanisms of arctic lake ecosystems: a limnocorral experiment. *Hydrobiologia*, **240**, 143–188.

Ogilvie, M.A. & St Joseph, A.K.M. (1976). Dark-bellied brent geese in Britain and Europe, 1955–76. *British Birds*, **69**, 422–439.

Oksanen, L., Fretwell, S.D., Arruda, J. & Niemala, P. (1981). Exploitation ecosystems in gradients of primary productivity. *American Naturalist*, **118**, 240–261.

Oksanen, L., Oksanen, T., Ekerholm, P., Moen, J., Lundberg, P., Schnieder, M. & Aunapuu, M. (1996). Structure and dynamics of arctic–subarctic grazing webs in relation to primary productivity. *Foodwebs: Integration of Patterns and Dynamics* (Ed. by G.A. Polis & K.O. Winemiller), pp. 231–242. Chapman & Hall, New York.

Oksanen, T. (1990). Exploitation ecosystems in heterogeneous habitat complexes. *Evolutionary Ecology*, **4**, 220–234.

Owen, M. (1990). The damage–conservation interface illustrated by geese. *Ibis*, **132**, 238–252.

Owen, M. & Black, J. (1991). Geese and their future fortune. *Ibis*, **133** (Suppl. 1), 28–35.

Owens, N.J.P. (1988). Natural variations in ^{15}N in the marine environment. *Advances in Marine Biology*, **24**, 389–451.

Pastor, J., Naiman, R.J. & Dewey, B. (1987). A hypothesis of the effects of moose and beaver foraging on soil carbon and nitrogen cycles, Isles Royale. *Alces*, **23**, 107–124.

Pastor, J., Naiman, R.J., Dewey, B. & McInnes, P. (1988). Moose, microbes and the boreal forest. *BioScience*, **38**, 770–777.

Peterson, B.J. & Fry, B. (1987). Stable isotopes in ecosystem studies. *Annual Review of Ecology and Systematics*, **18**, 293–320.

Polis, G.A. (1991). Complex trophic interactions in deserts: an empirical critique of food web theory. *American Naturalist*, **138**, 123–155.

Power, M.E. (1990). Effects of fish in river food webs. *Science*, **250**, 811–814.

Power, M.E. (1992). Top-down and bottom-up forces in food webs: do plants have primacy? *Ecology*, **73**, 733–746.

Power, M.E., Marks, J.C. & Parker, M.S. (1992). Variation in the vulnerability of prey to different predators: community-level consequences. *Ecology*, **73**, 2218–2223.

Prins, H.H.T., Ydenberg, R.C. & Drent, R.H. (1980). The interaction of brent geese *Branta bernicla* and sea plantain *Plantago maritima* during spring staging: field observations and experiments. *Acta Botanica Neerlandica*, **29**, 585–596.

Rannie, W.F. (1986). Summer air temperature and number of vascular species in arctic Canada. *Arctic*, **39**, 133–137.

Rosenheim, J.A., Wilhoit, L.R. & Armer, C.A. (1993). The influence of intraguild predation among generalist insect predators on the suppression of an herbivore population. *Oecologia*, **96**, 439–449.

Rublee, P.A. (1992). Community structure and bottom-up regulation of heterotrophic microplankton in arctic LTER lakes. *Hydrobiologia*, **240**, 133–141.

Ruess, R.W., Hik, D.S. & Jefferies, R.L. (1989). The role of lesser snow geese as nitrogen processors in a subarctic salt marsh. *Oecologia*, **79**, 23–29.

Ryan, J.K. (1977). Synthesis of energy flows and population dynamics of Truelove Lowland invertebrates. *Truelove Lowland, Devon Island, Canada: A High Arctic Ecosystem* (Ed. by L.C. Bliss), pp. 395–409. University of Alberta Press, Edmonton.

Schultz, A.M. (1968). A study of an ecosystem: the arctic tundra. *The Ecosystem Concept in Natural Resource Management* (Ed. by V.M. Dyne), pp. 77–93. Academic Press, New York.

Scott, P.A. (1992). Annual development of climatic summer in northern North America: accurate prediction of summer heat availability. *Climate Research*, **2**, 91–99.

Shaver, G.R., Chapin, F.S. III. & Gartner, B.L. (1986). Factors limiting seasonal growth and peak biomass accumulation in *Eriophorum vaginatum* in Alaskan tussock tundra. *Journal of Ecology*, **74**, 257–278.

Smith, T.J. & Odum, W.E. (1981). The effects of grazing by snow geese on coastal salt marshes. *Ecology*, **62**, 98–106.

Spiller, D.A. & Schoener, T.W. (1990). A terrestrial field experiment showing the impact of eliminating top predators on foliage damage. *Nature*, **347**, 469–472.

Sprules, W.G. & Bowerman, J.E. (1988). Omnivory and food chain length in zooplankton food webs. *Ecology*, **69**, 418–426.

Srivastava, D.S. & Jefferies, R.L. (1995a). The effects of salinity on the leaf and shoot demography of two arctic forage species. *Journal of Ecology*, **83**, 421–430.

Srivastava, D.S. & Jefferies, R.L. (1995b). Mosaics of vegetation and soil salinity: a consequence of goose foraging in an arctic salt marsh. *Canadian Journal of Botany*, **73**, 75–85.

Srivastava, D.S. & Jefferies, R.L. (1996). A positive feedback: herbivory, plant growth, salinity and the desertification of an arctic salt marsh. *Journal of Ecology*, **84**, 31–42.

Stirling, I., Archibald, R.W. & DeMaster, D. (1977). The distribution and abundance of seals in the eastern Beaufort Sea. *Journal of Fisheries Research Board of Canada*, **34**, 976–988.

Stirling, I., Kingsley, M. & Calvert, W. (1982). *The Distribution and Abundance of Seals in the Eastern Beaufort Sea, 1974–1979.* Canadian Wildlife Service Occasional Papers No. 47, 25 pp. Ottawa.

Stirling, I. & McEwan, E.H. (1975). The caloric value of whole ringed seals (*Phoca hispida*) in relation to polar bear (*Ursus maritimus*) ecology and hunting behaviour. *Canadian Journal of Zoology*, **53**, 1021–1027.

Strong, D.R. (1992). Are trophic cascades all wet? Differentiation and donor-control in speciose ecosystems. *Ecology*, **73**, 747–754.

Strong, D.R., Maron, J.L., Harrison, S., Connors, P.G., Jefferies, R.L. & Whipple, A. (1995). High mortality, fluctuations in numbers and heavy subterranean insect herbivory in bush lupine *Lupinus arboreus. Oecologia*, **104**, 85–92.

Tieszen, L.L., Boutton, T.W., Tesdahl, K.G. & Slade, N.A. (1983). Fractionation and turnover of stable carbon isotopes in animal tissues: implications for $\delta^{13}C$ analysis of diet. *Oecologia*, **57**, 32–37.

Walker, D.A. & Walker, M.D. (1991). History and pattern of disturbance in Alaskan arctic terrestrial ecosystems: a hierarchical approach to analysing landscape change. *Journal of Applied Ecology*, **28**, 244–276.

Welch, H.E., Bergmann, M.A., Siferd, T.D. & Amarualik, P.S. (1991). Seasonal development of ice algae near Chesterfield Inlet, N.W.T., Canada. *Canadian Journal of Fisheries and Aquatic Science*, **48**, 2395–2402.

Welch, H.E., Bergmann, M.A., Siferd, T.D., Martin, K.A., Curtis, M.F., Crawford, R.E., Conover, R.J. & Hop, H. (1992). Energy flow through the marine ecosystem of the Lancaster Sound region, Arctic Canada. *Arctic*, **45**, 343–357.

Westoby, M. (1979/80). Elements of a theory of vegetation dynamics in arid rangelands. *Israeli Journal of Botany*, **28**, 169–194.

Whalen, S.C. & Alexander, V. (1984). Diel variations in inorganic carbon and nitrogen uptake by phytoplankton in an arctic lake. *Journal of Plankton Research*, **6**, 571–590.

Whitfield, D.W.A. (1977). Energy budgets and ecological efficiencies on Truelove Lowland. *Truelove Lowland, Devon Island, Canada: A High Arctic Ecosystem* (Ed. by L.C. Bliss), pp. 607–620. University of Alberta Press, Edmonton.

Wilson, D. (1993). *Nitrogen Mineralization in a Grazed Sub-arctic Salt Marsh.* MSc Thesis, University of Toronto.

Zellmer, I.D., Claus, M.J., Hik, D.S. & Jefferies, R.L. (1993). Growth responses of arctic graminoids following grazing by captive lesser snow geese. *Oecologia*, **93**, 487–492.

9. Pathways and effects of contaminants in the Arctic

EILIV STEINNES

Department of Chemistry, Norwegian University of Science and Technology,
Trondheim, Norway

INTRODUCTION

During the last decades it has become increasingly evident that the Arctic is no longer a pristine area with respect to pollutants. Since the vulnerable arctic environment is particularly sensitive to perturbations from outside, it is important to follow closely the impact of pollutants in the different ecosystems in order to assess any ecological effects. In this chapter, pathways of contaminants that may have harmful effects even when present at low levels are reviewed, and possible ecological effects of these substances in the Arctic are discussed. The groups of substances concerned are man-made radionuclides, toxic metals and persistent organochlorine compounds.

Sources and pathways of contaminants of the Arctic may be classified as follows:
1 Local pollution sources within the Arctic.
2 Long-range transport:
    (a) airborne: gases, particles (aerosols);
    (b) via ocean currents;
    (c) with drifting ice.

In the following discussion, long-term transport via the atmosphere and ocean currents is emphasized. For a more detailed discussion of these matters the reader should refer to the proceedings of two recent conferences, notably 'Environmental Radioactivity in the Arctic and Antarctic' (Strand & Holm 1993) and 'Ecological Effects of Arctic Airborne Contaminants' (Landers 1995).

MAN-MADE RADIONUCLIDES

Ever since the first atmospheric nuclear weapons tests, man-made radionuclides have entered the polar regions from different sources. So far, the nuclides subject to most attention have been ^{137}Cs and ^{90}Sr, both of which are products of the fission of uranium and plutonium and have half-lives of the order of 30 years. Inventory estimates of still uncontained (i.e. potentially bioavailable) ^{137}Cs and ^{90}Sr are presented in Table 9.1. There is estimated to be more ^{137}Cs than ^{90}Sr, and amounts in the sea are estimated to be about five times higher than on land (Aarkrog 1993). It is interesting to note that the highest contribution to ^{137}Cs in the Arctic Ocean is neither nuclear weapons fallout nor USSR river discharges, but authorized

TABLE 9.1. Inventory estimates ($\times 10^{15}$ Bq) for uncontained radionuclides in the Arctic. (Data from Aarkrog 1993.)

	^{90}Sr	^{137}Cs
Sea		
Global fall-out (nuclear weapons)	2.6	4.1
Sellafield discharges	1–2	10–15
USSR river discharges	1–5	1–5
Runoff of global fall-out	1.5	0.5
Chernobyl accident	0	2–5
Total	6–11	17–30
Land		
Global fall-out	1.6	2.6

discharges from the Sellafield nuclear fuel reprocessing plant in the UK, which were at a maximum in 1975. The 1975 ^{137}Cs 'pulse' has been monitored around the coast of Scotland, across the North Sea, and along the coast of Norway until it reached the Barents Sea 6–7 years later (Kershaw & Baxter 1993).

However, there are also some other documented and possible future sources of radionuclides within the Arctic. The Thule accident in 1968, where nuclear weapons material was released, supplied about 10^{12} Bq ^{239}Pu–^{240}Pu to the local environment (Aarkrog 1993). Other potential sources not yet shown to contribute significantly include the Kosmomolets submarine that sank near the Bear Island in 1989, dumped submarine reactors in the Kara Sea, and ^{90}Sr-powered lighthouses along the Siberian coast. These sources could release large amounts of radioactivity in the future. Substantial amounts of long-lived radionuclides, particularly ^{90}Sr, may also be supplied to the Arctic Ocean from Siberian rivers during the years to come.

Marine food chains in the Arctic seem to have been contaminated by man-made radionuclides only to a very slight extent, in spite of the significant inventories listed above. The concentrations of ^{137}Cs, ^{90}Sr and ^{239}Pu–^{240}Pu observed in the organs of minke whale from the Arctic Ocean may serve as an example (Table 9.2).

TABLE 9.2. Radionuclide concentrations (Bq kg^{-1} wet wt) in organs of minke whale caught in the Arctic Ocean north of Norway in 1988 and 1992 (G.C. Christensen & E. Steinnes unpublished data). Number of samples are given in parentheses, and uncertainties listed are standard deviations.

	^{137}Cs (muscle)	^{90}Sr (bone)	^{239}Pu + ^{240}Pu (liver)
Norwegian Sea 1988	4.4 ± 1.6 (29)	15.2 ± 2.8 (5)	0.8 ± 0.7 (5)
Norwegian Sea 1992	3.4 ± 2.5 (4)	—	—
Barents Sea 1992	1.2 ± 0.3 (17)	4.7 ± 2.9 (5)	0.17 ± 0.15 (5)

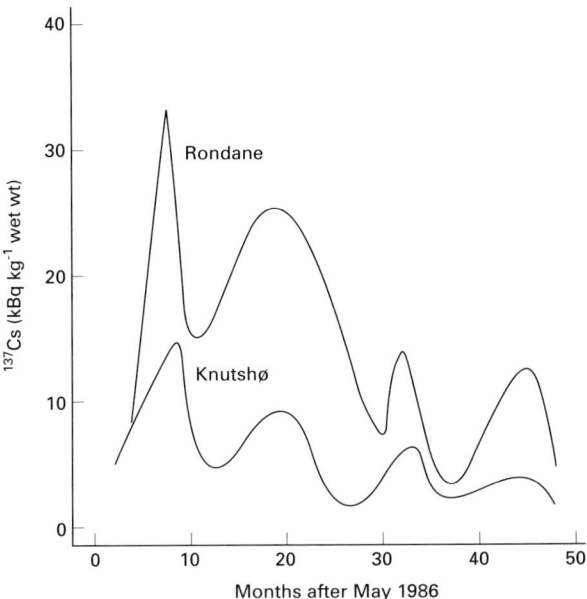

FIG. 9.1. Seasonal variation of ^{137}Cs in thigh muscle of reindeer in two mountain areas of central Norway after the Chernobyl accident. (After Bretten *et al.* 1992.)

These concentrations are orders of magnitude below what might have any significant ecological effect. Terrestrial mammals, however, have been much more exposed to man-made radioactivity (Aarkrog 1993). Reindeer, in particular, have been known to be affected because lichens, absorbing radionuclides efficiently from atmospheric deposition, form an important part of their diet (Lidén 1961). This was recognized as a circumpolar problem during the 1960s, but the main concern was the radiation dose to human populations consuming reindeer meat, rather than possible effects to the animals themselves. Even higher levels of ^{137}Cs were observed after the Chernobyl accident in reindeer populations in central Scandinavia (e.g. Bretten *et al.* 1992). Examples on the seasonal variation of ^{137}Cs in reindeer after the Chernobyl accident are shown in Fig. 9.1, illustrating the higher exposure during the winter when the animals feed exclusively on lichens. Even at these high exposure levels, however, effects at the population level have not been convincingly demonstrated.

HEAVY METALS

Heavy metals emitted to the atmosphere from various high-temperature anthropogenic processes constitute another group of contaminants that are transported to the Arctic and deposited there. Perhaps the most striking example is from studies of polar snow and ice (Boutron, Candelone & Hong 1994), demonstrating the strong increase of Pb deposition in Greenland during the last two centuries compared to preindustrial times. It has been known for some time that the aerosols thought to be involved in the formation of arctic haze (Rahn & McCaffrey 1980) also contain substantial amounts of some heavy metals. During the winter period with stagnant air, heavy metal concentrations in the Eurasian part of the Arctic may reach values that are of the same order of magnitude as air concentrations typically observed at much lower latitudes (Table 9.3). Some elements (V, Zn, Pb, As, Sb) are more typical of long-range transported aerosols than others (Cr, Ni, Cu), because they are more likely to occur in volatile forms at the source and hence be preferentially concentrated on small particles. The ratios between these elements in aerosols are remarkably similar from southern Scandinavia to the High Arctic (Table 9.3). There appears to be a predominant transport of pollution aerosols from source regions in Eurasia across the polar basin towards the North-American Arctic.

Even though air concentrations of metals in the Arctic sometimes reach appreciable levels, their deposition rates are probably very low because of the lack of precipitation during the winter season. No data, however, seem to exist on the bulk deposition of heavy metals in the High Arctic. Analyses of terrestrial moss from Svalbard indicate that the deposition of metals such as lead and cadmium there must be similar to, or lower than, what is observed in the northernmost part of mainland Norway. However, the use of mosses as biomonitors of atmospheric metal deposition in polar regions is associated with considerable difficulties (Steinnes 1995).

TABLE 9.3. Air concentrations ($ng\,m^{-3}$) of some heavy metals at Ny-Ålesund, Svalbard (Maenhaut et al. 1989) and at Birkenes, southern Norway (Amundsen et al. 1992).

| Element | Ny-Ålesund | | Birkenes‡ ($n = 160$) | Birkenes (1985) Ny-Ålesund (winter) ratio |
	Winter* ($n = 46$)	Summer† ($n = 13$)		
V	0.54	0.022	1.3	2.4
Zn	3.9	< 0.15	11	2.8
As	0.52	< 0.01	0.49	0.9
Sb	0.092	< 0.003	0.26	2.8
Pb	3.0	< 0.7	7.8	2.6
Cr	< 0.4	0.56	0.45	> 1.1
Ni	0.29	< 0.2	0.70	2.4
Cu	< 0.9	< 0.3	1.1	> 1.2

Year of sampling: *1983, 1984 and 1986; †1984; ‡1985–86.

While heavy metals supplied by long-range atmospheric transport are unlikely to cause any significant ecological problems in the Arctic there are local emission sources, particularly in the Russian Arctic, that cause excessive contamination on a local scale. The copper–nickel smelters on the Kola Peninsula are a well-known example. In Table 9.4 metal deposition levels in the western part of the Kola Peninsula (E. Steinnes & V. Nikonov unpublished data), as illustrated by concentrations in terrestrial mosses, are compared with similar data from a simultaneous study in Norway (Berg *et al.* 1995). In areas close to the smelters the fall-out levels of copper and nickel are extremely high, and coincide with strong effects of SO_2 emissions (see Chapter 10). Unlike the SO_2 deposition, however, the contamination of the terrestrial ecosystems by metals decreases rapidly with distance from the smelters because the predominant metals are emitted mainly with large particles that settle rapidly in the air. At distances of 100–150 km from the smelters the deposition of copper and nickel approaches regional background levels. It is difficult in these areas to distinguish ecological effects of heavy metals from those associated with foliar damage from SO_2 and soil acidification, but the combined effects on the ecosystems concerned are severe.

As with radionuclides, heavy metals discharged to the ocean at lower latitudes are likely to be transported towards the Arctic by ocean currents. Since these metals also occur naturally in sea water, however, the contribution from anthropogenic

TABLE 9.4. Concentrations of 13 elements (ppm dry wt) in moss samples (*n* = 55) from the Kola Peninsula (1993). Corresponding median values for moss samples from Norway taken in 1990 (*n* = 500) are also given for comparison. (E. Steinnes & V. Nikonov unpublished data.)

Element	Kola Peninsula			Norway median
	Median	Minimum	Maximum	
Lead	7.7*	3.3	33.2	9.3
Cadmium	0.14	0.01	1.42	0.13
Copper	**27.8†**	3.9	858	5.2
Zinc	29.8	9.8	83.4	36
Chromium	< 3.8	< 2.6	45.1	0.9
Nickel	**27.5**	2.5	765	1.6
Cobalt	**1.57**	0.17	13.7	0.25
Iron	**1340**	372	16 500	470
Manganese	457	62	1410	300
Vanadium	**5.4**	2.2	31.4	2.4
Arsenic	0.40	0.12	7.3	0.27
Barium	25	13	60	24
Strontium	13.4	4.8	63.0	12

* Italic, similar level as in Norway.
† Bold, substantially higher in Kola.

TABLE 9.5. Heavy metals (mean values) in minke whale liver (ppm wet wt) from the Norwegian Sea (E. Steinnes & B. Munro Jenssen unpublished data), West Greenland (Hansen *et al.* 1990) and the Antarctic Sea (Honda *et al.* 1987).

	N	Zn	Cu	Pb	Cd	Hg
Norwegian Sea	50	40.1	5.3	0.076	0.52	0.87
West Greenland	17	34.5	—	—	0.90	0.39
Antarctic Sea	135	36.6	4.3	0.102	40.3	0.049

activity is not easy to distinguish. So far, no estimates of ocean transport of heavy metals to the Arctic seem to be available. Heavy metal concentration levels in arctic marine mammals using minke whale as an example are presented in Table 9.5, where a comparison is also made with similar data from the Antarctic. Two striking differences are observed, a higher mercury level in the arctic whales and a much higher cadmium burden in the antarctic whales. While the difference in mercury levels might be associated with a generally higher contamination of the northern hemisphere, the difference in cadmium levels is most probably associated with differences in food habits, the antarctic whales primarily feeding on krill which concentrates cadmium efficiently from the sea water. In most cases, the metal levels observed in arctic marine organisms are probably derived from natural sources.

PERSISTENT ORGANOCHLORINE COMPOUNDS

The occurrence of persistent organochlorine compounds such as DDT in polar animals was observed more than 20 years ago. More recently, it has become clear that top predators in arctic food chains sometimes accumulate some of these compounds to very high concentration levels, which may be damaging to the individual and possibly cause effects even at the population level. This means that there must be an extensive and efficient transport of organochlorine compounds to the Arctic from lower latitudes where these substances are predominantly being used. While this transport in the past was often thought to occur via ocean currents, it is now most likely that the atmosphere is the main pathway. According to the so-called 'Global distillation hypothesis' (Ottar 1989; Wania & Mackay 1993), most of the persistent organochlorine compounds have a sufficiently high vapour pressure at ambient temperatures that they tend to partially volatilize from surfaces at lower latitudes and thus become available for atmospheric transport to colder areas. This process will eventually result in accumulation of organochlorines in polar regions, and will also lead to a fractionation of organochlorines according to differences in volatility.

The compounds – or groups of compounds – most regularly observed in the European Arctic are ΣDDT (DDT + metabolites), PCB (polychlorinated biphenyls),

HCB (hexachlorobenzene), and HCH (hexachlorocyclohexane), which is the sum of the isomers α-HCH and γ-HCH (lindane). Oehme, Haugen and Schlabach (1995) monitored air concentrations of several organochlorines at four stations ranging from Lista, southern Norway (58°N) to Ny-Ålesund, Svalbard (79°N). Their observations supported the idea inherent in the 'Global distillation hypothesis'. While the PCB congeners 138 and 153 occurred at about 10-fold higher concentrations at Lista compared to the more northerly latitudes, the HCB concentrations were quite similar at all stations, and α-HCH showed a steady increase with latitude. The α-HCH/γ-HCH ratio was about 10 times higher at Ny-Ålesund compared to Lista. A vast number of data are available for the content of organochlorine compounds in arctic animals, and a few examples will be mentioned here. Although most PCB congeners are generally less volatile than some of the other compounds mentioned above, PCB is of most serious concern because of its high toxicity and because it appears to accumulate more efficiently in the animals than most other organochlorine compounds. The levels observed, however, depend strongly on the position in the food chain. Considering birds in the Norwegian Arctic, the glaucous gull appears to be most at risk because it is a top predator. Gabrielsen *et al.* (1995) observed the following organochlorine levels (mg kg^{-1} wet wt) in glaucous gull liver: ΣPCB, 16.0 (mean), 1–32 (range); ΣDDT, 2.3, 0.9–6.3; HCB, 0.5, 0.3–1.0; ΣHCH, 0.02, not detected–0.14. Mehlum and Daelemans (1995) observed a similar PCB level in glaucous gull, while black guillemot and common eider showed PCB levels two orders of magnitude lower. Among marine mammals the polar bear appears to be the species exhibiting the highest levels, and presumably is most at risk (Skaare, Wiig & Bernhoft 1994). Purely terrestrial animals such as the caribou (Elkin & Bethke 1995) appear to accumulate organochlorine compounds only to a very low extent, while the arctic fox (Wang-Andersen *et al.* 1993) shows rather high PCB levels, presumably because of the element of marine animals in its diet.

CONCLUSIONS

The main conclusions from this paper may be summarized as follows:

1 Man-made radionuclides are generally present in measurable amounts in the Arctic, but the levels are far below what might represent any ecological risk.

2 Toxic metals are supplied in excessive amounts to the local environment at some industrial sites in the Low Arctic. In general, however, the metals supplied to arctic ecosystems from anthropogenic activities probably add little to the natural levels of these metals.

3 Organochlorine compounds are supplied to the Arctic in large amounts, predominantly by atmospheric transport, and are concentrated in the food chain. PCB levels in some predator species are very high, and may constitute a hazard to these species.

It is conceivable that the long-range transport of contaminants to the Arctic will continue in the foreseeable future. Monitoring of these substances is therefore a matter of great importance, and more effect studies should be performed in cases where exposure levels are particularly high.

REFERENCES

Aarkrog, A. (1993). Radioactivity in polar regions – main sources. *Environmental Radioactivity in the Arctic* (Ed. by P. Strand & E. Holm), pp. 15–34. Norwegian Radiation Protection Authority, Østerås, Norway.

Amundsen, C.E., Hanssen, J.E., Semb, A. & Steinnes, E. (1992). Long-range transport of trace elements to southern Norway. *Atmospheric Environment*, **26A**, 1309–1324.

Berg, T., Røyset, O., Steinnes, E. & Vadset, M. (1995). Atmospheric trace element deposition: principal component analysis of ICP-MS data from moss samples. *Environmental Pollution*, **88**, 67–77.

Boutron, C.F., Candelone, J.P. & Hong, S. (1994). Past and recent changes in the large-scale tropospheric cycles of lead and other heavy metals as documented in Antarctic and Greenland snow and ice: a review. *Geochimica et Cosmochimica Acta*, **58**, 3217–3235.

Bretten, S., Gaare, E., Skogland, T. & Steinnes, E. (1992). Investigations of radiocaesium in the natural terrestrial environment in Norway following the Chernobyl accident. *Analyst*, **117**, 501–503.

Elkin, B.T. & Bethke, R.W. (1995). Environmental contaminants in caribou in the Northwest Territories, Canada. *The Science of the Total Environment*, **160/161**, 307–321.

Gabrielsen, G.W., Skaare, J.U., Polder, A. & Bakken, V. (1995). Chlorinated hydrocarbons in glaucous gulls (*Larus hyperboreus*) in the southern part of Svalbard. *The Science of the Total Environment*, **160/161**, 337–346.

Hansen, C.T., Nielsen, C.O., Dietz, R. & Hansen, M.M. (1990). Zinc, cadmium, mercury and selenium in minke whales, belugas and narwhales from West Greenland. *Polar Biology*, **10**, 529–539.

Honda, K., Yamamoto, Y., Kato, H. & Tatsukawa, R. (1987). Heavy metal accumulations and their recent changes in southern minke whales *Balaenoptera acutorostrata*. *Archives of Environmental Contamination and Toxicology*, **16**, 209–216.

Kershaw, P.J. & Baxter, A.J. (1993). Sellafield as a source of radioactivity to the Barents Sea. *Environmental Radioactivity in the Arctic* (Ed. by P. Strand & E. Holm), pp. 161–174. Norwegian Radiation Protection Authority, Østerås.

Landers, D.H. (Ed.) (1995). Proceedings of the conference 'Ecological Effects of Arctic Airborne Contaminants', Reykjavik, Iceland, October 1993. *The Science of the Total Environment*, **160/161**.

Lidén, K. (1961). Caesium-137 burden in Swedish laplanders and reindeer. *Acta Radiologica*, **56**, 237–240.

Maenhaut, W., Cornille, P., Pacyna, J.M. & Vitols, V. (1989). Trace element composition and origin of the atmospheric aerosol in the Norwegian Arctic. *Atmospheric Environment*, **23**, 2551–2569.

Mehlum, F. & Daelemans, F.F. (1995). Ambient air levels of persistent organochlorines in spring 1992 at Spitsbergen and the Norwegian mainland: comparison with 1984 results and quality control measures. *The Science of the Total Environment*, **160/161**, 139–152.

Oehme, M., Haugen, J.E. & Schlabach, M. (1995). Ambient air levels of persistent organochlorines in spring 1992 at Spitsbergen and the Norwegian mainland: Comparison with 1987 results and quality control measures. *The Science of the Total Environment*, **160/161**, 139–152.

Ottar, B. (1989). Arctic air pollution: a Norwegian perspective. *Atmospheric Environment*, **23**, 2349–2356.

Rahn, K.A. & McCaffrey, R.J. (1980). On the origin and transport of the winter Arctic aerosol. *Annals of the New York Academy of Sciences*, **388**, 486–503.

Skaare, J.U., Wiig, Ø. & Bernhoft, A. (1994). *Organochlorines in Polar Bear: Concentration Levels and Effects*. Report No. 86. Norwegian Polar Institute, Oslo. [In Norwegian.]

Steinnes, E. (1995). A critical evaluation of the use of naturally growing moss to monitor the deposition of atmospheric metals. *The Science of the Total Environment*, **160/161**, 243–249.

Strand, P. & Holm, E. (Eds) (1993). *Environmental Radioactivity in the Arctic*, 432 pp. Norwegian Radiation Protection Authority, Østerås.

Wang-Andersen, G., Skaare, J.U., Presterud, P. & Steinnes, E. (1993). Levels and congener pattern of PCB in arctic fox, *Alopex lagopus*, in Svalbard. *Environmental Pollution*, **82**, 269–275.

Wania, F. & Mackay, D. (1993). Global fractionation and cold condensation of low volatility organochlorine compounds in polar regions. *Ambio*, **22**, 10–18.

10. Effects of acid deposition on arctic vegetation

SARAH J. WOODIN

Department of Plant and Soil Science, University of Aberdeen, Cruickshank Building, St Machar Drive, Aberdeen AB24 3UU, UK

INTRODUCTION

It is well recognized that the Arctic is subject to atmospheric pollution. 'Arctic haze' was first described almost 40 years ago (Mitchell 1956) and 20 years later studies linked it to pollutants in the atmosphere (reviewed by Barrie 1986a). Since then, aircraft studies, ground-based monitoring, and meteorological studies have greatly increased our knowledge and understanding of arctic haze. It is formed mainly by SO_4^{2-} aerosols, accompanied by SO_2 and other compounds including nitrogenous pollutants and heavy metals. Arctic haze is strongly seasonal, with the highest atmospheric concentrations of pollutants occurring in winter and early spring. For example, at Bjørnøya in the Norwegian Arctic, the maximum monthly mean concentration of SO_4^{2-}–S in arctic haze (averaged over 1978–86) is $0.85\,\mu g\,m^{-3}$, and this occurs in February–March (Ottar *et al.* 1986). This seasonal maximum is caused by strong north–south transport conditions bringing pollutants from mid-latitudes, by the stability of the arctic air mass and long atmospheric lifetime of pollutants in winter, and by low rates of pollutant deposition (Rahn *et al.* 1980; Barrie 1986a; Ottar *et al.* 1986).

For terrestrial ecosystems in the Arctic, rates of pollutant deposition are just as important as their atmospheric concentrations. Winter precipitation, in the form of snow, originates from high in the atmosphere above the polluted air, and thus has a low pollutant content (Semb, Braekkan & Joranger 1984). The rate of scavenging of pollutants as the snow falls through the polluted air is also low in arctic winter conditions and thus concentrations of pollutants in precipitation and rates of deposition are low (Ottar *et al.* 1986). In addition, rates of dry deposition to the surface of the snowpack are low, although deposition of fog onto surface snow may increase concentrations of some ionic species including SO_4^{2-} and NO_3^- (Davidson *et al.* 1993). In summer, the atmospheric concentrations of pollutants are much lower than in winter (e.g. monthly mean $c.$ $0.25\,\mu g\,m^{-3}$ SO_4^{2-}–S, calculated as above), but much of the precipitation falls as mist and fog. This originates at lower altitudes, within the polluted layer of the atmosphere, and is very efficient at scavenging pollutants from the air mass (Joranger & Semb 1989). Concentrations of pollutants in deposition are actually highest (e.g. monthly mean $c.$ $1\,mg\,l^{-1}\,SO_4^{2-}$–S, calculated as above) during April and May when air

pollutant concentrations are declining, but the predominant form of precipitation
is changing from snow to mist and scavenging efficiency is thus increasing (Ottar
et al. 1986).

Long-term trends in deposition of acid pollutants in the Arctic have been well
documented by ice-core analyses (e.g. Neftel *et al.* 1985). In a recent study of a
core from Greenland, Laj, Palais and Sigurdsson (1992) demonstrated an increase
in S deposition of more than 300% since before the Industrial Revolution and an
increase in N deposition of *c.* 100% since the middle of this century (Fig. 10.1).
The deposition of both S and N during the past 20 years is described as having
remained stable. Historically, S deposition is shown to have reflected S emissions
in Europe and North America. Sulphur emissions in western Europe, the former
USSR and North America are currently on a downward trend (e.g. Lövblad *et al.*

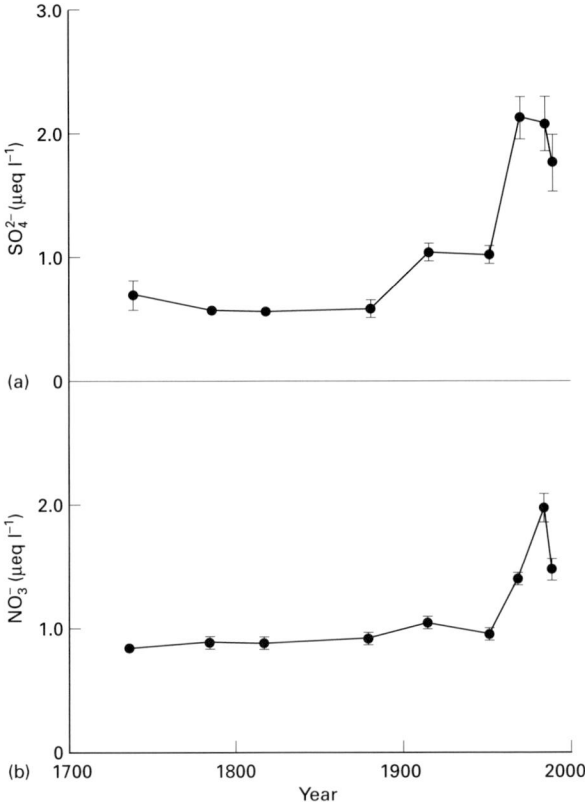

FIG. 10.1. Concentrations of (a) SO_4^{2-} and (b) NO_3^- (mean ± 2 SE) in a dated ice core from central
Greenland. (From data in Laj *et al.* 1992.)

1992; EMEP 1994), and thus S deposition in the Arctic is likely to decrease. The situation regarding N deposition is less clear. Even if all the nations signed up to the United Nations NO_x protocol meet their commitments, N deposition in arctic Scandinavia is predicted to decrease by only *c.* 10%, due in part to NO_x emissions in non-protocol countries increasing (Lövblad *et al.* 1992). However, NO_x emission reductions are much less advanced, and more difficult to achieve, than S emission reductions and if commitments are not met, or indeed if NH_3 emissions increase, then N deposition in the Arctic could increase. For this reason, particular attention is paid to the ecological effects of N deposition in this chapter.

CRITICAL LOADS FOR ARCTIC ECOSYSTEMS

Actual rates of deposition of S and N in the Arctic are low in comparison to deposition at more southerly latitudes. For example, wet deposition of non-marine N and S in the Norwegian Arctic is an order of magnitude less than that in southern Norway, with *c.* 0.14 and $1.6 \, g \, N \, m^{-2} year^{-1}$ in the Arctic and south respectively, and *c.* 0.19 and $1.5 \, g \, S \, m^{-2} year^{-1}$ (calculated from Joranger & Semb 1989). However, to assess the ecological significance of acid deposition in the Arctic, comparison should be made not with other regions, but with the sensitivity of arctic ecosystems to pollutants. Such comparison forms the basis of the 'critical loads' approach to assessment of the ecological effects of acid deposition. A critical load is defined as 'the quantitative estimate of an exposure to one or more pollutants below which significant harmful effects on specified sensitive elements of the environment do not occur according to present knowledge' (Nilsson & Greenfelt 1988).

The slow soil processes, low nutrient availability, short growing season and extreme climate typical of arctic ecosystems are likely to make them sensitive to further stresses such as S and N inputs, and thus they will have low critical loads for acid deposition. Inputs of pollutant N are known to change species composition in other ecosystems which have naturally low N availability and conservative N cycling (e.g. Hornung, Sutton & Wilson 1995). Critical-load studies are based on theoretical calculations of the biogeochemistry of plant–soil–water systems, and utilize modelled or extrapolated estimates of deposition rates. The actual definition of the relationships between water or soil chemistry and the condition of the organisms selected as indicators of 'harm' is also always open to discussion. Thus critical loads have to be viewed with some caution. However, it is interesting that critical-loads mapping of Svalbard has shown that in an extremely sensitive area of the north of the archipelago, with a critical load of $0–0.2 \, g \, S \, m^{-2} year^{-1}$ for surface waters, deposition already exceeds the critical load by a small amount (Lien, Heinriksen & Traaen 1993). The area concerned is equivalent to 5% of the ice-free area of Svalbard. Thus, there are grounds for concern about the effects of acid deposition even in remote areas of the Arctic receiving low rates of deposition.

ECOLOGICAL EFFECTS OF
MAJOR POLLUTANT SOURCES WITHIN THE ARCTIC

By studies of the size distribution of pollutant particles, their chemical composition and sulphur isotope ratio, and of meteorological pathways associated with individual pollution events, source regions contributing to the pollution of the Arctic have been identified and their contributions quantified (Table 10.1). These vary with altitude and season, but the contribution of the former USSR to the arctic pollutant load is notable, and particularly important because of its very large contribution at lower altitudes of the atmosphere, which are the source of most deposition (Ottar *et al.* 1986). Several source regions within arctic Russia have been identified, but the availability of data on their emissions varies (AMAP 1994). As recently as 1992, Russian authors described a region in north-west Siberia where 'annual thermal ejections to the atmosphere during gas torching attains the annual amount of solar radiation absorbed by soil and vegetation within the region'. However, despite this gross pollution, they state that 'unfortunately, there are no exact data on chemical pollution of the air in the region ... but it may cause acid precipitation within the neighbouring areas' (Vilchek & Bykova 1992). Thus, the need for better monitoring remains.

Pollutant emissions in one important source area within arctic Russia, the Kola Peninsula, are relatively well documented, partly due to the proximity of the area to northern Scandinavia. The fourth and fifth largest sources of S in Europe are sited within the Peninsula, these being the nickel smelters in the towns of Nickel and Monchegorsk, which each emit over $200000\,t\,S\,year^{-1}$ (Barrett & Protheroe 1994) and significant quantities of nickel and copper. The average SO_2 concentration in winter exceeds the International Union of Forest Research Organization's air-quality guideline for the protection of forest health ($25\,\mu g\,m^{-3}$, 6-month average) over an area of approximately $2000\,km^2$ around Nickel alone (e.g. Sivertsen *et al.* 1992).

TABLE 10.1. Annual SO_2 emissions (estimates, based on early 1980s) and percentage contributions to atmospheric SO_4^{2-} concentration at different altitudes in the atmosphere north of 72.5°N (modelled, based on June/July 1983) of areas which contribute pollutants to the arctic air mass. (Data abstracted from *Barrie 1986a and †Ottar *et al.* 1986.)

	Europe	Russia	North America	Far East
Annual SO_2 emissions (Mt)*	31	23	23.5	—
Modelled percentage contribution to atmospheric SO_4^{2-} concentration at†:				
1 m	23.5	74.9	1.4	0.2
1000 m	30.1	67.4	2.1	0.4
5000 m	64.2	10.0	23.1	2.7

TABLE 10.2. Pollution climate and condition of forest vegetation in zones of 'disturbed and transformed ecosystems' around metal smelters on the Kola Peninsula, as described by Kryuchkov (1990).

Ecosystem zone	I	II	III	IV	V
Pollution climate					
S deposition (g m^{-2} year^{-1})	25–30	3–5	2–3	1–2	<1
Mean SO$_2$ concentration (µg m^{-3})	200	90	80	70	50
Metal deposition Σ Ni, Cu, Mn, Zn (g m^{-2})	5–6	2–5	0.5–2	0.05–0.5	<0.05
Vegetation condition					
Mosses and lichens	None	None	Some ground lichens and mosses	60–70% cover of ground moss, unhealthy bog moss, some ground lichens	Ground moss and lichens healthy, small epiphytic lichens appear
Field layer	Some grasses and sedges in hollows only	Grasses and sedges more widespread	20–30% grass–shrub cover	60–80% herb–shrub cover	Healthy vegetation
Spruce condition					
Spruce health	Few prostrate trees in hollows only	Few live prostrate, many dead/dying standing trees	30–40% trees dead, live trees with 'flag' crowns	Live trees with 'flag' crowns	Healthy, but not in optimum condition
Maximum percentage of buds live	10	50	70	85	100
Needle retention (years)	1–2	1–2	2–9	5–10	9–13

The state of the vegetation on the Kola Peninsula has been documented by Russian workers for *c*. 20 years. In a summary paper, Kryuchkov (1990) describes the pollution climate and state of the vegetation in five zones representing different stages of 'degradation'. Information from this paper is summarized in Table 10.2. For an impression of the severity of pollution in the different zones, the S deposition should perhaps be compared with the critical load for sensitive ecosystems of $0.3\,\mathrm{g\,S\,m^{-2}\,year^{-1}}$ (Nilsson & Grennfelt 1988) and SO_2 concentrations with the critical level of SO_2 (annual mean concentration) for forests and natural vegetation of $20\,\mathrm{\mu g\,m^{-3}}$ (Umwelt Bundesamt 1993). In the most polluted zone, Zone 1, Kryuchkov (1990) notes an absence of soil and of lower plants. Vegetation, including a few prostrate spruce trees, appears to be limited to hollows which must in some way provide protection from pollutant exposure. It is clear that ecological damage is likely to occur even in Zone 5, where epiphytic lichens only just begin to appear and spruce needles are noted not to 'reach a biological age optimum'.

There have been several studies of the effects of emissions from the Monchegorsk smelter on the health and physiology of Scots pine (*Pinus sylvestris*) in the forests of the Kola Peninsula and adjacent Finnish Lapland. The S concentration in pine needles is elevated over a distance of approximately 50 km from the smelter (Raitio 1992) and sensitivity to frost damage is also increased over this distance (Sutinen 1992). The number of needle year classes retained by the trees is reduced, and the amount of defoliation increased, over a distance of about 100 km from the smelter (Raitio 1992; Salemaa, Lindgren & Jukola-Sulonen 1992) (Fig. 10.2).

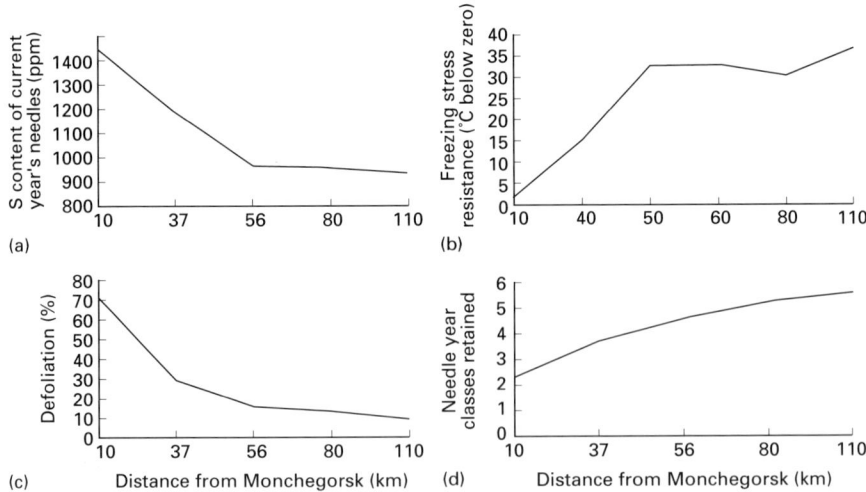

FIG. 10.2. Health of Scots pine with increasing distance from the Monchegorsk smelter, Kola Peninsula. (a) S content of current year's needles, (b) freezing stress resistance, (c) defoliation and (d) number of needle year classes retained. (After Raitio 1992; Sutinen 1992; Salemaa *et al.* 1992.)

Estimates of the total area of forest affected by the Kola Peninsula smelter emissions vary somewhat, but Kalabin (1992) gives a value of 128 000 km^2. The area over which the critical load of S deposition for sensitive soils and fresh waters (0.3 g S m^{-2}year^{-1}) is exceeded is similarly estimated at 150 000 km^{-2}, extending into arctic Norway and Finland (Tuovinen & Laurila 1992). In a study of fresh-water acidification, Moiseenko (1994) concludes that in the northern Kola region critical loads of S are exceeded for 48% of the lakes, with mountainous and remote tundra areas being most affected due to their high sensitivity. Thus, emissions from these massive pollution sources within the Arctic region not only have very severe local effects, but also affect forest and fresh-water ecosystems over a wide area of north-west Russia and northern Fennoscandia. As discussed above, these emissions also contribute by long-distance transport to the total pollution loading of the arctic air mass; consequent impacts at remote locations in the High Arctic are more difficult to identify directly, but should not be overlooked.

EFFECTS OF NITROGEN DEPOSITION ON ARCTIC SHRUBS

Low rates of mineralization in cold, and in some cases waterlogged, arctic soils result in arctic vegetation being N limited (Nadelhoffer *et al.* 1992; Shaver & Chapin 1980). It has been stated that an increase in N supply 'always increases plant growth' (Berendse & Jonasson, 1992), although Chapin *et al.* (see Chapter 4) argue that the long-term community response to increased nutrient availability is change in species composition without overall increase in biomass. In either case, it is apparent that increased N input to arctic terrestrial ecosystems through acid deposition is likely to cause change in the productivity and/or species composition of the vegetation. It could be argued that any such anthropogenic change in a natural system is 'harmful', this being the word used in the definition of a critical load. It is important to determine what rate of N deposition would cause detectable change in arctic vegetation, and how this compares to current rates of deposition. In other words, what is the critical load, and is it being exceeded?

There are few datasets for N deposition in the Arctic and many of those that do exist are incomplete. Table 10.3 provides a summary of data, calculated on the assumptions (based on data in Statens Forurensningstilsyn 1994) that deposition of NH$_4^+$–N equals deposition of NO$_3^-$–N and that dry deposition equals 50% of wet deposition (a conservative estimate). The data in Table 10.3 demonstrate that total N deposition at most arctic sites studied is within the range 0.1–0.5 g N m^{-2} year^{-1}, but at some it reaches 1 g N m^{-2}year^{-1}, this being due to pollutant deposition (see Nenonen 1991; Jaffe & Zukowski 1993).

There have been many N fertilization experiments in the Arctic. In a study at Ny-Ålesund, Spitsbergen, the effects of N deposition on the arctic dwarf shrubs *Salix polaris*, *Dryas octopetala* and *Cassiope tetragona* are under investigation

(Baddeley, Woodin & Alexander 1994). Background deposition at the site is
c. $0.1 g N m^{-2} year^{-1}$ (see Table 10.3), and experimental additions are at 1 and 5 g
$N m^{-2} year^{-1}$, representing current deposition rates at polluted arctic and British
sites respectively. Previous fertilization experiments on the same, and closely related,
species have shown them to conform to expected trends (see Chapter 4) with the
deciduous *Salix* showing the most positive response to fertilization, the evergreen
Cassiope showing the least positive (and sometimes negative) response, and *Dryas*,
which has an intermediate wintergreen strategy, showing an intermediate range of
responses (Table 10.4). However, the fertilization rates used in all these experiments
were greatly in excess of actual rates of N deposition in the Arctic and thus do not
help define the critical load.

TABLE 10.3. Estimates of total nitrogen deposition at sites in the Arctic. For incomplete datasets,
calculation of total N is based on the assumptions that $NO_3^--N = NH_4^+-N$ and dry deposition = 50%
wet deposition (see text).

Site	Estimated total N deposition ($g N m^{-2} year^{-1}$)	Reference on which estimate is based
Norwegian Arctic		
Spitsbergen	0.08–0.12	Statens Forurensningstilsyn 1994
Bjornoya	0.33	Joranger & Semb 1989
Hopen	0.39	Joranger & Semb 1989
Jan Mayen	0.43	Joranger & Semb 1989
Northern Norway		
Jergul	0.16	Joranger & Semb 1989
Svanvik	0.18–0.31	Statens Forurensningstilsyn 1994
Andoya	0.61	Cited in Malmer 1988
Northern Sweden		
Ricksgränsen	0.10	Cited in Malmer 1988
Stordalen	0.09	Cited in Malmer 1988
Arctic Russia		
Most areas	≤ 0.20	Cited in Nenonen 1991
Taymyr Peninsula	≤ 1.00	Cited in Nenonen 1991
Kola Peninsula	≤ 0.40	Cited in Nenonen 1991
Iceland	0.06–0.60	Jónsdóttir, Callaghan & Lee 1995
Greenland	0.20–0.50	Cited in Jaffe & Zukowski 1993
Alaska		
Northern	1.10 (winter only)	Jaffe & Zukowski 1993
Interior	0.64 (winter only)	Jaffe & Zukowski 1993

TABLE 10.4. Effects of fertilization at various rates of N application on the above-ground growth of *Salix* spp., *Dryas* spp. and *Cassiope tetragona* at arctic sites: +, positive effect; o, no effect; −, negative effect.

N fertilization rate (g N m^{-2} year^{-1} × no. of years of treatment)	Species response			Site	Reference
	Salix pulchra	*Dryas octopetala*	*Cassiope tetragona*		
25 × 1	+			Alaska, tussock tundra	Chapin & Shaver 1989§
10 × 5	+			Alaska, tussock tundra	Matthes-Sears *et al.* 1988§
10 × 2	+*	−		Alaska, fellfield	Fox 1992¶
25 × 2	+		−	Alaska, tussock tundra	Shaver & Chapin 1986¶
25 × 2	+		−	Alaska, tussock tundra	Shaver & Chapin 1986¶
25 × 1	o		o	Alaska, tussock tundra	Shaver & Chapin 1986¶
25 × 1	+†	−‡	−	Ellesmere Island, sedge meadow, mesic and dry-mesic heath	Henry, Freedman & Svoboda 1986
5 × 1	o†	−‡	o	Ellesmere Island, sedge meadow, mesic and dry-mesic heath	Henry, Freedman & Svoboda 1986
15 × 1, 10 × 2		+		Alaska, snow-bed	McGraw 1985§
15 × 1, 10 × 2		−		Alaska, fellfield	McGraw 1985§
17 × 1		+		N. Sweden, graminoid tundra	Jonasson 1992¶
5 × 1, 10 × 2			o	Spitsbergen, beach ridge	Havström, Callaghan & Jonasson 1993¶
5 × 1, 10 × 2			o	N. Sweden, fellfield	Havström, Callaghan & Jonasson 1993¶
5 × 1, 10 × 2			+	N. Sweden, subalpine heath	Havström, Callaghan & Jonasson 1993¶

* *Salix phlebophylla*, † *Salix arctica*, ‡ *Dryas integrifolia*.
§ P also added, ¶ PK also added.

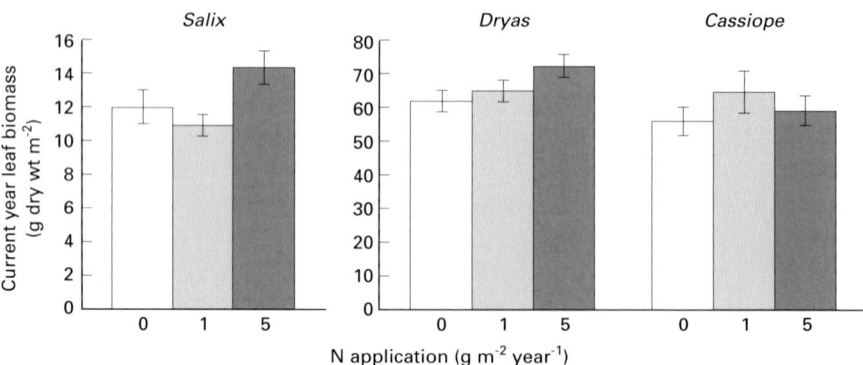

FIG. 10.3. Current year leaf biomass production (mean ± 1 SE) by *Salix polaris*, *Dryas octopetala* and *Cassiope tetragona* in the third year of N fertilization at 0, 1 or 5gm⁻²year⁻¹. Effects of N (ANOVA): *Salix* $P < 0.05$; *Dryas* and *Cassiope* $P > 0.05$.

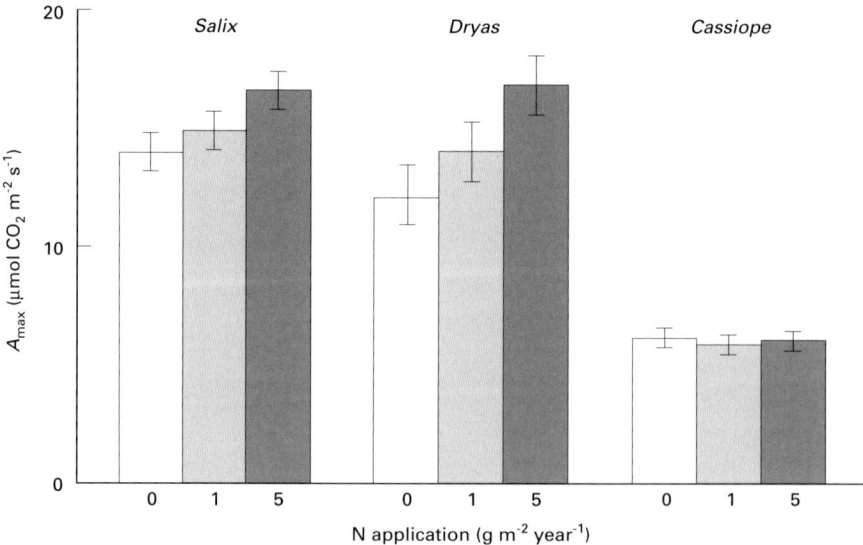

FIG. 10.4. Light-saturated rate of photosynthesis (A_{max}; mean ± 1 SE) in *Salix polaris*, *Dryas octopetala* and *Cassiope tetragona* in the second year of N fertilization at 0, 1 or 5 g N m⁻²year⁻¹.

In the experiment at Ny-Ålesund, species responsiveness followed the trend described above (Baddeley *et al.* 1994). Over 3 years, only *Salix* showed a significant increase in leaf biomass production, and only in response to 5 g N m⁻²year⁻¹ (Fig. 10.3). Photosynthetic parameters in both *Salix* and *Dryas* responded to fertilization, with intermediate responses to 1 g N m⁻²year⁻¹ apparent (Fig. 10.4 shows A_{max} as an example). N fertilization resulted in increased tissue N concentration

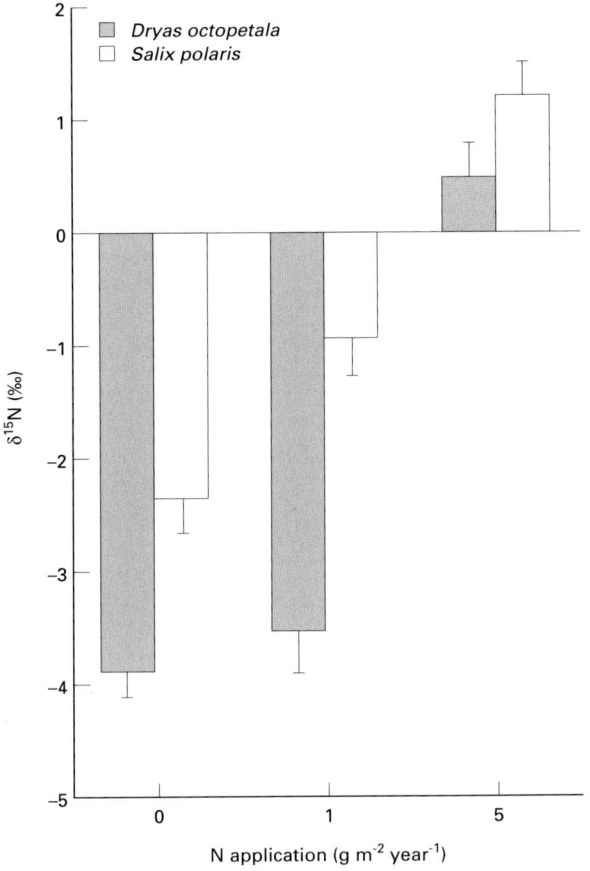

F IG. 10.5. Effect of 4 years' N fertilization on $\delta^{15}N$ in *Dryas octopetala* and *Salix polaris* leaf material (mean ± 1 SE). $\delta^{15}N$ of fertilizer (NH_4NO_3) = 2.6‰.

in all three species, again with an intermediate increase in plants receiving just 1 g N m^{-2} year^{-1}. Clear evidence that the added N has been taken up by the plants also comes from $\delta^{15}N$ data (Fig. 10.5) which shows the ^{15}N signature of the leaf material shifted towards that of the fertilizer N ($\delta^{15}N$ = 2.6‰) after 4 years of treatment (L. Högbom, I.J. Alexander & S.J. Woodin unpublished data). Below ground, the 5 g N m^{-2} year^{-1} treatment caused reductions in root biomass, but mycorrhizal infection of *Salix* and *Cassiope* roots was significantly reduced by just 1 g N m^{-2} year^{-1} (Fig. 10.6 shows data for *Salix*) (I.J Alexander, J.A. Baddeley & S.J. Woodin unpublished data).

The critical load of N for arctic heath has most recently been suggested to be within the range 0.5–1.5 g N m^{-2} year^{-1} (Hornung *et al.* 1995). On the basis that any anthropogenically induced change to vegetation is harmful, the data

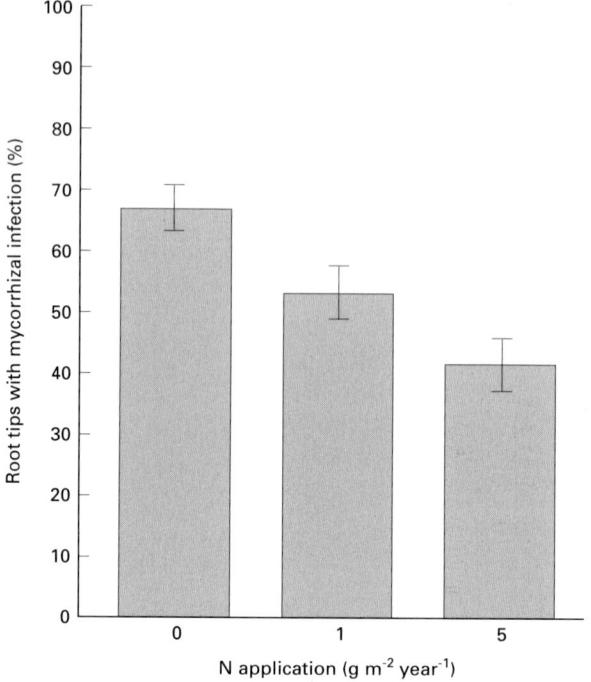

FIG. 10.6. Effect of 3 years' N fertilization on mycorrhizal infection of *Salix polaris* (mean ± 1 SE). Effect of N (ANOVA) $P < 0.001$.

described above verify for the first time that the critical load is indeed less than $1 \, g \, N \, m^{-2} year^{-1}$. Considering the deposition data in Table 10.3, it is likely that this critical load is already exceeded for some heathland ecosystems in the Arctic.

EFFECTS OF NITROGEN DEPOSITION ON ARCTIC BRYOPHYTES

Bryophytes attain their maximum relative importance in terms of biomass and production in tundra vegetation (Richardson 1981), and are influential in many ecological processes (see Chapter 3). Since most bryophytes assimilate inorganic N directly from the atmosphere rather than the soil, and they have a very low N requirement, they are potentially more sensitive to pollutant N deposition than are higher plants. Bryophytes in the Arctic are considered not to be N limited (Russell 1990), and thus any increase in N deposition may constitute an excess supply for some species.

The ability of an arctic bryophyte to utilize the N deposited to it from the atmosphere was illustrated in an experiment at Abisko in subarctic Sweden, in which *Sphagnum fuscum* was packed into mini-lysimeters and replaced into the bryophyte carpet. Precipitation falling, and that percolating through the *Sphagnum*, was analysed, demonstrating that all the NO_3^-–N and NH_4^+–N deposited from the atmosphere to the *Sphagnum* was immobilized by the moss (Woodin & Lee 1987a). In similar work in Iceland, Jónsdóttir, Callaghan and Lee (1995) found that *Racomitrium lanuginosum* immobilized all the N (as NO_3^-, NH_4^+ and NH_4NO_3) deposited to it, even after 4 years of N treatment at an average rate of $4\,g\,N\,m^{-2}$ year^{-1}, suggesting that bryophytes may buffer the ecosystem against effects of increased N deposition, at least until some of it is released through decomposition.

A study of the substrate-inducible nitrate assimilating enzyme, nitrate reductase, has also demonstrated the close coupling of *Sphagnum fuscum* to its atmospheric N supply, with enzyme activity being induced in response to the NO_3^-–N deposited in each precipitation event (Fig. 10.7). However, repeated spraying of the bryophyte carpet with 1 mM NO_3^- rapidly caused a loss of enzyme activity (Fig. 10.8), showing that excess N deposition disrupts the physiology of *Sphagnum* (see Woodin & Lee 1987b). These data demonstrate extreme sensitivity to N deposition, but they do not enable the calculation of a critical load.

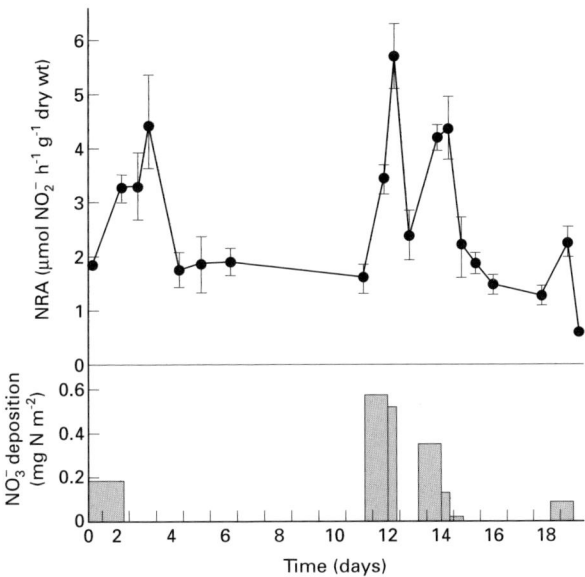

FIG. 10.7. Nitrate reductase activity (NRA; mean ± 1 SE) induced in *Sphagnum fuscum* by nitrate deposition in precipitation at Abisko, northern Sweden. (After Lee & Woodin 1988.)

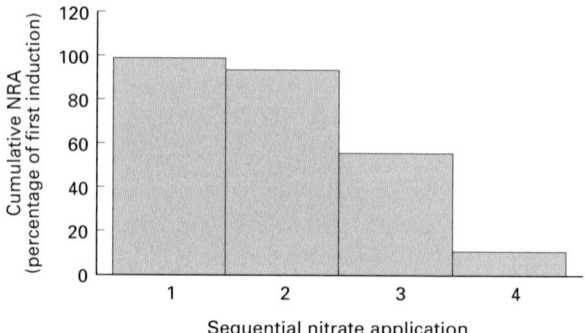

FIG. 10.8. Total nitrate reductase activity in *Sphagnum fuscum* induced by each of four successive 1 mM nitrate applications to a mire surface at Abisko, northern Sweden (cumulative nitrate reductase activity (NRA) expressed as percentage of total activity induced by first application, $n = 16$). Enzyme activity returned to a constitutive level between each application. (After Lee, Woodin & Press 1986 with kind permission from Kluwer Academic Publishers.)

Racomitrium showed only a very small positive growth response to the addition of $4 \, g \, N \, m^{-2} year^{-1}$ for 4 years (Jónsdóttir *et al.* 1995). On Spitsbergen, a fertilizer treatment containing $5 \, g \, N \, m^{-2} year^{-1}$ (applied for 5 years) caused 'relatively rapid' establishment of a bryophyte mat on denuded areas (Klokk & Rønning 1987). In another experiment at Abisko, Potter *et al.* (1995) found a decrease in overall bryophyte cover, with a significant decrease in growth of *Hylocomium splendens* and an increase in growth of *Polytrichum commune* in response to $10 \, g \, N \, m^{-2} year^{-1}$ for 3 years. Positive and negative growth responses of different bryophyte species have been reported to follow N fertilization in a few other arctic experiments (e.g. Chapin & Shaver 1985; Jonasson 1992), but these have been at even higher rates of N deposition, and thus are not useful for defining the critical load. A study of *Sphagnum* species at sites throughout Scandinavia, including subarctic Norway and Sweden, demonstrated increases in tissue N in proportion to atmospheric N deposition at deposition rates greater than $0.7 \, g \, N \, m^{-2} year^{-1}$, and this was suggested to be due to an excess of N in relation to the productivity of the moss layer as a result of anthropogenic emissions (Malmer 1988). Long-term fertilization experiments on bryophytes using realistic N application rates are required to demonstrate whether pollutant N deposition in the Arctic is currently sufficient to cause changes in the bryophyte component of the vegetation, and to define the rate at which it begins to do so.

EFFECTS OF POLLUTED SNOWMELT
ON ARCTIC BRYOPHYTES

Snowmelt provides an important input of nitrogen to tundra ecosystems (e.g.

Barsdate & Alexander 1975) and it has been suggested that the flush of growth shown by many tundra bryophyte species in early summer is due in part to the availability of nutrients from snowmelt (Russell 1990).

Solutes in the snowpack, including pollutants, are released at high concentrations in the initial stages of thaw (Tranter *et al.* 1986), and episodic acidification of rivers on the Kola Peninsula has already been related to the release of pollutants in the first melt waters (Moiseenko 1994). The potential for ecological damage in the Arctic by the release of pollutants from snow at high concentrations has been suggested (e.g. Körner & Fisher 1982; Barrie 1986a; Jaffe & Zukowski 1993), but not previously investigated. An investigation in an arctic environment in Scotland, the top of the Cairngorm plateau, is of direct relevance to this question. The snow on Cairngorm is more polluted than that in the Arctic, but although the maximum concentrations of NO_3^- and SO_4^{2-} in snow on Cairngorm are at least an order of magnitude greater, the mean values are at the upper end of the range values reported for arctic snow (Table 10.5).

The species investigated in the Cairngorm study was *Kiaeria starkei* which, in Scotland, is strongly associated with areas of the longest snow-lie duration within late snow-beds (Woolgrove & Woodin 1994). The first question addressed was whether the bryophyte actually assimilates pollutants from snowmelt, particularly when still under the snow. *Kiaeria* collected from under snow was packed into mini-lysimeters with columns of snow placed over them and allowed to melt slowly (at 2–4°C) so that the melt water percolated through the bryophyte mat. The snow itself provided a NO_3^- 'dose' equivalent to 0.04 g N m^{-2} to the *Kiaeria* mat. Additional NO_3^- was sprayed on to the surface of some snow columns to provide doses ranging up to 57 g N m^{-2}. The *Kiaeria* immobilized virtually all the NO_3^- from the snowmelt; even at the highest dose less than 1% of the NO_3^- applied passed through the bryophyte mat. It was also demonstrated that *Kiaeria* is capable

TABLE 10.5. Concentrations of NO_3^-–N and SO_4^{2-}–S (mean or range) in snow samples from arctic sites.

Site(s)	NO_3^-–N (μg l^{-1})	SO_4^{2-}–S (μg l^{-1})	Reference
Spitsbergen	< 14–56	32–640	Semb, Braekkan & Joranger 1984
Northern Alaska	155	613	Jaffe & Zukowski 1993
Interior Alaska	36	60	Jaffe & Zukowski 1993
Barrow, Alaska	26		Barsdate & Alexander 1975
Greenland	21	47	Davidson *et al.* 1993
NW Territory, Canada	70–98	112–288	Cited in Barrie 1986b
Cairngorm, Scotland	1–3248, mean 131	19–4928, mean 885	Tranter *et al.* 1986

of nitrate reductase activity, and thus nitrate assimilation, in the environmental conditions which prevail under the snowpack (Woolgrove & Woodin 1996a). Thus bryophytes do have the potential to accumulate pollutants from snowmelt, even while they are still under snow cover.

The second question addressed was whether the concentrations of pollutants in snowmelt are sufficient to be physiologically damaging to bryophytes. Membrane integrity in *Kiaeria* was monitored by measurement of potassium leakage from cells. Artificial snowmelt solutions containing SO_4^{2-} and NO_3^- at the maximum, mean and minimum concentrations measured on Cairngorm (based on Tranter *et al.* 1986) all caused a rapid increase in potassium leakage, and thus membrane damage, in *Kiaeria* while it was being treated (Fig. 10.9). Four weeks after the end of a week of pollutant treatment, potassium leakage in plants which had received

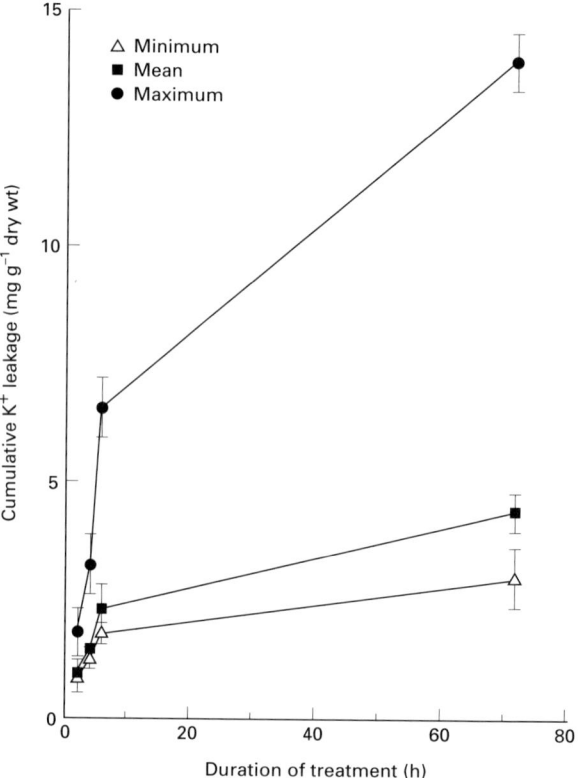

FIG. 10.9. Potassium leakage from *Kiaeria starkei* (cumulative; mean ± 95% confidence interval) caused by timed exposure to simulated snowmelt containing SO_4^{2-} and NO_3^- at the minimum, mean and maximum concentrations recorded on Cairngorm by Tranter *et al.* (1986).

the mean and maximum concentrations of pollutants was 150% and 250% that in control plants respectively (Woolgrove & Woodin 1996b). Thus, the physiological damage caused by realistic concentrations of pollutants in snowmelt is significant and sustained and, for a bryophyte which only has a few weeks growing season available, may significantly reduce its annual productivity.

The potential for pollutants in snowmelt to cause damage to underlying bryophytes has been clearly demonstrated; the question remains as to whether this actually occurs in the Arctic. Since mean pollutant concentrations on Cairngorm, which do cause damage, are similar to the highest concentrations reported for the Arctic it is likely that in some areas of the Arctic such damage does occur.

CONCLUSIONS

The fact that air pollution occurs in the Arctic has long been accepted, and arctic air quality has been the subject of several major research programmes (e.g. Ottar *et al.* 1986; Davidson & Schnell 1993). Actual deposition rates of air pollutants are less well documented but, generally speaking, are relatively low in the Arctic, and this has perhaps led to a perception that acid deposition is not really a problem to arctic ecosystems. Two relatively recent developments give us cause to examine the situation in a new light. One is the increased access to data on pollutant emissions within Arctic regions of the former USSR, which makes it very apparent that wide areas of subarctic and tundra vegetation are actually subject to significant rates of acid deposition. The second is the development of the concept of critical loads, in which pollutant exposure is considered in relation to ecosystem tolerance. This approach suggests that the high sensitivity of arctic ecosystems will lead to change or damage being caused by small amounts of pollutants.

Studies on the Kola Peninsula and in northern Fennoscandia have now clearly demonstrated that in addition to very severe ecological damage over areas of hundreds to thousands of square kilometres, there is evidence of exceedence of critical loads in relation to tree vitality, water and soil chemistry and frequency of sensitive lichens, in an area of over $100000\,km^2$ of arctic vegetation. It is likely that the same situation exists in and around other source areas in arctic Russia for which data are less available, such as Norilsk and the Taimyr Peninsula (AMAP 1994).

The ecological damage described above, although very extensive, is 'localized' in the sense that it can be directly related to adjacent pollutant sources. It is more difficult to identify whether pollutants transported long distances are causing damage at remote locations within the Arctic. Considering N deposition, this chapter has demonstrated that rates of deposition already occurring at some locations may be sufficient to affect dwarf shrubs and their mycorrhizal symbionts within arctic heath ecosystems. Arctic bryophytes have also been shown to be highly sensitive to N deposition, but the quantification of a critical load for damage is still required. The potential threat to this important component of arctic vegetation from polluted snowmelt has been demonstrated, although studies of both snowmelt chemistry and bryophyte physiology in the Arctic are required to confirm whether such damage does occur.

There is sufficient evidence that acid deposition in the Arctic has the potential to cause change within vegetation communities to merit concern and spur further investigation of both pollutant deposition and ecosystem sensitivity. The extent to which the Arctic should be protected from air pollution can only be an international political decision, but it should be informed by ecological information through the critical-loads approach.

REFERENCES

AMAP (1994). *Minutes from the Sixth Meeting of the Arctic Monitoring and Assessment Programme Working Group.* AMAP, Oslo.

Baddeley, J.A., Woodin, S.J. & Alexander, I.J. (1994). Effects of increased nitrogen and phosphorus availability on the photosynthesis and nutrient relations of three arctic dwarf shrubs from Svalbard. *Functional Ecology*, **8**, 676–685.

Barrett, M. & Protheroe, R. (1994). *Sulphur Emission from Large Point Sources in Europe.* Swedish NGO. Secretariat on Acid Rain, Goteborg.

Barrie, L.A. (1986a). Arctic air pollution: an overview of current knowledge. *Atmospheric Environment*, **20**, 643–663.

Barrie, L.A. (1986b). Background pollution in the Arctic air mass and its relevance to North American acid rain studies. *Water, Air and Soil Pollution*, **30**, 765–777.

Barsdate, R.J. & Alexander, V. (1975). The nitrogen balance of Arctic tundra: pathways, rates and environmental implications. *Journal of Environmental Quality*, **4**, 111–117.

Berendse, F. & Jonasson, S. (1992). Nutrient use and nutrient cycling in northern ecosystems. *Arctic Ecosystems in a Changing Climate: An Ecophysiological Perspective* (Ed. by F.S. Chapin, R.L. Jefferies, J.F. Reynolds, G.R. Shaver & J. Svoboda), pp. 337–356. Academic Press, San Diego.

Chapin, F.S. & Shaver, G.R. (1985). Individualistic growth response of tundra plant species to environmental manipulations in the field. *Ecology*, **66**, 564–576.

Chapin, F.S. & Shaver, G.R. (1989). Differences in growth and nutrient use among arctic plant growth forms. *Functional Ecology*, **3**, 73–80.

Davidson, C.I., Jaffrezo, J.-I., Mosher, B.W., Dibb, J.E., Borys, R.D., Bodhaine, B.A., Rasmussen, R.A., Boutron, C.F., Gorlach, U., Cachier, H., Ducret, J., Colin, J.-L., Heidam, N.Z., Kemp, K. & Hillamo, R. (1993). Chemical constituents in the air and snow at Dye 3, Greenland – I. Seasonal variations. *Atmospheric Environment*, **27A**, 2709–2722.

Davidson, C.I. & Schnell, R.C. (1993). Arctic air, snow and ice chemistry. *Atmospheric Environment*, Special Issue **17/18**.

EMEP (1994). *Transboundary Acidifying Pollution in Europe: Calculated Fields and Budgets 1985–93.* EMEP MSC-W Report 1/94. Norwegian Meteorological Institute, Oslo.

Fox, J.F. (1992). Responses of diversity and growth-form dominance to fertility in Alaskan tundra fellfield communities. *Arctic and Alpine Research*, **24**, 233–237.

Grennfelt, P. & Thornelof, E. (1992). *Critical Loads for Nitrogen.* Nordic Council of Ministers, Copenhagen.

Havström, M., Callaghan, T.V. & Jonasson, S. (1993). Differential growth responses of *Cassiope tetragona*, an arctic dwarf-shrub, to environmental perturbations among three contrasting high- and subarctic sites. *Oikos*, **66**, 389–402.

Henry, G.H.R., Freedman, B. & Svoboda, J. (1986). Effects of fertilization on three tundra plant communities of a polar desert oasis. *Canadian Journal of Botany*, **64**, 2502–2507.

Hornung, M., Sutton, M.A. & Wilson, R.B. (Eds) (1995). *Mapping and Modelling of Critical Loads for Nitrogen – a Workshop Report.* Institute of Terrestrial Ecology, Edinburgh.

Jaffe, D.A. & Zukowski, M.D. (1993). Nitrate deposition to the Alaskan snowpack. *Atmospheric Environment*, **27A**, 2935–2941.

Jonasson, S. (1992). Plant response to fertilisation and species removal in tundra related to community structure and clonality. *Oikos*, **63**, 420–429.

Jónsdóttir, I.S., Callaghan, T.V. & Lee, J.A. (1995). Fate of added nitrogen in a moss–sedge Arctic community and effects of increased nitrogen deposition. *The Science of the Total Environment*, **160/161**, 677–685.

Joranger, E. & Semb, A. (1989). Major ions and scavenging of sulphate in the Norwegian Arctic. *Atmospheric Environment*, **23**, 2463–2469.

Kalabin, G. (1992). Implementation of the ecological imperative strategy when accessing the natural resources of the northern regions. *Symposium on the State of the Environment and Environmental Monitoring in Northern Fennoscandia and the Kola Peninsula* (Ed. by E. Tikkanen, M. Varmola & T. Katermaa), pp. 365–367. Arctic Centre Publications, Rovaniemi.

Klokk, T. & Rønning, O.I. (1987). Revegetation experiments at Ny-Ålesund, Spitsbergen, Svalbard. *Arctic and Alpine Research*, **19**, 549–553.

Körner, R.M. & Fisher, D. (1982). Acid snow in the Canadian high Arctic. *Nature*, **295**, 137–140.

Kryuchkov, V. (1990). Extreme anthropogenic load and the state of the north taiga ecosystem. *Effects of Air Pollution and Acidification in Combination with Climatic Factors on Forests, Soils and Waters in Northern Fennoscandia* (Ed. by K. Kinnunen & M. Varmola), pp. 197–205. Nordic Council of Ministers, Copenhagen.

Laj, P., Palais, J.M. & Sigurdsson, H. (1992). Changing sources of impurities to the Greenland ice sheet over the last 250 years. *Atmospheric Environment*, **26A**, 2627–2640.

Lee, J.A. & Woodin, S.J. (1988). Vegetation structure and the interception of acidic deposition by ombrotrophic mires. *Vegetation Structure in Relation to Carbon and Nutrient Economy* (Ed. by J.T.A. Verhoeven, G.W. Heil & M.J.A. Werger), pp. 137–147. Academic Publishing, The Hague.

Lee, J.A., Woodin, S.J. & Press, M.C. (1986). Nitrogen assimilation in an ecological context. *Fundamental, Ecological and Agricultural Aspects of Nitrogen Metabolism in Higher Plants* (Ed. by H. Lambers, J.J. Neetson & I. Stulen), pp. 331–346. Martinus Nijhoff, Dordrecht.

Lien, L., Henriksen, A. & Traaen, T.S. (1993). *Critical Loads of Acidity for Surface Waters – Svalbard.* Norwegian Institute for Water Research, Oslo.

Lövblad, G., Amann, M., Andersen, B., Hovmand, M., Joffre, S. & Pedersen, U. (1992). Deposition of sulfur and nitrogen in the Nordic countries: present and future. *Ambio*, **21**, 339–347.

Malmer, N. (1988). Patterns in the growth and the accumulation of inorganic constituents in the *Sphagnum* cover on ombrotrophic bogs in Scandinavia. *Oikos*, **53**, 105–120.

Matthes-Sears, U., Mattes-Sears, W.C., Hastings, S.J. & Oechel, W.C. (1988). The effects of topography and nutrient status on the biomass, vegetative characteristics and gas exchange of two deciduous shrubs on an arctic tundra slope. *Arctic and Alpine Research*, **20**, 343–351.

McGraw, J.B. (1985). Experimental ecology of *Dryas octopetala* ecotypes. III Environmental factors and plant growth. *Arctic and Alpine Research*, **17**, 229–239.

Mitchell, M. (1956). Visual range in the polar regions with particular reference to the Alaskan Arctic. *Journal of Atmospheric and Terrestrial Physics*, Special Supplement, 195–211.

Moiseenko, T. (1994). Acidification and critical loads in surface waters: Kola, northern Russia. *Ambio*, **23**, 418–424.

Nadelhoffer, K.J., Giblin, A.E., Shaver, G.R. & Likens, A.E. (1992). Microbial processes and plant nutrient availability in Arctic soils. *Arctic Ecosystems in a Changing Climate: An Ecophysiological Perspective* (Ed. by F.S. Chapin, R.L. Jefferies, J.F. Reynolds, G.R. Shaver & J. Svoboda), pp. 281–300. Academic Press, San Diego.

Neftel, A., Beer, J., Oeschger, H., Zürcher, F. & Finkel, R.C. (1985). Sulphate and nitrate concentrations in snow from south Greenland 1895–1978. *Nature*, **314**, 611–613.

Nenonen, M. (1991). Report on acidification in the Arctic countries: man-made acidification in a world of natural extremes. *The State of the Arctic Environment*. Arctic Centre Publications, Rovaniemi.

Nilsson, I. & Grennfelt, P. (1988). *Critical Loads for Sulphur and Nitrogen*. Nordic Council of Ministers, Copenhagen.

Ottar, B., Gotaas, Y., Hov, O., Ivresen, T., Joranger, E., Oehme, M., Pacyna, J., Semb, A., Thomas, W. & Vitols, V. (1986). *Air Pollutants in the Arctic*. NILU, Lillestrom.

Potter, J.A., Press, M.C., Callaghan, T.V. & Lee, J.A. (1995). Growth responses of *Polytrichum commune* and *Hylocomium splendens* to simulated environmental change in the sub-arctic. *New Phytologist*, **131**, 533–541.

Rahn, K.A., Joranger, E., Semb, A. & Conway, T.J. (1980). High winter concentrations of SO_2 in the Norwegian Arctic and transport from Eurasia. *Nature*, **287**, 824–825.

Raitio, H. (1992). The foliar chemical composition of Scots pines in Finnish Lapland and on the Kola Peninsula. *Symposium on the State of the Environment and Environmental Monitoring in Northern Fennoscandia and the Kola Peninsula* (Ed. by E. Tikkanen, M. Varmola & T. Katermaa), pp. 226–231. Arctic Centre Publications, Rovaniemi.

Richardson, D.H.S. (1981). *The Biology of Mosses*. Blackwell Scientific Publications, Oxford.

Russell, S. (1990). Bryophyte production and decomposition in tundra ecosystems. *Biological Journal of the Linnean Society*, **104**, 3–22.

Salemaa, M., Lindgren, M. & Jukola-Sulonen, E.-L. (1992). Vitality of Scots pine stands at different distances from emission sources in the Kola Peninsula. *Symposium on the State of the Environment and Environmental Monitoring in Northern Fennoscandia and the Kola Peninsula* (Ed. by E. Tikkanen, M. Varmola & T. Katermaa), pp. 323–324. Arctic Centre Publications, Rovaniemi.

Semb, A., Braekkan, R. &, Joranger, E. (1984). Major ions in Spitsbergen snow samples. *Geophysical Research Letters*, **11**, 445–448.

Shaver, G.R. & Chapin, F.S. (1980). Response to fertilization by various plant growth forms in an Alaskan tundra: nutrient accumulation and growth. *Ecology*, **61**, 662–675.

Shaver, G.R. & Chapin, F.S. (1986). Effects of fertiliser on production and biomass of tussock tundra, Alaska. *Arctic and Alpine Research*, **18**, 261–268.

Sivertsen, B., Makarova, T., Hagen, L.O. & Baklanov, A.A. (1992). *Air Pollution in the Border Areas of Norway Russia*. Norwegian Institute for Air Research, Lillestrom.

Statens Forurensningstilsyn (1994). *Overåvking av Langtransportert Forurenset Luft og Nedbør. Årsrapport 1993*. Statlig Program for Forurensningsovervåking Rapport 583/94, Oslo.

Sutinen, M.-L. (1992). The effects of air pollution on the seasonal changes in the frost hardiness of the needles of *Pinus sylvestris* L. *Symposium on the State of the Environment and Environmental Monitoring in Northern Fennoscandia and the Kola Peninsula* (Ed. by E. Tikkanen, M. Varmola & T. Katermaa), pp. 232–234. Arctic Centre Publications, Rovaniemi.

Tranter, M., Brimblecombe, P., Davies, T.D., Vincent, C.E., Abrahams, P.W. & Blackwood, I. (1986). The composition of snowfall, snowpack and meltwater in the Scottish Highlands – evidence for preferential elution. *Atmospheric Environment*, **20**, 517–525.

Tuovinen, J.-P. & Laurila, T. (1992). Key aspects of sulphur pollution in northernmost Europe. *Symposium on the State of the Environment and Environmental Monitoring in Northern Fennoscandia and the Kola Peninsula* (Ed. by E. Tikkanen, M. Varmola & T. Katermaa), pp. 37–40. Arctic Centre Publications, Rovaniemi.

Umwelt Bundesamt (1993). *Manual on Methodologies and Criteria for Mapping Critical Loads/Levels and Geographical Areas where they are Exceeded*. Federal Environmental Agency, Berlin.

Vilchek, G.E. & Bykova, O.Y. (1992). The origin of regional ecological problems within the northern Tyumen Oblast, Russia. *Arctic and Alpine Research*, **24**, 99–107.

Woodin, S.J. & Lee, J.A. (1987a). The fate of some components of acidic deposition in ombrotrophic mires. *Environmental Pollution*, **45**, 61–72.

Woodin, S.J. & Lee, J.A. (1987b). The effects of nitrate, ammonium and temperature on nitrate reductase activity in *Sphagnum* species. *New Phytologist*, **105**, 103–115.

Woolgrove, C.E. & Woodin, S.J. (1994). Relationships between the duration of snowlie and the distribution of bryophyte communities within snowbeds in Scotland. *Journal of Bryology*, **18**, 253–260.

Woolgrove, C.E. & Woodin, S.J. (1996a). Effects of pollutants in snowmelt on *Kiaeria starkei*, a characteristic species of late snowbed dominated vegetation. *New Phytologist*, **133**, 519–529.

Woolgrove, C.E. & Woodin, S.J. (1996b). Ecophysiology of a snowbed bryophyte, *Kiaeria starkei*, during snowmelt and uptake of nitrate from polluted meltwater. *Canadian Journal of Botany*, **74**, 1095–1103.

11. Effects of enhanced UV-B radiation on subarctic vegetation

LARS OLOF BJÖRN*, TERRY V. CALLAGHAN§,
CAROLA GEHRKE†¶, DYLAN GWYNN-JONES§,
BJÖRN HOLMGREN‡¶, ULF JOHANSON*
AND MATS SONESSON†¶

*Section of Plant Physiology, Lund University, Box 117, S-221 00 Lund,
Sweden,
†Section of Plant Ecology, Lund University, S-223 62 Lund, Sweden,
‡Department of Meteorology, Uppsala University, Box 516, S-751 20 Uppsala,
Sweden,
§Department of Animal and Plant Sciences, University of Sheffield,
Sheffield S10 2TN, UK and
¶Abisko Scientific Research Station, Royal Swedish Academy of Sciences,
S-981 07 Abisko, Sweden

INTRODUCTION

Stratospheric ozone depletion over Antarctica (the 'ozone hole') has been in the headlines for a number of years. The biological effects of the resulting increase in ultraviolet-B (UV-B) radiation in the sea and on the Antarctic continent have been the subject of a recent monograph (Weiler & Penhale 1994). Although less dramatic, and not given the same public attention, there has also been a continuing depletion of stratospheric ozone in the Arctic.

To compare levels or exposures of ultraviolet radiation in a biological context when the spectral composition of the radiation is not constant, one has to take into account that different spectral components have different efficiencies; that is, different impacts per unit power or unit energy. This is done by using a weighting function, the value of which varies with wavelength. The spectral irradiance at each wavelength is multiplied by the value of the weighting function for that wavelength, and the product is integrated over wavelength (in practice, the integral is approximated by the sum of a finite number of products). One of the more common weighting functions in use is the absorption spectrum of DNA, as DNA is one of the most important molecular targets and the efficiency of radiation to alter DNA is, in the first approximation (there are deviations), proportional to its absorption by DNA.

For much of the time the annual average amount of stratospheric ozone over Scandinavia is now below the natural level, due to anthropogenic air pollution. Over the period 1979–93 the decrease in stratospheric ozone has been equivalent to a 10% increase in annual DNA-damaging radiation at a latitude of 65°N (Madronich *et al.* 1995). In the spring the relative increase is greater than the annual increase. However, it is probable that the ultraviolet radiation level is still close to natural, due to the compensating effect of increased sulphate aerosol and tropospheric ozone. For obvious reasons we do not wish to depend on this 'protecting' effect in the future. Much further decrease is not currently predicted; rather a slow return to more normal levels is foreseen to start within a few years (Madronich *et al.* 1995).

The most authoritative recent general overview of the effects of increased UV-B is UNEP's assessment of environmental effects of ozone depletion (UNEP 1995). Other general treatises are those of Young *et al.* (1993), Biggs and Joyner (1993), two SCOPE reports (SCOPE 1992, 1993), and, most recently, Rozema *et al.* (1996). Special topics have been covered by Caldwell and Flint (1994), Cullen and Neale (1994), Strid, Chow and Anderson (1994), Teramura and Sullivan (1994) and Teramura and Ziska (1994).

The special problems associated with arctic ozone depletion have been discussed in a recent IASC report (IASC 1995). The natural ultraviolet radiation level in the Arctic is low. On the other hand, organisms living there are not adapted to high UV-B, and the relative change in radiation level is likely to be greater there than at lower latitudes. A comparison of plants from different latitudes (Robberecht, Caldwell & Billings 1980; Barnes, Flint & Caldwell 1987) has shown that the epidermis of arctic plants offers little protection to UV-B.

A further reason to take special interest in the effects of UV-B at high latitudes is the recent finding by Takeuchi, Ikeda and Kasahara (1993) that growth inhibition by UV-B is more pronounced at a lower than at a higher temperature (although both temperatures in their study, 20 and 25°C, must be regarded as 'very high' in an arctic or subarctic perspective). Also most plants in the Arctic and Subarctic are long-lived, and do not often reproduce sexually. For this reason it is possible that ultraviolet damage could accumulate year after year to a larger extent than is the case with vegetation from lower latitudes.

The experiments outlined here are part of a larger group of projects dealing with effects of UV-B radiation on plants in the subarctic, in which the authors are involved. The field experiments described here are carried out in northern Sweden at Abisko Scientific Research Station of the Royal Swedish Academy of Sciences. Ozone depletion is simulated by supplementing the daylight intercepted by a natural subarctic heath ecosystem and by a bog with artificial ultraviolet radiation. The objective of this chapter is to give an overview of the project and the results obtained so far.

METHODS

Ozone column

To monitor the changes in the ozone layer, ozone values ('column ozone', meaning the total ozone from the ground to the top of the atmosphere) measured south (Vindeln) and north (Tromsø) of Abisko (68.3°N, 18.8°E) were compared (Björn & Holmgren 1996) with a computer model (Björn & Murphy 1985) based on measurements several decades ago. Usually linear interpolation between Vindeln and Tromsø was used to obtain an ozone value for Abisko, which is close to the line connecting these observation points. On a few days, when the value for Tromsø or Vindeln was lacking, observations at Sodankylä in Finland were used. The interpolation was then done between either Vindeln or Tromsø and Sodankylä to the point closest to Abisko, and this interpolated value used as the measured ozone value. The station at Vindeln (64.2°N, 19.6°E) is run by the Swedish Meteorological and Hydrological Institute, using a Dobson instrument; that at Tromsø (69.7°N, 19.0°E) by Nordlysobservatoriet, University of Tromsø, using a Dobson instrument; and that at Sodankylä (67.4°N, 26.6°E) by the Finnish Meteorological Institute, using a Brewer instrument.

Daylight monitoring, daylight models and computer programs

Proper administration of ultraviolet radiation in such a way that it corresponds to a certain level of ozone depletion depends both on measurements and computer modelling.

A computer program devised by Björn and Murphy (1985, see Björn & Teramura 1993 for a listing) makes it possible to compute the spectral irradiance throughout the UV-A and UV-B range provided that there are no clouds and no fog, and the ozone column agrees with the long-term average for the particular place and time. Initially this program alone was used for planning the experiments. Most of the UV-B treatments have been designed so that they would correspond to 15% ozone depletion (in addition to that actually taking place), based on clear skies. However, in reality many of the days at Abisko are more or less cloudy. This means that ambient UV-B is lower than the irradiance predicted by the model, and the UV-B enhancement in reality corresponds to a higher depletion level than the nominal 15%. For modelling photosynthetically active radiation (PAR), a modification of the model by Bird and Riordan (1986, see Björn 1994 for a program list) was used. There was excellent agreement for the time course of irradiance over the day for both PAR and UV-B when modelled and measured values were compared on a clear day, 5 July 1994. PAR and UV-B have different time courses over the day, with UV-B more confined to the hours around noon than is PAR.

From summer 1994 UV-B irradiance has been continuously monitored (from

March to October) at Abisko (68.3°N, 18.8°E, 360 m asl), along with PAR irradiance, and several other radiation-related parameters (UV-A, sun-minutes per hour, solar radiation from ultraviolet to 3000 nm and long-wavelength infrared thermal radiation from the Earth's atmosphere). UV-B is monitored by a broad-band sensor similar to the Robertson–Berger instrument, and also by a UV-B-sensitive photodiode. This allows correction for cloudiness when computing the levels of ozone depletion corresponding to the irradiation programs imposed.

Radiation measurements to date have been taken with 'cosine corrected' sensors, which record vectorial irradiance; that is, the radiation falling on a horizontal surface. This is not proportional to the radiation falling on a three-dimensional plant, and for this reason a measuring method has been devised by which fluence rate (scalar irradiance), that is, the radiation falling on a sphere of unit cross-section, can be estimated using cosine corrected sensors (Björn 1995).

UV-lamp arrangement

UV-B was generated by 40 W fluorescent lamps (Philips TL12 in the first experiments, later Q-Panel UVB-313). The 40 W lamps were arranged in parallel at a distance of 0.5 m from one another in groups of six, 1.5 m above ground. The centres of the central two lamps were covered with aluminium foil, to obtain a more uniform distribution of radiation (Björn & Teramura 1993). The lamps were connected to time switches in such a way that half the lamps in a group were switched on first, then the rest, then the first half were switched off, and then the rest. In this way a step-wise change of irradiance over the day was obtained. Half of the lamp groups were equipped with cellulose acetate filters that transmit UV-A and UV-B (but stop UV-C) radiation to provide treatments, while the other (control) groups were equipped with window glass that stops UV-B (as well as UV-C and some UV-A) but transmits most of the UV-A.

To compute the daily UV-B addition required, one needs to define a 'weighting function' describing the relative sensitivity of the plants to ultraviolet radiation of different wavelengths. We used the UV-B model mentioned above in combination with a weighting function devised by Thimijan, Carns and Campbell (1978), that resembles the Caldwell generalized plant spectrum (Caldwell 1971). The mathematical function used for the spectrum is $\exp\{-[(265-\lambda)/21]^2\}/\exp\{-(35/21)^2\}$, where λ is wavelength in nanometres. The ambient clear-sky radiation was computed for every nm and every hour of the day using the program, weighted and added up to yield the daily exposure by daylight. The radiation from the lamps was measured using an Optronic model 742 spectroradiometer and weighted in a similar fashion. Occasional spectral measurements were also taken of ambient UV-B.

It should be recognized that the weighting function used is somewhat arbitrary, and we have done several experiments to determine action spectra for ultraviolet effects on plants (Bornman, Björn & Åkerlund 1984; Negash & Björn 1986; Negash 1987; Cen & Björn 1994). However, it is assumed that the errors due to inappropriate weighting are not of such a magnitude as to invalidate the results, since the spectrum of the administered artificial radiation is rather similar to the difference spectrum between 'depleted' and 'normal' daylight (see Björn & Teramura 1993).

Complementary to the field experiments, experiments with plants grown in greenhouses and growth chambers have been carried out.

Plant material

In most experiments only the plants naturally growing on the ground below the lamps were investigated. These include the deciduous dwarf shrubs, *Vaccinium myrtillus* and *V. uliginosum*, the evergreen dwarf shrubs *V. vitis-idaea* and *Empetrum hermaphroditum* and the mosses *Hylocomium splendens*, *Polytrichum commune* and *Sphagnum fuscum*. Indoor UV-B exposures (greenhouse and growth chamber) were carried out with mosses and lichens (*Hylocomium splendens*, *Polytrichum commune*, *Cladonia arbuscula*, *Cetraria islandica* and *Stereocaulon paschale*). Leaf litter of *V. uliginosum* was collected for a laboratory experiment on litter decomposition, while the decomposition of leaf litter from *V. myrtillus* was studied *in situ* using the litter cup method.

RESULTS

Ozone column during the summer 1994

No ozone depletion at all could be detected during the summer of 1994. The values for Abisko interpolated from recent measurements at Vindeln and Tromsø are scattered around the model prediction, and their measured/modelled ratio is in fact 1.01 when averaged over the whole period (Fig. 11.1). For comparison, during the years 1988–93 the depletions during the summer half years at Norrköping (further south in Sweden) average 2.3% as compared to summer values measured at Uppsala during 1951–66, and the present trend at Norrköping is $0.8 \pm 0.3\%$ depletion per year (Josefsson 1996).

Radiation climate

By taking the ratio of the monitored and the modelled PAR values and the same for the UV-B radiation values, 'cloud transmission factors' for PAR and UV-B were computed. The factors differ for PAR and UV-B, but are strongly correlated ($r^2 = 0.84$, Fig. 11.2), and this may open a possibility for estimation of UV-B at places where it is not monitored, if PAR values are available. The data shown here

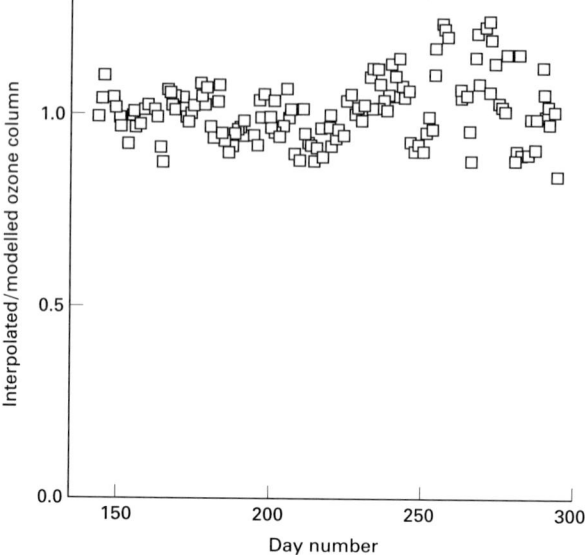

FIG. 11.1. The ratio of interpolated ozone column to modelled ozone column as a function of the time of the year.

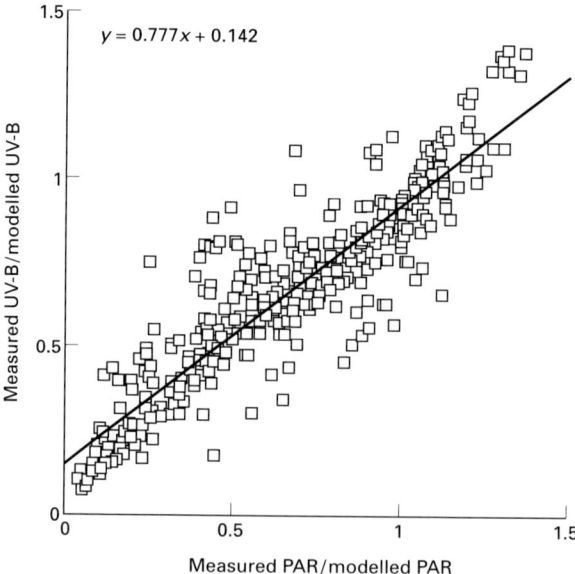

FIG. 11.2. The ratio between measured and modelled UV-B (cloud transmission factor) plotted vs. the corresponding ratio for photosynthetically active radiation (PAR). Both the PAR and the UV-B data were averaged hourly, and only values from Greenwich mean time (UTC) 7–15 were used. The linear regression coefficient (r^2) is 0.84.

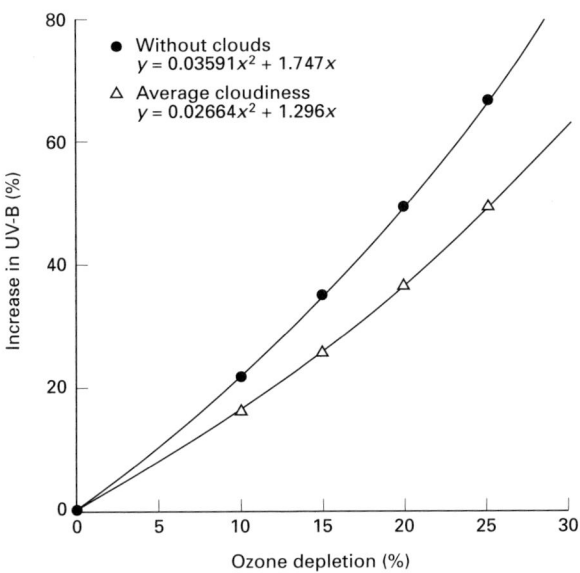

Fɪɢ. 11.3. The relation between ozone depletion and increase in plant-weighted UV-B for Abisko summer conditions. The upper graph is for clear skies, the lower one is for skies with normal cloud cover.

were collected during the period 19 June to 14 August 1994, and the relation between the 'cloud transmission factors' for PAR and UV-B might well be different during the spring. This is under investigation.

Integration of modelled and monitored UV-B over much of the summer (19 June–14 August) gave a real UV-B exposure that was 74.2% of that modelled for clear skies. By comparison with the relationship between ozone levels and UV-B exposure given by the model, it was determined from this that the treatment exposure nominally corresponding to 15% ozone depletion in fact, taking the effect of clouds into account, corresponds to 19% depletion, and the nominal 20% depletion corresponds to 25% with clouds (Fig. 11.3).

Biological effects of increased UV-B radiation

Only a brief outline of our results is given here. For details the reader is referred to other reports (Johanson *et al.* 1995a,b; Gehrke *et al.* 1995, 1996; Sonesson, Callaghan & Björn 1995; Yu & Björn 1996).

Growth and morphology of dwarf shrubs

Plants of *V. vitis-idaea* exposed to enhanced UV-B developed thicker leaves than

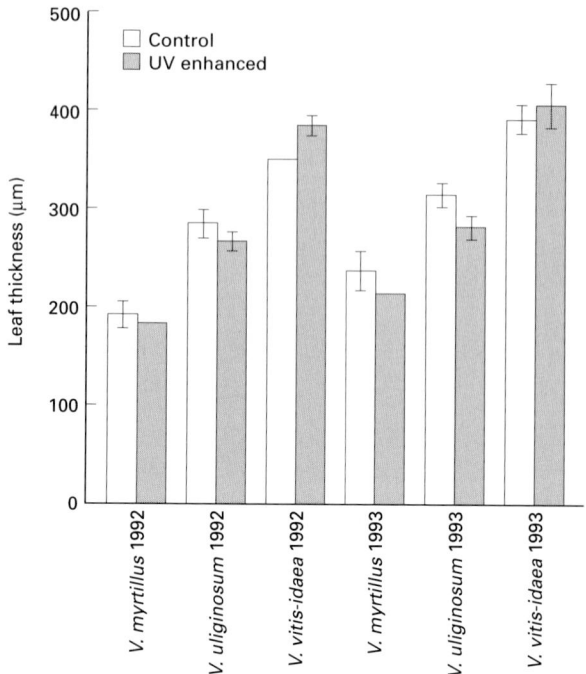

Fɪɢ. 11.4. Effects of enhanced UV-B radiation on leaf thickness of dwarf shrubs. UV-B enhancement results in thinner leaves of the deciduous species *Vaccinium myrtillus* and *V. uliginosum*, and thicker leaves of the evergreen species *V. vitis-idaea*. Vertical bars show standard errors.

control plants (by 4–9% depending on year), while those of the deciduous dwarf shrubs, *V. myrtillus* and *V. uliginosum,* became thinner (by 4–10% depending on year and species) as a consequence of UV-B enhancement (Fig. 11.4). There were significant effects on leaf dry weight and leaf area only for *V. uliginosum,* for which enhanced ultraviolet radiation caused increases of 29% in both parameters. The thinning of the deciduous leaves was not in line with what is usually observed, whereas the thickening of evergreen leaves conformed to earlier studies. Changes in leaf thickness would be expected to alter the light climate within individual leaves and also within whole plant canopies. There is also evidence that it affects herbivory: more area is devoured of thinner leaves to maintain the same input of leaf volume.

The relative longitudinal shoot growth (i.e. shoot growth under enhanced UV-B as divided by growth of the same shoot during the year before application of enhanced UV-B) was reduced in *E. hermaphroditum* by 14% after 1 year and by 33% after 2 years. For the *Vaccinium* species no significant effects on longitudinal growth of shoots were found after 1 year of irradiation, but after 2 years there was a reduction of 11% for *V. myrtillus*, of 10% for *V. uliginosum* and of 27% for *V. vitis-idaea* (Johanson *et al.* 1995b). Such reductions in shoot growth in response to enhanced UV-B radiation are commonly found (Sullivan & Teramura 1992).

These results suggest that the evergreen plants accumulate damage to a greater degree than deciduous plants, which annually discard damaged leaves. Although UV-B impacts on leaf and shoot growth were evident, preliminary analyses of changes in the cover of species within the dwarf-shrub community fail to show any impact at this scale. However, the sensitivity of cover measurements is less than that of the more precise measurements of leaf thickness and stem growth. The results so far suggest that subtle differences in shoot architecture and competition (Barnes *et al.* 1988) should eventually occur.

Mosses

Quite unexpectedly, growth of *Hylocomium splendens* was strongly stimulated by UV-B, provided the moss received additional water (Gehrke *et al.* 1996). The stimulation of annual growth in length by enhanced UV-B (15% calculated for cloudless skies) was, in three successive growing seasons in the field, 15%, 31% and 27%, respectively. The stimulation of growth in length by UV-B was also observed in a greenhouse experiment, when water supply was sufficient. For *H. splendens* in the field not receiving water in addition to that naturally available, however, there was either no effect (first year of irradiation), or an inhibition of growth in length (by 25% and 18% during the second and third year, respectively). Preliminary results from a field irradiation of *Sphagnum fuscum* indicate that during the first year of the experiment, longitudinal growth was reduced by about 20% (Gehrke *et al.* 1996).

Impacts of enhanced UV-B on mosses are potentially particularly important in the Arctic where mosses, together with lichens, are dominant components of ecosystems, affecting herbivores, energy and water balance, plant nutrient availability and even trace greenhouse gas emissions from the peat-forming keystone genus *Sphagnum*. The counterintuitive increase in annual growth is, as yet, unexplained. The response of *Sphagnum*, although yet to be reported in detail (C. Gehrke personal communication), is potentially particularly important because of the role of this genus in sequestering atmospheric CO_2 in many northern ecosystems. The only other results apparently published so far on ultraviolet effects on mosses is the finding of trends of increasing flavenoid (UV-B protecting pigment) with decreasing stratospheric ozone in the Antarctic (Markham *et al.* 1990), indicating that UV-B is a stress factor for these mosses, as for the *Hylocomium* not receiving extra water and for the *Sphagnum*.

Lichens

Thalli collected at Abisko were compared to thalli of the same species (*Cladonia arbuscula, Cetraria islandica* and *Stereocaulon paschale*) collected in southern Sweden. Experiments were carried out under 350, 600 and 1000 ppm CO_2. UV-B exposure increased the photochemical quantum yield of photosystem II as measured with pulse modulated fluorimetry, $(F'_m - F_t)/F'_m$, under low CO_2 (except late in the season, i.e. August), but not under the highest concentration. The quantum yield was greater for lichens from Abisko than for those from southern Sweden (Sonesson *et al.* 1995).

These results are again counterintuitive in that lichens from the southern site, with higher UV-B, should be better adapted than those from the northern site. Comparable results on lichens have not been reported, although Heide-Jørgensen and Johnsen (1995) suggest that high UV-B has damaged lichens in north-west Greenland, and it is currently impossible to generalize from these data. The impacts of enhanced UV-B on lichens from the Subarctic are potentially important because of the critical role of the lichens in providing winter forage for reindeer.

UV-B effects on litter quality and decomposition

Plant growth in northern latitudes is generally constrained by low nutrient availability due to low decomposition rates limited by low microbial activity in cold soils and often poor resource quality. As UV-B is known to influence secondary chemistry in plant tissues, enhanced UV-B radiation could potentially reduce decomposition rates, plant nutrient availability and primary and secondary production (Moorhead & Callaghan 1994).

During the first 8 days the CO_2 release from 'UV-B litter' in microcosm experiments was only 65% of the control rate, but the release rate later became similar to the control rate. During 62 days of decomposition the mass loss was 83% of the control rate. Irradiation during decomposition also had an inhibiting effect on decomposition. In decomposition experiments in the field (12 months) the decomposition rate of 'UV-B litter' was 91% of that of the control.

It is likely that increased UV-B, by slowing down decomposition, would decrease the availability of plant nutrients, which is limiting plant growth in this ecosystem. However, no difference was found in ammonium nitrogen, phosphorus or potassium in leachate from 'UV-B litter' and control litter.

Exposure of *V. uliginosum* to enhanced UV-B changed the chemical composition, causing a decrease in α-cellulose and cellulose/lignin ratio and increase in tannins (Gehrke *et al.* 1995), making the litter more resistant to decomposition. UV-B exposure during decomposition decreased the proportion of lignin and decreases fungal colonization and total microbial respiration. Of three fungal species investigated, *Mucor hiemalis* and *Truncatella truncata* were more UV-B sensitive than was *Penicillium brevicompactum* (Gehrke *et al.* 1995).

Leaf litter from treated (UV-B supplemented) *V. myrtillus* plants is decomposed more slowly by micro-organisms than is litter from control plants (Gehrke *et al.* 1995).

These experimental results are in line with the model output and predictions made by Moorhead and Callaghan (1994).

CONCLUSIONS

To our knowledge this is the first experiment aimed at determining the impact of enhanced UV-B radiation on natural ecosystems. The methodology is technically difficult yet critical to understand in order to interpret ecosystem responses. These responses are subtle and sometimes surprising, vary from species to species and cannot be predicted without detailed experimentation. There is some evidence that stem growth inhibition in perennial plants increases over time. Although it is too early to make any detailed statements, one can envisage that not only direct UV-B effects on certain plant species are of importance, but also complex ecological interactions. The earliest results to be recorded were on individual organisms or plant parts which can be measured with accuracy; the higher level responses of populations and communities, although potentially important because of those responses identified on individuals, are likely to become measurable only after some longer time interval.

ACKNOWLEDGEMENTS

The investigation was financially supported by The Swedish Environmental Protection Agency, Astra-Draco AB, the European Commission and Abisko Scientific Research Station. We thank the Swedish Meteorological and Hydrological Institute, Nordlyslaboratoriet and the Finnish Meteorological Institute for providing us with ozone data.

REFERENCES

Barnes, P.W., Flint, S.D. & Caldwell, M.M. (1987). Photosynthesis damage and protective pigments in plants from a latitudinal arctic/alpine gradient exposed to supplemental UV-B radiation in the field. *Arctic and Alpine Research*, **19**, 21–27.
Barnes, P.W., Jordan, P.W., Gold, W.G., Flint, S.D. & Caldwell, M.M. (1988). Competition, morphology and canopy structure in wheat (*Triticum aestivum* L.) and wild oat (*Avena fatua* L.) exposed to enhanced ultraviolet-B radiation. *Functional Ecology*, **2**, 319–330.

252 L.O.Björn *et al.*

Biggs, R.H. & Joyner, M.E.B. (Eds) (1993). *Stratospheric Ozone Depletion/UV-B Radiation in the Biosphere.* NATO ASI Series, **118**. Springer, Berlin.

Bird, R.E. & Riordan, C. (1986). Simple solar spectral model for direct and diffuse irradiance on horizontal and tilted planes at the earth's surface for cloudless atmospheres. *Journal of Climate & Applied Meteorology*, **25**, 87–97.

Björn, L.O. (1994). Modelling the light environment. *Photomorphogenesis in Plants* (Ed. by R.E. Kendrick & G.H.M. Kronenberg), pp. 537–551, Kluwer Academic, Dordrecht.

Björn, L.O. (1995). Estimation of fluence rate from irradiance measurements with a cosine corrected sensor. *Journal of Photochemistry & Photobiology B: Biology*, **29**, 179–183.

Björn, L.O. & Holmgren, B. (1996). Monitoring and modelling of the radiation climate at Abisko. *Ecological Bulletin*, **45**, 204–209.

Björn, L.O. & Murphy, T.M. (1985). Computer calculation of solar ultraviolet radiation at ground level. *Physiologie Végétale*, **23**, 555–561.

Björn, L.O. & Teramura, A.H. (1993). Light sources for UV-B photobiology. *Environmental UV Photobiology* (Ed. by A.R. Young, L.O. Björn, J. Moan & W. Nultsch), pp. 41–71. Plenum Press, New York.

Bornman, J.F., Björn, L.O. & Åkerlund, H.-E. (1984). Action spectrum for inhibition by ultraviolet radiation of photosystem II activity in spinach thylakoids. *Photobiochemistry & Photobiophysics*, **8**, 305–313.

Caldwell, M.M. (1971). Solar UV irradiation and growth and development of higher plants. *Photophysiology*, Vol. 6 (Ed. by A.C. Giese), pp. 131–177. Academic Press, New York.

Caldwell, M.M. & Flint, S.D. (1994). Stratospheric ozone reduction, solar UV-B radiation and terrestrial ecosystems. *Climatic Change*, **28**, 375–394.

Cen, Y-P. & Björn, L.O. (1994). Action spectra for enhancement of ultraweak luminescence by UV radiation (270–340 nm) in leaves of *Brassica napus*. *Journal of Photochemistry & Photobiology B: Biology*, **22**, 125–129.

Cullen, J.J. & Neale, P.J. (1994). Ultraviolet radiation, ozone depletion, and marine photosynthesis. *Photosynthesis Research*, **39**, 303–320.

Gehrke, C., Johanson, U., Callaghan, T.V., Chadwick, D. & Robinson, C.H. (1995). The impact of enhanced ultraviolet-B radiation on litter quality and decomposition processes in *Vaccinium* leaves from the Subarctic. *Oikos*, **72**, 213–222.

Gehrke, C., Johanson, U., Gwynn-Jones, D., Björn, L.O., Callaghan, T.V. & Lee, J.A. (1996). Effects of enhanced ultraviolet-B radiation on terrestrial subarctic ecosystems and implications for interactions with increased atmospheric CO$_2$. *Ecological Bulletin*, **45**, 192–203.

Heide-Jørgensen, H.S. & Johnsen, I. (1995). Analysis of surface structures of *Cladonia mitis* podetia in historic and recent collections from Greenland. *Canadian Journal of Botany*, **73**, 457–464.

IASC (1995). *Effects of Increased Ultraviolet Radiation in the Arctic – An Interdisciplinary Report on the State of Knowledge and Research Needed.* IASC (International Arctic Science Committee) Report No. 2. IASC Secretariat, Oslo.

Johanson, U., Gehrke, C., Björn, L.O., Callaghan, T.V. & Sonesson, M. (1995a). The effects of enhanced UV-B radiation on a subarctic heath ecosystem. *Ambio*, **24**, 106–111.

Johanson, U., Gehrke, C., Björn, L.O. & Callaghan, T.V. (1995b). The effects of enhanced UV-B radiation on the growth of dwarf shrubs in a subarctic heathland. *Functional Ecology*, **9**, 713–719.

Josefsson, W. (1996). *Measurements of Total Ozone: National Environmental Monitoring 1993/94.* Swedish Environmental Protection Agency, Stockholm.

Madronich, S., McKenzie, R.L., Caldwell, M.M. & Björn, L.O. (1995). Changes in ultraviolet radiation reaching the earth's surface. *Ambio*, **24**, 143–151.

Markham, K.R., Franke, A., Given, D.R. & Browsney, P. (1990). Historical antarctic ozone level trends from herbarium specimen flavonoids. *Bulletin of the Liason Group for Polyfenols*, **15**, 230–235.

Moorhead, D.L. & Callaghan, T.V. (1994). Effects of increasing ultraviolet-B radiation on decomposition of soil organic matter dynamics: a synthesis and modelling study. *Biology and Fertility of Soils,* **18,** 19–26.

Negash, L. (1987). Wavelength-dependence of stomatal closure by ultraviolet radiation in attached leaves of *Eragrostis tef:* action spectra under backgrounds of red and blue lights. *Plant Physiology and Biochemistry,* **25,** 753–760.

Negash, L. & Björn, L.O. (1986). Stomatal closure by ultraviolet radiation. *Physiologia Plantarum,* **66,** 360–364.

Robberecht, R., Caldwell, M.M. & Billings, W.D. (1980). Leaf ultraviolet optical properties along a latitudinal gradient in the arctic-alpine life zone. *Ecology,* **61,** 612–619.

Rozema, J., Gieskes, W.W.C., van de Geijn, S.C., Nolan, C. & de Boois, H. (1996). *UV-B and Biosphere.* Kluwer, Dordrecht.

SCOPE (1992). *Effects of Increased Ultraviolet Radiation on Biological Systems.* Scientific Committee on Problems on the Environment (SCOPE), Paris.

SCOPE (1993). *Effects of Increased Ultraviolet Radiation on Global Ecosystems.* Scientific Committee on Problems on the Environment (SCOPE), Paris.

Sonesson, M., Callaghan, T.V. & Björn, L.O. (1995). Short-term effects of enhanced UV-B and CO_2 on lichens at different latitudes. *Lichenology,* **27,** 547–557.

Strid, Å., Chow, W.S. & Anderson, J.M. (1994). UV-B damage and protection at the molecular level in plants. *Photosynthesis Research,* **39,** 475–489.

Sullivan, J.H. & Teramura, A.H. (1992). The effects of ultraviolet-B radiation on loblolly pine. 2. Growth of field-grown seedlings. *Trees (Berl.),* **6,** 115–120.

Takeuchi, Y., Ikeda, S. & Kasahara, H. (1993). Dependence on wavelength and temperature of growth inhibition induced by UV-B irradiation. *Plant Cell Physiology,* **34,** 913–917.

Teramura, A.H. & Sullivan, J.H. (1994). Effects of UV-B radiation on photosynthesis and growth of terrestrial plants. *Photosynthesis Research,* **39,** 463–473.

Teramura, A.H. & Ziska, L.H. (1994). Ultraviolet-B radiation. *Photosynthesis and the Environment* (Ed. by N.R. Baker). Kluwer Academic, Dordrecht.

Thimijan, R.W., Carns, H.R. & Campbell, L.E. (1978). *Final Report. Radiation Sources and Related Environmental Control for Biological and Climatic Effects: UV Research (BACER).* United States Environmental Protection Agency, Washington, DC.

UNEP (1995). *Environmental Effects of Ozone Depletion: 1994 Assessment* (Ed. by J.C. van der Leun, X. Tang & M. Tevini). United Nations Environment Programme. *Ambio,* **24,** 137–196 (special issue).

Weiler, C.S. & Penhale, P.A. (Eds) (1994). *Ultraviolet Radiation in Antarctica: Measurements and Biological Effects.* American Geophysical Union, Washington, DC.

Young, A.R., Björn, L.O., Moan, J. & Nultsch, W. (Eds) (1993). *Environmental UV Photobiology.* Plenum Press, New York.

Yu, S.-G. & Björn, L.O. (1996). Differences in UV-B sensitivity between PSII from grana lamellae and stroma lamellae. *Journal of Photochemistry and Photobiology. B: Biology,* **34,** 35–38.

12. Effects of CO_2
and climate change
on arctic ecosystems

WALTER C.OECHEL, ANDREA C.COOK, STEVEN
J.HASTINGS AND GEORGE L.VOURLITIS
Global Change Research Group, San Diego State University,
San Diego, CA 92182, USA

INTRODUCTION

The increase in the concentration of radiatively active gases such as CO_2 has the potential to affect temperature and climate on a global scale (Gates *et al.* 1992; Watson *et al.* 1992; Kattenberg *et al.* 1996). Northern latitudes are considered more sensitive to radiative forcing, due to potentially greater changes in atmospheric and surface properties. The atmosphere over polar regions contains relatively little water vapour (Hinzman & Kane 1992), and increased cloudiness, which is predicted for high-latitude regions in a double-ambient CO_2 climate, may affect the surface temperature significantly (Mitchell *et al.* 1989; Gates *et al.* 1992; Chapman & Walsh 1993; Groisman *et al.* 1994). Reductions in the distribution of polar ice and snow cover also are expected in CO_2-induced climate change scenarios, and would likely represent a positive feedback on the increase in surface temperature by reducing surface albedo (Mitchell *et al.* 1989; Gates *et al.* 1992; Hinzman & Kane 1992; Groisman *et al.* 1994).

General circulation models (GCMs) based on a doubling of atmospheric CO_2 predict mean surface temperature increases of approximately 2–5°C during the winter and spring months, while summer temperature increases are expected to be more modest (e.g. 1–3°C: Gates *et al.* 1992; Meehl *et al.* 1993; Kattenberg *et al.* 1996). Observations of high-latitude temperature trends are consistent with simulations based on atmospheric forcing by increased trace-gas concentrations and aerosols (Mitchell *et al.* 1995; Kattenberg *et al.* 1996). Thermal profiles of permafrost indicate a temperature rise of 2–4°C across the north slope of Alaska and throughout northern Canada during the last few decades (Lachenbruch & Marshall 1986; Beltrami & Mareshal 1991). Northern-latitude weather records indicate a similar increase in annual surface temperature (Karl *et al.* 1991; Chapman & Walsh 1993; Oechel *et al.* 1993). The annual increase is due primarily to warmer winter and spring temperatures, which have increased on the order of 4.5°C over the past three decades, while summer and fall temperatures have increased by approximately 2°C (Chapman & Walsh 1993).

Changes in energy balance will undoubtedly lead to changes in the hydrology of northern ecosystems (Kane *et al.* 1990, 1992; Gates *et al.* 1992; Hinzman & Kane 1992; Waelbroeck 1993; Groisman *et al.* 1994). Precipitation is expected to increase during the summer months by on average 7% (−10 to +20%; Kattenberg *et al.* 1996); however, evapotranspiration is predicted to increase as well (Gates *et al.* 1992; Hinzman & Kane 1992), and warmer winter and spring temperatures may result in an earlier spring thaw, which may lead to further reductions in soil water during the summer growing season. For example, a 3°C rise in air temperature is predicted to increase the length of the growing season by 22%, soil surface temperature by 40% and evaporation rates by nearly 35% (Fig. 12.1a). The increased evaporation and warmer temperatures are subsequently predicted to decrease soil water content in the organic and mineral soil layers significantly (Fig. 12.1b). Assuming a temperature increase of 0.08°C year^{-1} (or 2.4°C over three decades), Waelbroeck (1993) predicted soil water content to decrease by 15 and 70% for organic and mineral soil layers, respectively (Fig. 12.1b). It must be noted, however, that the predicted change in northern latitude soil moisture is quite variable and, in many cases, soil moisture is predicted to increase in response to a doubling of atmospheric CO_2 (Mitchell *et al.* 1995; Kattenberg *et al.* 1996), due to increased winter precipitation and runoff. Regardless of the direction of change, variations in soil moisture are likely to have profound impacts on the net CO_2 flux of northern latitude ecosystems, as CO_2 exchange is strongly linked to soil water content (Oechel *et al.* 1993, 1995; Oechel & Vourlitis 1994, 1995).

Recent observations of the trend in growing season length, surface albedo and soil water content are consistent with model predictions described above. An increasing trend in growing season length over the past 65 years has been observed in several regions throughout Alaska (Sharratt 1992). These observations correspond with reductions in the extent of snow and ice cover throughout North America, Europe and Asia (Groisman *et al.* 1994). Similarly, increases in mean annual precipitation for North America and the former Soviet Union, on the order of those predicted by greenhouse gas driven GCMs, have been observed over the past 120 years (Bradley *et al.* 1987), with much of the increase observed during the winter and spring months. However, recent data indicate that soil water content in the upper 10 cm of soil has decreased by approximately 50% over the past two decades at the US International Biological Program (IBP) research site II near Barrow, Alaska (Oechel *et al.* 1995).

The changes in the hydrology and soil thermal regime of arctic soils will ultimately determine the fate of vast stores of soil C present in the active layer and permafrost. Northern ecosystems (tundra and boreal forest) represent only about 14% of the total global land area; however, they contain approximately 25–33% of the total global soil C pool in the permafrost and seasonally thawed soil active layer (Miller *et al.* 1983; Billings 1987; Gorham 1991; Schlesinger 1991). Arctic

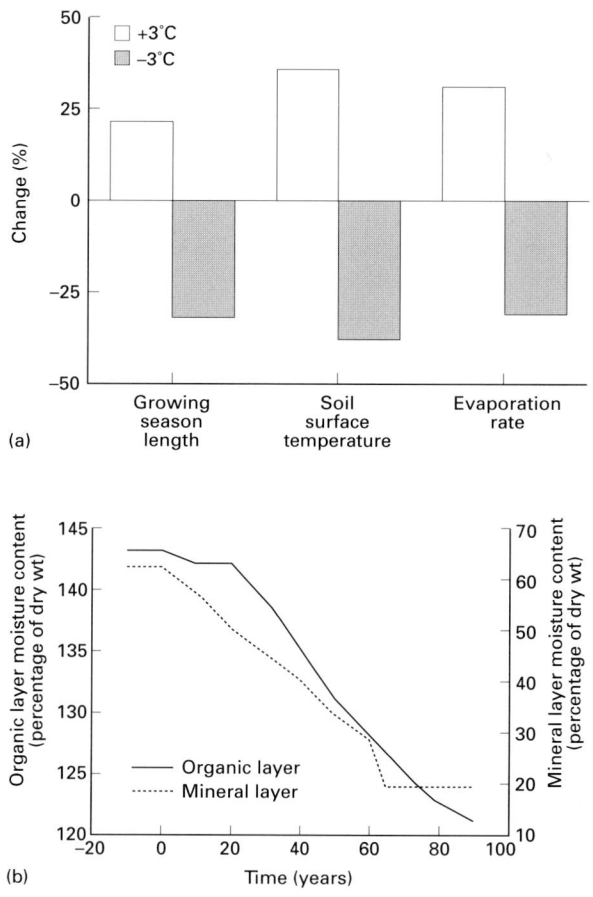

FIG. 12.1. (a) The percentage of change in growing season length, soil surface temperature and evaporation rate in response to a 3°C increase or decrease in air temperature. (b) The simulated effect of a 0.08°C year^{-1} rise in air temperature on soil moisture content in the organic (solid line) and mineral (dashed line) soil layers. (Data modified from Waelbroeck 1993.)

tundra ecosystems alone contain approximately 12% of the global soil C pool even though they only make up about 5% of the total land area (Miller *et al.* 1983; Billings 1987; Schlesinger 1991).

The response of high-latitude ecosystems to increased temperature and associated climate change will largely depend on how plants and ecosystems respond to elevated CO$_2$; how soil microbial processes and plant productivity are affected by changes in temperature and hydrology; and how community dynamics are affected by changing thermal, nutrient and competitive regimes (Oechel &

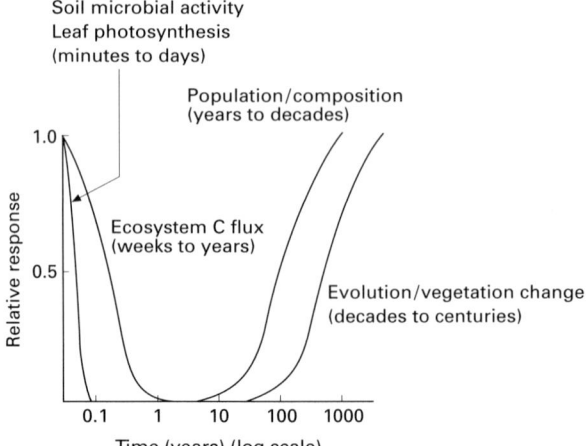

Fɪɢ. 12.2. Hypothetical relative response times for adjustment to an environmental perturbation for changes at the level of leaf photosynthesis and soil microbial activity, ecosystem CO_2 flux, population structure, community composition, vegetation and evolution. (After Oechel & Vourlitis 1994.)

Billings 1992; Shaver *et al.* 1992; Oechel *et al.* 1993, 1994; Oechel & Vourlitis 1994; Parsons *et al.* 1994; Wookey *et al.* 1994). However, any discussion regarding ecosystem response to elevated CO_2 and climate change must consider the varying time scales in which the various ecosystem components respond to external forcing (Pastor & Post 1993). For example, soil microbial activity and leaf-level processes may respond to perturbations on time scales that are on the order of minutes to days (Fig. 12.2). Ecosystem-level processes, however, respond to external forcing on time scales of weeks to years, while the response times of plant populations and communities are even longer still (Fig. 12.2). The actual trajectory of the biotic response to climate change may be non-linear (Pastor & Post 1993), indicating the importance of conducting process-level research over the short, intermediate and long term. Therefore, a comprehensive understanding of the effects of elevated CO_2 and climate change on plant- and ecosystem-level function requires observations over a variety of temporal scales.

DIRECT EFFECTS OF ELEVATED CO_2

Increases in atmospheric CO_2 have the potential to increase photosynthetic rates and biomass production on a global scale (Idso *et al.* 1991). Over short time scales, atmospheric CO_2 concentration appears to universally limit photosynthesis rates (Stitt 1991; Drake 1992; Strain 1992). However, the long-term photosynthetic response to elevated atmospheric CO_2 concentration is much more variable, due to genetic (Graham *et al.* 1991; Oechel & Billings 1992), anatomical (Woodward 1987), growth, demographic, community, and ecosystem-level controls and feed-backs (Körner *et al.* 1996). These adjustments can be homeostatic and act to maintain ecosystem photosynthetic rates at similar levels, or they can further exaggerate the effect of elevated CO_2 on photosynthesis by increasing the magnitude of the plant or ecosystem response.

Early simulation models indicate that ecosystem C accumulation would increase minimally in a doubled CO_2 environment over the next 50 years, and by 3–4% under elevated CO_2 and a 4°C increase in temperature (Miller *et al.* 1983). In a more recent simulation, however, C accumulation under elevated CO_2 alone is expected to increase by 12% over the next 50 years (Rastetter *et al.* 1992). The cause for the increase in C accumulation is thought to be due to greater N-use efficiency of plants exposed to elevated CO_2, so more C should be sequestered per unit N mineralized from decomposition (Oechel & Billings 1992; Rastetter *et al.* 1992; Shaver *et al.* 1992). Other simulations, however (e.g. Terrestrial Ecosystem Model), indicate that there will be no stimulation of net primary productivity (NPP) by CO_2 without concomitant increases in nutrient (especially N) availability (Melillo *et al.* 1993).

Initial exposure to elevated CO_2 results in significant increases in photosynthesis rates of many arctic species; however, this stimulation appears to be relatively short lived (Oberbauer *et al.* 1986; Tissue & Oechel 1987). *In situ* studies of *Eriophorum vaginatum* (Oechel *et al.* 1992), the tussock forming sedge which is dominant in Alaskan moist tussock tundra, indicate that homeostatic adjustment of photosynthesis can occur in as little as 3 weeks (Tissue & Oechel 1987). Plants grown in ambient CO_2 exhibited significantly higher photosynthesis rates than plants grown in elevated CO_2 when measured over a variety of CO_2 concentrations, indicating that a loss in photosynthetic potential occurred (Fig. 12.3). This rapid loss of photosynthetic potential is further illustrated by the fact that the photo-synthetic rates of ambient and double-ambient grown plants were identical when plants were measured at their growth CO_2 concentrations (Fig. 12.3). The mecha-nism for this rapid homeostatic adjustment in photosynthetic rates appears to be due to sink limitations (Tissue & Oechel 1987), which are caused by nutrient limitations to growth (Chapin & Shaver 1985). These sink limitations may cause excess carbohydrate that is produced during the initial photosynthetic stimulation to accumulate, which can feed back to limit the amount and activity of Rubisco (ribulose-1,5-bisphosphate carboxylase/oxygenase), resulting in a concomitant reduction in photosynthesis (Tissue & Oechel 1987; Arp 1991).

Results from a 3-year *in situ* ecosystem-level study at Toolik Lake, Alaska are qualitatively similar to results from short-term leaf-level studies; however, due to the longer response time of ecosystems to external forcing (Fig. 12.2), complete homeostatic adjustment of ecosystem C assimilation occurred over a 3-year period

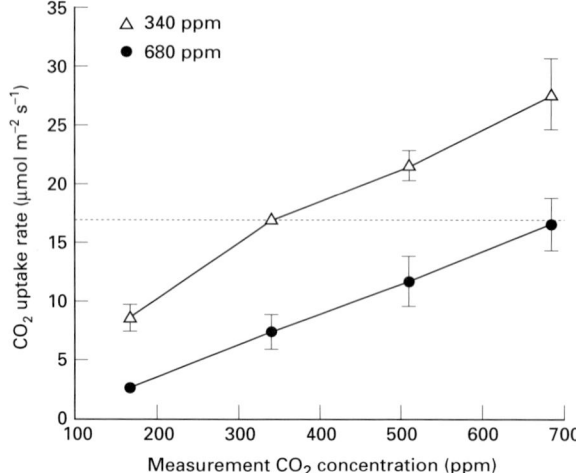

FIG. 12.3. The net photosynthetic response of *Eriophorum vaginatum* plants grown at ambient CO_2 (340 ppm) and double-ambient CO_2 (680 ppm) to a range of CO_2 concentrations. The horizontal line indicates complete homeostatic adjustment of photosynthesis rates of plants grown and measured at 680 ppm. (After Tissue & Oechel 1987.)

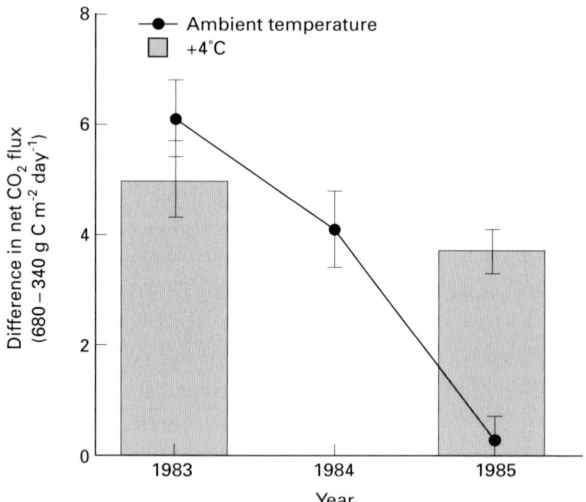

FIG. 12.4. Difference in net CO_2 uptake of moist tussock tundra plots exposed to 680 ppm CO_2 concentrations (circles) and 680 ppm CO_2 and a 4°C increase in air temperature (bars) relative to plots exposed to 340 ppm CO_2 and ambient temperature during a 3-year *in situ* CO_2 exposure experiment at Toolik Lake, Alaska. (Data are means ± 1 SE, $n = 3$, and are modified from Oechel *et al.* 1994.)

(Fig. 12.4). During 1983, which was the first year of the study, plots exposed to double ambient CO_2 (680 ppm) accumulated nearly $6 \, g \, C \, m^{-2} day^{-1}$ more than ambient-grown (340 ppm) plots (Fig. 12.4). Körner *et al.* (1996) report that net CO_2 accumulation increased by approximately 41% in *Carex curvula*-dominated alpine tundra ecosystems after 2 years of exposure to elevated CO_2. Changes in plant growth (both above and below ground) due to CO_2 exposure were minimal, indicating that the excess carbohydrate produced in elevated CO_2 plots is being incorporated in organic matter (Körner *et al.* 1996). These results are consistent with the results reported by Oechel *et al.* (1994) after 2 years of elevated CO_2 exposure (approximately 110% greater uptake by *E. vaginatum*-dominated plots exposed to double ambient CO_2; Oechel *et al.* 1994; Oechel & Vourlitis 1996), further illustrating that over the short term, the effects of elevated CO_2 on whole ecosystem C accumulation can be substantial.

By the end of the third year of the *in situ* study by Oechel *et al.* (1994), the initial stimulation in net ecosystem CO_2 uptake decreased entirely (Fig. 12.4). The rapid ecosystem-level down regulation of C assimilation is thought to be due to the over-riding effects of nutrient limitation which cause reduced growth and sink formation (Grulke *et al.* 1990; Oechel *et al.* 1994). In the Arctic, plants and whole ecosystems are likely to be limited by nutrients (Shaver & Chapin 1980; Chapin & Shaver 1985; Parsons *et al.* 1994; Wookey *et al.* 1994) and, as a result, the direct effects of CO_2 over the medium term are likely to be minimal (Grulke *et al.* 1990; Oechel & Billings 1992; Oechel *et al.* 1994).

When plants were exposed to double-ambient CO_2 and a 4°C increase in air temperature, the initial stimulation in net CO_2 flux was found to persist (Fig. 12.4). Plants exposed to 680 ppm CO_2 and a 4°C increase in ambient temperatures continued to incorporate as much as $4 \, g \, C \, m^{-2} day^{-1}$ more than plants exposed to ambient CO_2 and temperature (Fig. 12.4). Unfortunately, the mechanism for the continued stimulation is largely unknown, due to the inability to determine the full-factorial ecosystem-level response to variations in both CO_2 and temperature (Oechel & Vourlitis 1994; Oechel *et al.* 1994). It is likely that the elevated temperature may have stimulated plant respiration and growth which would in turn stimulate plant sink strength (Limbach *et al.* 1982; Kummerow & Ellis 1984; Chapin & Shaver 1985). In a 2-year study of subarctic evergreen and deciduous plant species in Sweden, elevated temperature was found to significantly stimulate growth of shoot and stem biomass, and was generally found to increase leaf area per stem (Parsons *et al.* 1994). Similarly, the development and production of reproductive structures was significantly increased following exposure to elevated temperature (Wookey *et al.* 1994). Elevated temperature may also stimulate soil respiration and nutrient mineralization and availability (Nadelhoffer *et al.* 1991). So, although the mechanism responsible for the apparent interaction between elevated temperature and CO_2 is largely unknown

(Oechel & Vourlitis 1994; Oechel *et al.* 1994), elevated temperature can lead to increased growth, reproductive development and soil respiration rates, which would provide a sink for the excess carbohydrate produced under elevated CO_2 (Tissue & Oechel 1987; Grulke *et al.* 1990; Oechel & Vourlitis 1994, 1996; Oechel *et al.* 1994).

Our knowledge of the long-term (decade to century) response to elevated CO_2 stems largely from observations of low and high altitude plants (e.g. Körner & Diemer 1987; Körner 1993) and from recent work near cold, CO_2-emitting springs in Iceland (A.C. Cook & W.C. Oechel unpublished data). Elevational studies (Körner & Diemer 1987) are based on the observation that low-altitude plants experience higher partial pressures of CO_2 compared to high-elevation plants. In general, estimates by Körner and Diemer (1987) indicate that an additional 100 ppm of atmospheric CO_2 would increase net photosynthesis by approximately 21 and 31% in low- and high-elevation plants, respectively, while an additional 200 ppm would result in an even smaller relative photosynthetic gain. These results suggest that the reduction in photosynthetic capacity observed under short-term exposure to elevated CO_2 is apparent even after centuries.

Results from cold CO_2 springs are similar (A.C. Cook & W.C. Oechel unpublished data). One spring that has recently been studied is near Olafsvik, Iceland, where historical evidence indicates that plants growing adjacent to the spring have been exposed to elevated levels of CO_2 for at least 150 years. Detailed analysis of the spring chemistry revealed that the elemental composition of the spring is similar to that of ground water, with extremely low background levels of nutrients and contaminants. Therefore, natural CO_2-emitting springs, such as the spring near Olafsvik, may represent an ideal natural laboratory for studying the long-term effects of elevated CO_2 on plant and ecosystem function. Photosynthesis measurements made during the 1994 field season of the dominant C_3 grass, *Nardus stricta*, indicate a reduction in the photosynthetic capacity of high-CO_2-grown plants (Fig. 12.5), due primarily to reductions in Rubisco content, final Rubisco activity and chlorophyll concentration (A.C. Cook *et al.* unpublished data). These results are similar to the short-term studies by Tissue and Oechel (1987) and suggest that even after a century or more of elevated CO_2 exposure, homeostatic adjustment of leaf-level photosynthesis persists. Preliminary results suggest that there is homeostatic adjustment of net ecosystem CO_2 flux as well (A.C. Cook & W.C. Oechel unpublished data).

In light of these results, it is unlikely that C accumulation will increase markedly over the intermediate and long term due to the direct effects of CO_2 alone. Because of sink limitations, an increase in nutrient availability will have to occur to allow the sink required for the utilization of the additional carbohydrates produced under elevated CO_2 (Miller *et al.* 1983; Billings *et al.* 1984; Hilbert *et al.* 1987; Tissue & Oechel 1987; Grulke *et al.* 1990; Melillo *et al.* 1993).

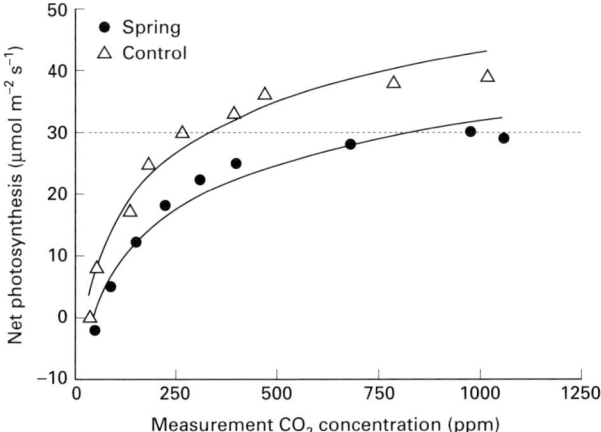

FIG. 12.5. The net photosynthetic response of *Nardus stricta* plants growing adjacent (circles) and away (triangles) from a CO$_2$-emitting spring near Olafsvik, Iceland to a range of CO$_2$ concentrations. Plants growing adjacent to the spring are estimated to have been exposed to CO$_2$ levels of 750 ppm for over a century or more, while plants growing away from the spring are estimated to have been exposed to the global ambient CO$_2$ concentration over the same time period. The horizontal line indicates complete homeostatic adjustment of photosynthesis rates of plants grown and measured at elevated CO$_2$. Photosynthetic rates were measured using a null-balance, temperature-controlled cuvette. Spring and control plants were exposed to saturating light (*c.* 1500 μmol m^{-2}s^{-1}) using cool-beam projector lamps, while leaf temperature was maintained at 10°C at a vapour pressure of 40–50 Pa. (Data from A.C. Cook & W.C. Oechel unpublished.)

EFFECTS OF CLIMATE CHANGE ON NET ECOSYSTEM CO$_2$ FLUX

The potential affects of high-latitude warming on surface energy and water balance (Fig. 12.1) will likely feed back on the C balance of arctic ecosystems. In the past, net primary productivity in northern ecosystems exceeded heterotrophic respiration due to the predominance of cold, wet soils (Schell 1983; Gorham 1991; Oechel & Billings 1992; Marion & Oechel 1993). Estimates based on harvest methods, photosynthesis and respiration measurements, and aerodynamic CO$_2$ flux estimates made primarily during the IBP effort of the early 1970s indicate that moist tussock tundra was accumulating approximately 25 g C m^{-2} year^{-1}, while coastal wet sedge ecosystems were estimated to be accumulating on average 70 g C m^{-2} year^{-1} (Coyne & Kelley 1975; Chapin *et al.* 1980; Miller *et al.* 1983). Recent plot-level measurements of CO$_2$ flux made over the past 13 years in arctic Alaska using passive, closed-system chamber sampling techniques (Vourlitis *et al.* 1993) and temperature-controlled, null-balance greenhouses (Grulke *et al.* 1990; Oechel *et al.* 1992) indicate that many tundra ecosystems are now sources of CO$_2$ to the atmosphere (Grulke *et al.* 1990; Oechel *et al.* 1993, 1995; Oechel & Vourlitis

1994). Tussock tundra ecosystems are currently losing on average $112\,\mathrm{g\,C\,m^{-2}}$ year^{-1} during the growing season. Similarly, wet sedge ecosystems are currently either slight sinks for atmospheric CO_2, or roughly in balance (Oechel *et al.* 1993, 1995; Oechel & Vourlitis 1994, 1995). An additional $24\,\mathrm{g\,C\,m^{-2}\,year^{-1}}$ may be lost from lakes and streams (Kling *et al.* 1991), and $19-68.5\,\mathrm{g\,C\,m^{-2}}$ during the cold season (October–May; Zimov *et al.* 1993; W.C. Oechel *et al.* unpublished data), thus increasing the actual amount of C lost from arctic ecosystems annually. Given the large C stores in arctic soils, this loss of C may represent a significant positive feedback to atmospheric CO_2 concentration and concomitant global change.

To illustrate the magnitude of this change, measurements of net CO_2 exchange were made in Barrow, Alaska in July 1991, and during the 1992 and 1993 growing season at the IBP site used by Coyne and Kelley (1975). Using aerodynamic methods from a 10-m tall tower, Coyne and Kelley (1975) estimated that the moist meadow and polygonized wet coastal tundra ecosystems accumulated approximately $25\,\mathrm{g\,C\,m^{-2}\,year^{-1}}$ during the 1971 field season (Fig. 12.6). During the 1992 growing season, these same ecosystems were net sources of CO_2 to the atmosphere of $1.3\,\mathrm{g\,C\,m^{-2}\,year^{-1}}$ (Fig. 12.6). A comparison of the seasonal trends in net CO_2 flux indicates that the current ecosystems were even stronger sinks during the early part of the growing season (June through mid-July). Around mid-July, which is considered the mid-season peak in ecosystem productivity, the sink strength of the current ecosystem began to decrease substantially (Fig. 12.6).

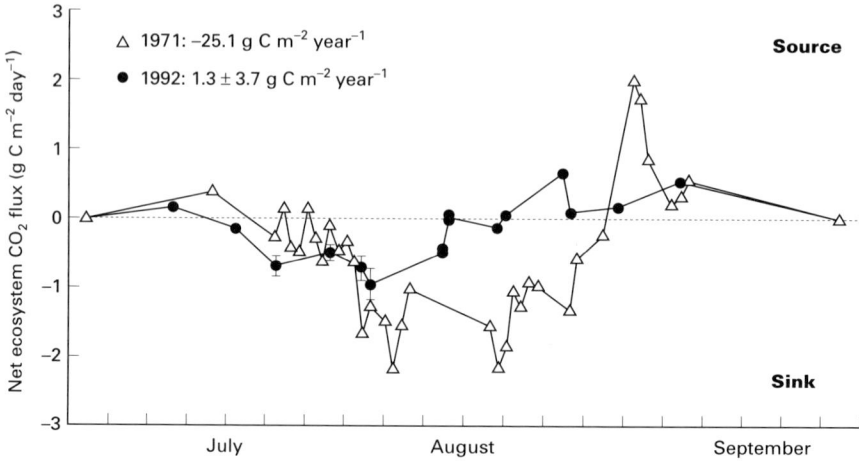

FIG. 12.6. Daily net CO_2 flux during the 1971 (triangles) and 1992 (circles) growing seasons at the US IBP site-II near Barrow, Alaska. Seasonal totals are calculated by integrating the daily flux estimates over a 100-day-long season. (After Oechel *et al.* 1995. Data for 1971 are from Coyne & Kelley 1975, and data from 1992 are from Oechel *et al.* 1995.)

By early August, the current ecosystem began losing more CO_2 from whole ecosystem respiration (ER) than it was accumulating by gross ecosystem productivity (GEP), while during 1971, these same ecosystems were significant sinks well into mid-August (Fig. 12.6). Similar trends were observed in July 1991 and during the 1993 field season using both chamber and aerodynamic techniques (W.C. Oechel *et al.* unpublished data; Harazono *et al.* 1995), indicating that the change in net CO_2 flux is persistent from year to year. Analysis of temperature data indicates that summer (June–August) air temperatures have increased only 1°C over the past two decades (Fig. 12.7a), however, comparison of soil water content reported by Gersper *et al.* (1980) for the 1971 growing season and data collected by Oechel *et al.* (1995) during 1992 indicates that soil moisture status has decreased by an estimated 50% during the same period (Fig. 12.7b).

These studies suggest that the change in net C balance of arctic ecosystems is due primarily to a change in water balance associated with the recently re-ported increase in high-latitude surface temperature (Oechel *et al.* 1993, 1995; Oechel & Vourlitis 1994, 1995). These observations are supported by a variety of laboratory and *in situ* experiments (Billings *et al.* 1982; Silvola 1986; Freeman *et al.* 1993; Funk *et al.* 1994). In laboratory manipulations with wet sedge tundra microcosms subjected to a 10-cm decrease in water-table depth, net ecosystem CO_2 incorporation decreased by $212\,g\,C\,m^{-2}$, resulting in a net CO_2 efflux of $84\,g\,C\,m^{-2}$ (Billings *et al.* 1982). Other experiments where water-table depth is lowered, or when soil cores are subjected to drying cycles, indicate that the rate of

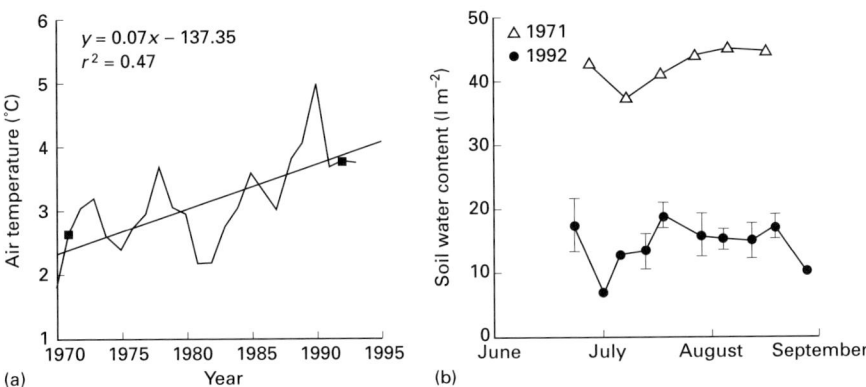

FIG. 12.7. (a) Summer (June–August) temperature trend for the period 1970–93 for Barrow, Alaska. (Data provided by the US National Weather Service.) Regression equation is calculated from least-squares linear regression. Closed boxes indicate the mean summer temperature for the years 1971 and 1992. (b) Average volumetric soil water content for the 1971 (triangles) and 1992 (circles) growing seasons for the US IBP site-II near Barrow, Alaska. (After Oechel *et al.* 1995. Data from 1971 are from Gersper *et al.* 1980, and data from 1992 are from Oechel *et al.* 1995.)

CO_2 efflux is 2–5 times greater during drained conditions (Hogg *et al.* 1992; Glenn *et al.* 1993; Freeman *et al.* 1993; Funk *et al.* 1994).

Similarly, draining of northern peatlands has been shown to convert these ecosystems from a CO_2 sink to approximately $25\,g\,C\,m^{-2}year^{-1}$ to a source of about $150\,g\,C\,m^{-2}year^{-1}$ to the atmosphere (Silvola 1986). Recent *in situ* experiments in wet sedge tundra ecosystems, where water-table depth is lowered on average by 10 cm and temperature is elevated on the order of 0.5–2°C over the course of the growing season, indicate that a lowering of the water-table results in a seasonal (June–September) loss of between 25 and $45\,g\,C\,m^{-2}$ in drained plots, while control plots were found to be net sinks of approximately $25\,g\,C\,m^{-2}\,season^{-1}$ (W.C. Oechel *et al.* unpublished data). The change in the direction of net CO_2 flux in drained plots was found to be due to significantly greater ER relative to GEP (W.C. Oechel *et al.* unpublished data). For example, ER nearly doubled in response to the 10-cm reduction in water-table depth, while GEP of drained plots was comparable to plots exposed to ambient fluctuations in water-table depth. This result is indicative of the differing response times of the various biotic components to external forcing (Fig. 12.2), as soil microbial activity was found to respond more rapidly to the decrease in water-table depth relative to GEP.

Elevated temperature, in and of itself, was found to be statistically less important in controlling net CO_2 flux than water table, although elevated temperature appeared to accentuate the amount of net CO_2 loss in drained plots (W.C. Oechel *et al.* unpublished data). This may be due to the fact that the temperature sensitivity of soil micro-organisms appears to vary according to the soil moisture status. In general, decomposition rates are slow and independent of temperature under conditions of low soil moisture content ($\leqslant 20\%$ of dry mass), while temperature sensitivity increases until soil moisture content reaches approximately 200% (Nadelhoffer *et al.* 1991). Under saturated conditions, however, the temperature sensitivity of soil microbial populations is low, and respiration rates appear to be limited more by poor soil aeration (Billings *et al.* 1982; Nadelhoffer *et al.* 1991; Moorhead & Reynolds 1993; Funk *et al.* 1994).

Increased nutrient mineralization is expected as arctic soils become warmer and drier (Marion & Miller 1982; Marion & Black 1987; Nadelhoffer *et al.* 1991). In laboratory incubations of tussock and wet sedge tundra soils, N mineralization increased by approximately 45% with a 12°C increase in soil temperature (3–15°C; Nadelhoffer *et al.* 1991). Because essentially all of the N taken up by arctic plants is supplied by mineralization of organic matter, increased mineralization is expected to increase plant growth (Shaver *et al.* 1992). Results from fertilization experiments of arctic plants indicate that increased nutrient availability acts to stimulate tissue production significantly (Shaver & Chapin 1980; Chapin & Shaver 1985; Parsons *et al.* 1994; Wookey *et al.* 1994). Surprisingly, however, the relationship between rates of soil decomposition and mineralization appears to be relatively weak (Nadelhoffer *et al.* 1991; Moorhead & Reynolds 1993). Although rates of microbial respiration appear to increase exponentially with elevated temperature, rates of N mineralization, especially in nutrient poor soils, appear to be insensitive to small or intermediate variations in soil temperature and soil moisture status due to rapid immobilization of mineralized N by soil micro-organisms (Flanagan & Bunnell 1980; Marion & Miller 1982; Marion & Black 1987; Nadelhoffer *et al.* 1991; Moorhead & Reynolds 1993). In simulations where soil moisture, temperature, and growing season length are varied over a range of 800%, 4.5°C, and 90 days, respectively, immobilization accounted for nearly all the N mineralized (Moorhead & Reynolds 1993).

This phenomenon, coupled with the longer response times of whole ecosystem-level processes to climatic forcing, might easily explain why a concomitant increase in NPP has not been observed with the recently reported change in net CO_2 flux (Grulke *et al.* 1990; Oechel *et al.* 1993, 1995; Oechel & Vourlitis 1994, 1995). Presumably, any additional N mineralized due to the reduction in the water-table depth and elevated temperature is rapidly immobilized by soil micro-organisms, resulting in an available N pool that is insufficient to significantly stimulate GEP.

Although the current data indicate that some arctic ecosystems may be losing significant amounts of C as CO_2 to the atmosphere annually, palaeoecological

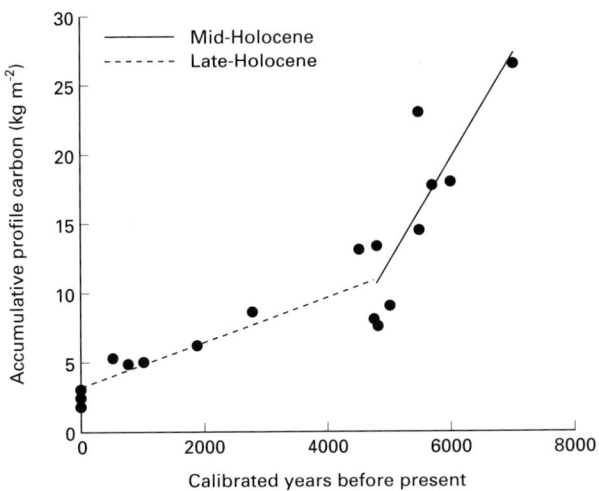

FIG. 12.8. Accumulative soil C measured from the top of soil cores of the soil profile at the West Dock wet-sedge tundra study site in Prudhoe Bay, Alaska as a function of calibrated years BP. Calibrated time scale is the radiocarbon date calibrated against dendrochronologically dated wood. The dashed line indicates the soil C accumulation rate calculated for the late-Holocene period (1.2 ± 0.3 g C m^{-2} year^{-1}), while the solid line indicates the C accumulation rate for the mid-Holocene period (6.7 ± 1.6 g C m^{-2} year^{-1}). (After Marion & Oechel 1993.)

studies indicate that high-latitude warming may, over the long term, result in increased C accumulation in arctic ecosystems. For example, C accumulation rates during the warmer mid-Holocene (6900–4800 years before present (BP)) were on average 5.5 times greater than C accumulation rates during the cooler late-Holocene (4800–400 BP: Fig. 12.8). The greater rate of C sequestration observed during the mid-Holocene is thought to be due to differences in species composition, although it is impossible to determine if the increased C accumulation observed in the middle Holocene was due to climate and/or vegetation change (Marion & Oechel 1993), or to the normal progression of soil formation following deglaciation of the Laurentide Ice Sheet (Harden *et al.* 1992). However, pollen records indicate that the vegetation in central and northern Alaska during the warmer mid-Holocene was composed primarily of a more productive *Picea, Betula* and *Alnus*-dominated forest, while during the cooler late-Holocene, the vegetation became more similar to the modern flora (Billings 1987). If the expected vegetation shift from arctic to more boreal forest species occurs in response to high-latitude warming (Neilson 1993; Smith & Shugart 1993), net accumulation of atmospheric CO_2 is expected to increase, making the Arctic a net sink over the long term (Marion & Oechel 1993).

SUMMARY AND CONCLUSIONS

Although elevated CO_2 has the potential to increase leaf-level photosynthesis and whole-ecosystem productivity (Idso *et al.* 1991), the evidence accumulating indicates that the effects of elevated CO_2 in nutrient-limited ecosystems, such as the Arctic, are minimal over the intermediate and long term (Tissue & Oechel 1987; Grulke *et al.* 1990; Oechel *et al.* 1994; Oechel & Vourlitis 1996; A.C. Cook & W.C. Oechel unpublished data). Complete homeostatic adjustment of leaf-level photosynthetic rates under elevated CO_2 were observed after only 3 weeks (Tissue & Oechel 1987) and, due to the longer response time of ecosystems relative to leaf biochemical processes, comparable homeostatic adjustment of net ecosystem CO_2 flux was observed within 3 years of elevated CO_2 exposure (Oechel *et al.* 1994). This rapid loss of leaf and ecosystem photosynthetic potential is due primarily to the lack of a suitable sink for the utilization of the excess carbohydrate produced (Tissue & Oechel 1987; Grulke *et al.* 1990; Oechel *et al.* 1994). Although this pattern may be modified by increased temperature (Oechel *et al.* 1994), the homeostatic adjustment observed over the short and intermediate term appears to persist even after a century or more of elevated CO_2 exposure (A.C. Cook & W.C. Oechel unpublished data). It is, therefore, unlikely that ecosystem C sequestration will be affected significantly by the direct effects of elevated CO_2 alone.

Climate change is likely to be amplified in high-latitude regions due to alterations in albedo, atmospheric composition and permafrost (Gates *et al.* 1992; Manabe & Stouffer 1993; Meehl *et al.* 1993; Kattenberg *et al.* 1996). Northern ecosystems, in turn, are sensitive to climate change due to the large soil C stocks, presence of permafrost, and the importance of hydrology in controlling C sequestration (Oechel & Billings 1992; Shaver *et al.* 1992). Whether the Arctic will represent a positive or negative feedback to global climate change is largely a function of how drastic the temperature and hydrologic change is; the response of vegetation and soils to the changes in hydrology associated with an increase in temperature; and the time scales required for each level of biological organization to respond to such changes (Oechel & Billings 1992; Shaver *et al.* 1992; Oechel *et al.* 1993; Smith & Shugart 1993; Oechel & Vourlitis 1994; Wookey *et al.* 1994).

The climate appears to be changing in northern regions (Lachenbruch & Marshall 1986; Chapman & Walsh 1993), and this change has apparently resulted in altered ecosystem function where arctic ecosystems are now net sources of CO_2 to the atmosphere (Grulke *et al.* 1990; Oechel *et al.* 1993, 1995; Oechel & Vourlitis 1994, 1995). Although this change is thought to be transient (Oechel *et al.* 1993; Marion & Oechel 1993; Smith & Shugart 1993; Oechel & Vourlitis 1994), there is currently no evidence that this situation will end in the near future, as plant productivity appears to be relatively unresponsive to the observed warming. Changes in the species composition of high-latitude ecosystems due to the northward migration of vegetation is expected (Landhäusser & Wein 1993; Smith & Shugart 1993; Neilson 1994; Wookey *et al.* 1994). However, the time scale for this migration is extremely long compared to the relatively instantaneous response of soil micro-organisms and the extant vegetation (Smith & Shugart 1993).

ACKNOWLEDGEMENTS

This research was supported by the Office of Health and Environmental Research of the US Department of Energy (DE-FG03-86-ER60479) and by the Office of Polar Programs, Arctic Systems Science, Land-Atmosphere-Ice Interactions Program of the US National Science Foundation (OPP 9318527). Constructive comments on earlier versions of the manuscript by Kathleen Turner and two anonymous reviewers are greatly appreciated.

REFERENCES

Arp, W.J. (1991). Effects of source–sink relations on photosynthetic acclimation to elevated CO_2. *Plant, Cell and Environment*, **14**, 869–875.

Beltrami, H. & Mareshal, J.C. (1991). Recent warming in eastern Canada inferred from geothermal measurements. *Geophysical Research Letters*, **18**, 605–608.

Billings, W.D. (1987). Carbon balance of Alaskan tundra and taiga ecosystems: Past, present, and future. *Quaternary Science Reviews*, **6**, 165–177.

Billings, W.D., Luken, J.O., Mortensen, D.A. & Peterson, K.M. (1982). Arctic tundra: A source or sink for atmospheric carbon dioxide in a changing environment? *Oecologia*, **53**, 7–11.

Billings, W.D., Peterson, K.M., Luken, J.D. & Mortensen, D.A. (1984). Interaction of increasing atmospheric carbon-dioxide and soil nitrogen in the carbon balance of tundra microcosms. *Oecologia*, **65**, 26–29.

Bradley, R.S., Diaz, H.F., Eischeid, J.K., Jones. P.D., Kelly, P.M. & Goodess, C.M. (1987). Precipitation fluctuations over northern hemisphere land areas since the mid-19th century. *Science*, **237**, 171–175.

Chapin, F.S. III, Miller, P.C., Billings, W.D. & Coyne, P.I. (1980). Carbon and nutrient budgets and their control in coastal tundra. *An Arctic Ecosystem: The Coastal Tundra at Barrow, Alaska* (Ed. by J. Brown, P.C. Miller, L.L. Tieszen & F.L. Bunnell), pp. 458–482. Dowden, Hutchinson & Ross, Stroudsburg, Penn.

Chapin, F.S. III & Shaver, G. (1985). Individualistic growth response of tundra plant species to environmental manipulations in the field. *Ecology*, **66**, 564–576.

Chapman, W.L. & Walsh, J.E. (1993). Recent variations of sea ice and air temperature in high latitudes. *Bulletin American Meteorological Society*, **74**, 33–47.

Coyne, P.I. & Kelley, J.J. (1975). CO_2 exchange over the Alaskan arctic tundra: meteorological assessment by an aerodynamic method. *Journal of Applied Ecology*, **12**, 587–611.

Drake, B.G. (1992). The impact of rising CO_2 on ecosystem production. *Water, Air, and Soil Pollution*, **64**, 25–44.

Flanagan, P.W. & Bunnell, F.L. (1980). Microflora activities and decomposition. *An Arctic Ecosystem: The Coastal Tundra at Barrow, Alaska* (Ed. by J. Brown, P.C. Miller, L.L. Tieszen, & F.L. Bunnell), pp. 243–287. Dowden, Hutchinson & Ross, Stroudsburg, Penn.

Freeman, C., Lock, M.A. & Reynolds, B. (1993). Fluxes of CO_2, CH_4 and N_2O from a Welsh peatland following simulation of water table draw-down: potential feedback to climate change. *Biogeochemistry*, **19**, 51–60.

Funk, D.W., Pullman, E.R., Peterson, K.M., Crill, P.M. & Billings, W.D. (1994). Influence of water table on carbon dioxide, carbon monoxide, and methane fluxes from taiga bog microcosms. *Global Biogeochemical Cycles*, **8**, 271–278.

Gates, W.L., Mitchell, J.F.B., Boer, G.J., Cubasch, U. & Meleshko, V.P. (1992). Climate modeling, climate prediction and model validation. *Climate Change 1992: The Supplemental Report to the IPCC Scientific Assessment* (Ed. by J.T. Houghton, B.A. Callander & S.K. Varney), pp. 97–135. Cambridge University Press, Cambridge.

Gersper, P.L., Alexander, V., Barkley, S.A., Barsdate, R.J. & Flint, P.S. (1980). The soils and their nutrients. *An Arctic Ecosystem. The Coastal Tundra at Barrow, Alaska* (Ed. by J. Brown, P.C. Miller, L.L. Tieszen & F.L. Bunnel), pp. 219–254. Dowden, Hutchinson & Ross, Stroudsburg, Penn.

Glenn, S., Heyes, A. & Moore, T. (1993). Carbon dioxide and methane fluxes from drained peat soils, southern Quebec. *Global Biogeochemical Cycles*, **7**, 247–257.

Gorham, E. (1991). Northern peatlands: role in the carbon cycle and probable responses to climate warming. *Ecological Applications*, **1**, 182–195.

Graham, R.L., Turner, M.G. & Dale, V.H. (1991). How increasing CO_2 and climate change affect forests. *BioScience*, **40**, 575–587.

Groisman, P.Y., Karl, T.R. & Knight, R.W. (1994). Observed impact of snow cover on the heat balance and the rise of continental spring temperatures. *Science*, **263**, 198–200.

Grulke, N.E., Riechers, G.H., Oechel, W.C., Hjelm, U. & Jaeger, C. (1990). Carbon balance in tussock tundra under ambient and elevated atmospheric CO_2. *Oecologia*, **83**, 485–494.

Harazono, Y., Yoshimoto, M., Miyata, A., Vourlitis, G.L. & Oechel, W.C. (1995). Micro-meteorological data and their characteristics over the arctic tundra at Barrow, Alaska during the summer of 1993. *Miscellaneous Publication of the National Institute of Agro-Environmental Sciences*, **16**, 1–215.

Harden, J.W., Sunquist, E.T., Stallard, R.F. & Mark, R.K. (1992). Dynamics of soil carbon during deglaciation of the Laurentide ice sheet. *Science*, **258**, 1921–1924.

Hilbert, D.W., Prudhomme, T.I. & Oechel, W.C. (1987). Response of tussock tundra to elevated carbon dioxide regimes: analysis of ecosystem flux through nonlinear modelling. *Oecologia*, **72**, 466–472.

Hinzman, L.D. & Kane, D.L. (1992). Potential response of an arctic watershed during a period of global warming. *Journal of Geophysical Research*, **97**, 2811–2820.

Hogg, E.H., Lieffers, V.J. & Wein, R.W. (1992). Potential carbon losses from peat profiles: effects of temperature, drought cycles, and fire. *Ecological Applications*, **2**, 298–306.

Idso, S.B., Kimball, B.A. & Allen, S.G. (1991). Net photosynthesis of sour orange trees maintained in atmospheres of ambient and elevated CO_2 concentration. *Agricultural and Forest Meteorology*, **54**, 95–101.

Kane, D.L., Gieck, R.E. & Hinzman, L.D. (1990). Evapotranspiration from a small arctic watershed. *Nordic Hydrology*, **21**, 253–272.

Kane, D.L., Hinzman, L.D., Woo, M.K. & Everett, K.R. (1992). Arctic hydrology and climate change. *Physiological Ecology of Arctic Plants: Implications for Climate Change* (Ed. by F.S. Chapin III, R. Jeffries, G. Shaver, J. Reynolds & J. Svoboda), pp. 243–287. Academic Press, New York.

Karl, T.R., Kukla, G., Razuvayev, V.N., Changery, M.J., Quayle, R.G., Heim, R.R., Easterling, D.R. & Fu, C.B. (1991). Global warming: evidence for asymmetric diurnal temperature change. *Geophysical Research Letters*, **18**, 2253–2256.

Kattenberg, A., Giorgi, F., Grassl, H., Meehl, G.A., Mitchell, J.F.B., Stouffer, R.J., Tokioka, T., Weaver, A.J. & Wigley, T.M.L. (1996). Climate models – projections of future climate. *Climate Change 1995: The Science of Climate Change* (Ed. by J.T. Houghton, L.G. Meira Filho, B.A. Callender, H. Harris, A. Kattenberg & K. Maskell), pp. 285–358. Cambridge University Press, Cambridge.

Kling, G.W., Kipphut, G.W. & Miller, M.C. (1991). Arctic lakes and streams as conduits to the atmosphere: implications for tundra carbon budgets. *Science*, **251**, 298–301.

Körner, C. (1993). CO_2 fertilization: the great uncertainty in future vegetation development. *Vegetation Dynamics and Global Change* (Ed. by A.M. Soloman & H.H. Shugart), pp. 53–70. Chapman & Hall, New York.

Körner, C. & Diemer, M. (1987). *In situ* photosynthetic responses to light, temperature, and carbon dioxide in herbaceous plants from low and high altitude. *Functional Ecology*, **1**, 179–194.

Körner, C., Diemer, M., Schäppi, B. & Zimmerman, L. (1996). Response of alpine vegetation to elevated CO_2. *Terrestrial Ecosystem Response to Elevated Carbon Dioxide* (Ed. by G. Koch), pp. 177–196. Academic Press, San Diego.

Kummerow, J. & Ellis, B.A. (1984). Temperature effect on biomass production and root/shoot biomass ratios in two arctic sedges under controlled environmental conditions. *Canadian Journal of Botany*, **62**, 2150–2153.

Lachenbruch, A.H. & Marshall, B.V. (1986). Changing climate: geothermal evidence from permafrost in the Alaskan Arctic. *Science*, **234**, 689–696.

Landhäusser, S.M. & Wein, R.W. (1993). Postfire vegetation recovery and tree establishment at the arctic treeline: climate-change–vegetation-regeneration hypotheses. *Journal of Ecology*, **81**, 665–672.

Limbach, W.E., Oechel, W.C. & Lowell, W. (1982). Photosynthetic and respiratory responses to temperature and light of three Alaskan tundra growth forms. *Holarctic Ecology*, **5**, 150–157.

Manabe, S. & Stouffer, R.J. (1993). Century-scale effects of increased atmospheric CO_2 on the ocean–atmosphere system. *Nature*, **364**, 215–218.

Marion, G.M. & Black, C.H. (1987). The effect of time and temperature on nitrogen mineralization in arctic tundra soils. *Soil Science Society of America Journal*, **51**, 1501–1507.

Marion, G.M. & Miller, P.C. (1982). Nitrogen mineralization in a tussock tundra soil. *Arctic and Alpine Research*, **14**, 287–293.

Marion, G.M. & Oechel, W.C. (1993). Mid- to late-Holocene carbon balance in arctic Alaska and its implications for future global warming. *The Holocene*, **3**, 193–200.

Meehl, G.A., Washington, W.M. & Karl, T.R. (1993). Low-frequency variability and CO_2 transient climate change: Part 1. Time averaged differences. *Climate Dynamics*, **8**, 117–133.

Melillo, J.M., McGuire, A.D., Kicklighter, D.W., Moore, B., Vorosmarty, C.J. & Schloss, A.L. (1993). Global climate change and terrestrial net primary production. *Nature*, **363**, 234–240.

Miller, P.C., Kendall, R. & Oechel, W.C. (1983). Simulating carbon accumulation in northern ecosystems. *Simulation*, **40**, 119–131.

Mitchell, J.F.B., Johns, T.C., Gregory, J.M. & Tett, S.B.F. (1995). Climate response to increasing levels of greenhouse gases and sulphate aerosols. *Nature*, **376**, 501–504.

Mitchell, J.F.B., Senior, C.A. & Ingram, W.J. (1989). CO_2 and climate: a missing feedback? *Nature*, **341**, 132–134.

Moorhead, D.L. & Reynolds, J.F. (1993). Effects of climate change on decomposition in arctic tussock tundra: a modeling synthesis. *Arctic and Alpine Research*, **25**, 403–412.

Nadelhoffer, K.J., Giblin, A.E., Shaver, G.R. & Laundre, J.A. (1991). Effects of temperature and substrate quality on element mineralization in six arctic soils. *Ecology*, **72**, 242–253.

Neilson, R.P. (1994). Vegetation redistribution: a possible biosphere source of CO_2 during climatic change. *Water, Soil and Air Pollution*, **70**, 659–673.

Oberbauer, S.F., Sionit, N., Hastings, S.J. & Oechel, W.C. (1986). Effects of CO_2 enrichment and nutrition on growth photosynthesis, and nutrient concentration of Alaskan tundra plant species. *Canadian Journal of Botany*, **64**, 2993–2998.

Oechel, W.C. & Billings, W.D. (1992). Anticipated effects of global change on carbon balance of arctic plants and ecosystems. *Physiological Ecology of Arctic Plants: Implications for Climate Change* (Ed. by F.S. Chapin III, R. Jeffries, G. Shaver, J. Reynolds & J. Svoboda), pp. 139–168. Academic Press, New York.

Oechel, W.C., Cowles, S., Grulke, N., Hastings, S.J., Lawrence, B., Prudhomme, T., Riechers, G., Strain, B., Tissue, D. & Vourlitis, G. (1994). Transient nature of CO_2 fertilization in arctic tundra. *Nature*, **371**, 500–503.

Oechel, W.C., Hastings, S.J., Vourlitis, G.L., Jenkins, M., Riechers, G. & Grulke, N. (1993). Recent change of Arctic tundra ecosystems from a net carbon dioxide sink to a source. *Nature*, **361**, 520–523.

Oechel, W.C., Riechers, G., Lawrence, W.T., Prudhomme, T.I., Grulke, N. & Hastings, S.J. (1992). 'CO_2LT' an automated, null-balance system for studying the effects of elevated CO_2 and global climate change on unmanaged ecosystems. *Functional Ecology*, **6**, 86–100.

Oechel, W.C. & Vourlitis, G.L. (1994). The effects of climate change on land–atmosphere feedbacks in arctic tundra regions. *Trends in Ecology and Evolution*, **9**, 324–329.

Oechel, W.C. & Vourlitis, G.L. (1995). Effects of global change on carbon storage in cold soils. *Advances in Soil Science: Soils and Global Change* (Ed. by R. Lal, J. Kimbel, E. Levine & B.A. Stewart), pp. 117–129. Lewis, Boca Raton, Fla.

Oechel, W.C. & Vourlitis, G.L. (1996). Direct effects of elevated CO_2 on arctic plant and ecosystem function. *Terrestrial Ecosystem Response to Elevated Carbon Dioxide* (Ed. by G. Koch), pp. 163–176. Academic Press, San Diego.

Oechel, W.C., Vourlitis, G.L., Hastings, S.J. & Bochkarev, S.A. (1995). Change in carbon dioxide flux at the U.S. Tundra International Biological Program Sites at Barrow, AK. *Ecological Applications*, **5**, 846–855.

Parsons, A.N., Welker, J.M., Wookey, P.A., Press, M.C., Callaghan, T.V. & Lee, J.A. (1994). Growth responses of four sub-Arctic dwarf shrubs to simulated environmental change. *Journal of Ecology*, **82**, 307–318.

Pastor, J. & Post, W.M. (1993). Linear regressions do not predict transient responses of eastern North American forests to CO_2-induced climate change. *Climate Change*, **23**, 111–119.

Rastetter, E.B., McKane, R.B., Shaver, G.R. & Melillo, J.M. (1992). Changes in C storage by terrestrial ecosystems: how C–N interactions restrict responses to CO_2 and temperature. *Water, Air, and Soil Pollution*, **64**, 327–344.

Schell, D.M. (1983). Carbon-13 and carbon-14 abundances in Alaskan aquatic organisms: delayed production from peat in arctic food webs. *Science*, **219**, 1068–1071.

Schlesinger, W.H. (1991). *Biogeochemistry: An Analysis of Global Change.* Academic Press, San Diego.

Sharratt, B.S. (1992). Growing season trends in the Alaskan climate record. *Arctic*, **45**, 124–127.

Shaver, G.R., Billings, W.D., Chapin, F.S. III, Giblin, A.E., Nadelhoffer, K.J., Oechel, W.C. & Rastetter, E.B. (1992). Global change and the carbon balance of arctic ecosystems. *BioScience*, **42**, 433–441.

Shaver, G.R. & Chapin, F.S. III. (1980). Response to fertilization by various plant growth forms in an Alaskan tundra: nutrient accumulation and growth. *Ecology*, **61**, 662–675.

Silvola, J. (1986). Carbon dioxide dynamics in mines reclaimed for forestry in eastern Finland. *Annales Botanici Fennici*, **23**, 59–67.

Smith, T.M. & Shugart, H.H. (1993). The transient response of terrestrial carbon storage to a perturbed climate. *Nature*, **361**, 523–526.

Stitt, M. (1991). Rising CO$_2$ levels and their potential significance for carbon flow in photosynthetic cells. *Plant, Cell and Environment*, **14**, 741–762.

Strain, B.R. (1992). Field measurements of CO$_2$ enhancement and change in natural vegetation. *Water, Air and Soil Pollution*, **64**, 45–60.

Tissue, D.T. & Oechel, W.C. (1987). Response of *Eriophorum vaginatum* to elevated CO$_2$ and temperature in the Alaskan tussock tundra. *Ecology*, **68**, 401–410.

Vourlitis, G.L., Oechel, W.C., Hastings, S.J. & Jenkins, M.A. (1993). A system for measuring *in situ* CO$_2$ and CH$_4$ flux in unmanaged ecosystems: an arctic example. *Functional Ecology*, **7**, 369–379.

Waelbroeck, C. (1993). Climate–soil processes in the presence of permafrost: a systems modeling approach. *Ecological Modeling*, **69**, 185–225.

Watson, R.T., Meira Filho, L.G., Sanhueza, E. & Janetos, A. (1992). Sources and sinks. *Climate Change 1992: The Supplemental Report to the IPCC Scientific Assessment* (Ed. by J.T. Houghton, B.A. Callander & S.K. Varney), pp. 25–46. Cambridge University Press, Cambridge.

Woodward, F.L. (1987). Stomatal numbers are sensitive to increases in CO$_2$ from pre-industrial levels. *Nature*, **327**, 617–618.

Wookey, P.A., Welker, J.M., Parsons, A.N., Press, M.C., Callaghan, T.V. & Lee, J.A. (1994). Differential growth, allocation and photosynthetic responses of *Polygonum viviparum* to simulated environmental change at a high arctic polar semi-desert, *Oikos*, **70**, 131–139.

Zimov, S.A., Zimova, G.M., Davidov, S.P., Davidova, A.I., Boropaev, Y.V., Boropaeva, Z.V., Prosiannikov, S.F., Prosiannikova, O.V., Semiletova, I.V. & Semiletov, I.P. (1993). Winter biotic activity and production of CO$_2$ in Siberian soils: a factor in the greenhouse effect. *Journal of Geophysical Research*, **98**, 5017–5023.

Index